DISEASES OF FRUIT CROPS

Diseases of Fruit Crops

R.S. Singh

Former Professor and Head
Department of Plant Pathology
G.B. Pant University
Pantnagar
India

OXFORD & IBH PUBLISHING CO. PVT. LTD.
New Delhi

DISEASES OF FRUIT CROPS

Oxford & IBH Publishing Company Pvt. Ltd.
113-B Shahpur Jat
Asian Games Village side
New Delhi 110 049, India
E-mail: oxford@oxford-ibh.in

Last reprinted 2016

ISBN 81-204-1441-1

Printed at Chaman Enterprises, New Delhi.

??-R6-05

Preface

Compendia dealing with diseases of fruits and fruit trees grown in temperate climate countries have been published but there are few single volumes dealing with the majority of fruit crops and their diseases. The need for a book giving details of symptoms of a disease observed and described in different regions of the world, identification and description of pathogen(s) associated with the disease in these regions, their relationships with the local environments and approaches to management of these diseases, particularly non-chemical where possible, has been felt by students and teachers of plant pathology. The objective of this book is to present a comprehensive account of economically important diseases of major fruit trees and fruits of the subtropical regions including the temperate climate areas of these regions. Attempt has been made to incorporate the latest references from throughout the world. There has been significant work on fruit diseases in India particularly on apple, citrus and peach, but the references are scattered in journals. This book has tried to bring together these references and put them along with the references from other countries.

The author is grateful to his erstwhile colleagues at the Pantnagar Agricultural University for help in various ways. He is particularly thankful to Dr. H.B. Singh of Central Institute of Medicinal and Aromatic Plants and to Dr. Ram Kishun of Central Institute of Sub-tropical Horticulture for providing valuable photographs of many diseases.

October, 2000   **R.S. Singh**

Contents

Preface   v

Chapter 1: CITRUS DISEASES   1
    Gummosis and root-rot of citrus   1
    Anthracnose, die-back or wither tip   9
    Armillaria root-rot of citrus trees   12
    Post-harvest decay of citrus fruits   15
    Citrus canker   28
    Citrus greening disease   37
    Citrus tristeza virus   39
    Xyloporosis or cachexia   46
    Psorosis or scaly bark   47
    Citrus exocortis viroid   49
    Citrus root nematode   50
    Spreading decline of citrus   55
    Nutritional disorders in citrus   56
    References   62

Chapter 2: DISEASES OF POME AND STONE FRUITS   73
    Apple scab   73
    Powdery mildew of apple   91
    Peach leaf curl   96
    Collar rot of apple trees   99
    White root rot of apple   103
    Brown rot of pome and stone fruits   106
    Botryosphaeria white rot, root rot and stem brown   111
    Botryosphaeria black rot and twig canker   113
    Bitter rot of apple   115
    Post-harvest decay of pome and stone fruits   117
    Pink mold rot   121
    Mucor rot of pome and stone fruits   122

Fire blight of apple and pear   125
Crown gall disease   131
Bacterial canker and gummosis   135
References   140

Chapter 3: MANGO DISEASES   153
Bacterial leaf spots, blight and canker   153
Powdery mildew of mango   157
Anthracnose and die-back   159
Twig blight, die-back, and leaf blight   163
Grey blight   166
Red rust   167
Post-harvest fungal rots of fruits   167
Black tip or mango necrosis   172
Mango malformation   173
Giant or leafy mistletoes   176
References   178

Chapter 4: BANANA DISEASES   183
Fusarium wilt of banana (Panama disease)   183
Cercospora leaf spots or Sigatoka   189
Post-harvest decay of banana fruits   193
Bacterial wilt or Moko disease   197
Bunchy top of banana   201
Root and rhizome rot of banana   204
References   207

Chapter 5: DISEASES OF GRAPEVINES AND GRAPES   212
Downy mildew   212
Powdery mildew   219
Bunch rot   227
Black rot   230
Anthracnose   233
References   230

Chapter 6: GUAVA DISEASES   242
Anthracnose and fruit rots   242
Grey blight, leaf spots and scab   245
Guava wilt   246
Zinc deficiency   249
References   251

Chapter 7: DISEASES OF PAPAYA   253
Stem and foot rot   253

Phytophthora root rot 254
Algal fruit spots 257
Post-harvest fruit rots 257
Papaya mosaic (ringspot) 259
Leaf curl of papaya 262
References 263

Chapter 8: STRAWBERRY DISEASES **265**
Anthracnose of strawberry 265
Mycosphaerella leaf spots 270
Red stele 272
Black root rot complex 274
Fungal rots of strawberry fruits 275
References 278

Chapter 9: DISEASES OF COCONUT PALM **281**
Bud rot 281
Stem bleeding disease 283
Root (wilt) disease 284
Cadang-cadang 285
Leaf spots of coconut 287
Red ring disease 287
References 289

Chapter 10: MISCELLANEOUS TROPICAL FRUITS **290**
Heart or stem rot and wilt of pineapple 290
Basal rot, leaf rot and fruit rot of pineapple 291
Bacterial leaf spots of pomegranate 292
Aspergillus rot of pomegranate 292
Leaf spots of litchi 293
Fruit rot of litchi 294
Die-back, twig blight and canker of loquat 294
Decline of cashewnut trees 295
Anthracnose of cashewnut 296
Pestalotia and other leaf spots of sapodilla plum 297
References 297

Index **298**

About the Author **311**

Citrus Diseases

■ GUMMOSIS AND ROOT ROT

Species of *Phytophthora* cause the most serious soil-borne diseases of citrus throughout the world (205). Losses occur from damping off of seedlings in the seed beds, root and crown rot in nurseries, and from diseases variously known as gummosis, brown rot gummosis, brown rot, trunk rot, foot rot, collar rot, root rot, fibrous root rot, leaf fall and fruit rot. Serious losses can occur in orchards with trees on susceptible rootstock such as rough lemon (*Citrus jambheri*), or in plantings on resistant rootstock where the graft union is at or below the soil surface exposing scion tissues to the pathogen. If the disease is not checked in time the entire tree may be destroyed. The brown rot of fruits occurs in orchards causing fruit drop and in storage it causes post-harvest decay of fruits.

Symptoms

Phytophthora spp. can cause damping off of newly germinated seedlings of *Citrus* spp. Typical symptoms result when soil- or seed-borne fungus penetrates the stem just above the soil line and causes the seedlings to topple. Seed rot and pre-emergence damping off may also occur. There is rapid killing when there is abundant soil moisture and temperatures are favourable for the fungus. Once the leaves have emerged the seedlings become resistant. These symptoms are similar to damping off of any plant caused by *Pythium* or *Rhizoctonia* (205).

Foot rot and gummosis are the most serious diseases of citrus caused by *Phytophthora* spp. Primary infection by *Phytophthora* normally occurs on the bark at the base of the trunk near the ground level, producing lesions on the trunk and the crown roots. It spreads around the trunk girdling it and killing the tree. The bark and the wood both are affected. The infection can spread upward and down the roots, often causing fibrous root rot. The root damage is especially serious on susceptible rootstock in nurseries. Fibrous root rot may occur even in bearing trees where the root damage causes tree decline and yield losses (178). In some tolerant rootstocks the fibrous root rot does not affect the fruit yield.

The main symptom of gummosis is oozing of gum from the affected parts on the trunk. Infected bark remains firm with small longitudinal cracks through which abundant amber-coloured gum exudation occurs. Citrus gum is water soluble. During rainy season the gum is washed down or gets mixed with soil near the ground level hence this symptom may not be clear. During summer gum deposits dry and stick to the bark making the symptom of gummosis very clear.

The root rot (69) and fibrous root rot symptoms are not seen in the early stages of the disease. However, the root rot destroys a major portion of the root system before well-marked symptoms are seen on the aerial parts. The effect of trunk and root infection is ultimately drying up of the tree. Before death the tree flowers profusely but fruits are small and drop before maturity.

Leaves of the affected trees show symptoms of nutritional deficiency. The veins turn yellow and there is premature leaf fall. In mandarin oranges, infection of leaves by *P. palmivora* is common in heavy rainfall areas. Water-soaked spots appear on the lower leaves and by the time the spots spread over the entire lamina the leaf drops. This usually results in heavy defoliation. The infection reaches the unripe, ripening and ripe fruits and produces water-soaked spots on the skin. This is brown rot of fruits (73, 207). The infection of fruits can occur directly when water splashed inoculum reaches the fruits near ground level. All the fruits on the tree may gradually become affected under humid conditions. Such fruits become soft and white fungus growth develops on the skin. Ultimately the fruit drops. The fungus continues to grow on fallen fruits.

The fruits which do not show symptoms and are still on the tree are harvested and packed. If fruits are untreated, the brown rot spreads from fruit to fruit by contact (112). In a few days of storage these fruits have a characteristic pungent, rancid odour. Brown rot epidemics are usually restricted to areas where heavy rainfall coincides with the early stages of fruit maturity (236). All cultivars are affected especially the lemons.

The causal organisms

Three species of *Phytophthora*, viz., *P. palmivora*, *P. parasitica* (*P. nicotianae* var. *parasitica*) and *P. citrophthora* attack citrus. In south India the main cause of gummosis and root rot is recognized as *P. palmivora* (170) although *P. parasitica* is also common as incitant of fruit rot in Karnataka (180). In other parts of India also *P. parasitica* was reportedly associated with gummosis. In the USA *P. parasitica* and *P. citrophthora* are the main pathogens of root rot of citrus. In addition, *P. hibernalis* and *P. syringae* attack citrus fruits to a limited extent in areas with cool moist weather. *P. citricola* is reported to attack citrus in some tropical areas (21).

Phytophthora nicotianae B. de Haan var. *parasitica* (Dastur) Waterhouse

The hyphae are tough, irregularly wide upto 9 microns, but without typical hyphal swellings. Sporangiophores are more slender than the mycelial hyphae, irregularly or sympodially branched, the sympodia being close in moist air. Sporangia are papillate, and occasionally have more than one apex. They are broadly ovoid, ellipsoid, obpyriform to spherical, not noticeably narrowed at the apex, occasionally lateral or intercalary. They measure 38-50 × 30-40 (av. 40 x 38) μm and are deciduous with very short (2-5 microns) pedicel. Oogonia are usually produced in single cultures though often very scarcely or not until some weeks, but readily produced in dual cultures with opposite strains (the A_1 and A_2 mating types). Antheridia are amphigynous, spherical or oval, and 12-16 × 18 μm in size. The oogonial diameter is usually 22-29 μm, rarely 31 μm. Oogonia become rough, thick walled and yellowish brown with age. Oospores are markedly aplerotic and 18-20 (sometimes 20-25) μm in diameter. The oospore wall is about 2 μm thick (232). Chlamydospores are abundantly produced. They are less than 25 to 60 μm in diameter, with 3-4 μm thick wall. Minimum, optimum and maximum temperatures for growth are 10°C, 30-32°C and 47°C, respectively. *P. parasitica* has a broad host range. Virulence of isolates from tomato and other non-citrus hosts towards citrus is low while all isolates of the species are pathogenic on tomato (133).

Phytophthora citrophthora (Smith and Smith) Leonian

Hyphae are fairly coarse and upto 7 μm wide. Sporangiophores are delicate, short, scarcely widening at the base of the sporangium, irregularly branched and with a swelling at the point of branching. Sporangia are rather scanty on some agar media and very variable in shape and size in water; often with more than 1 apices and papilla having 5 μm deep apical thickening. Sporangia are deciduous with a 10-12 μm long pedicel. The average size of sporangia is 40-45 × 27, often 50-55 × 30 or even up to 90 × 60 μm. Chlamydospores may be abundant, few or absent. Most isolates do not form chlamydospores. Their average diameter is 28 μm with 1.5-2 μm thick wall. Sexual reproduction is not seen (233). Minimum, optimum and maximum temperatures for growth are 5°C, 24-28°C and 32°C, respectively.

Phytophthora palmivora (Butler) Butler

The mycelium is intercellular with haustoria. Hyphae are large and often swollen at regular intervals. They are upto 7 μm wide. Sporangiophores are simple or branched with inverted pear shaped, rarely round and always terminal sporangia which measure 38-72 × 33-42 (av. 50 × 35) μm. Zoospores are large, 8-10 μm in diameter when encysted. Oospores are spherical, 35-45 μm in diameter with 4 μm thick wall. They produce secondary sporangium on germination.

Disease cycle

P. parasitica is a parasite but a poor saprophyte in soil. It causes noticeable yield losses only when high populations of the fungus are present in the soil. Tsao (215) and Tsao and Bricker (218, 219) had carried out extensive studies on the behaviour of the fungus in soil. The growth in soil is restricted due to antagonism including intense colonization of hyphae by bacteria leading to rapid lysis and breakdown of cytoplasm (125, 215). The entire mycelium is soon converted into oospores and chlamydospores which serve as survival structures. Oospores occur in low numbers in soil and probably are resistant to desiccation and cold temperature. The formation of oospore and perpetuation of the fungus through these structures is governed by the presence of mating types in the plant population (181, 182). The oospores mature more slowly than chlamydospores but once matured they germinate in response to nutrients from roots.

Chlamydospores are common and the most important source of soil survival of *P. parasitica* (197). These structures are commonly formed when soil moisture is limiting, conditions are cool or where the host roots are not actively growing and producing susceptible tissues. Formation of chlamydospores can also be stimulated by poor aeration and high carbon dioxide concentration in the soil atmosphere (213). They can survive in soil for several months under unfavourable conditions. In the presence of host roots, nutrients, aeration, optimum temperature and moisture these resting structures germinate by a germ tube and by sporangia which liberate motile zoospores. Germination of chlamydospores is optimum in well-aerated, moist environments when temperatures are favourable for root growth. Root exudates promote their germination (60). Chlamydospores cause more uniform infection than zoospores (90).

Kuch and Khew (115) noticed maximum populations of propagules of *P. palmivora* at 0.5-15 cm depth of soil and very low populations at 30-45 cm. Optimum soil moisture for survival of propagules was 25-45 per cent WHC at pH 6.5-7.0. The survival could be for upto 18 months in natural soil. Sporangia survive and remain infective after passage through the alimentary canals of two snail species. Ingestion of oospores of *P. palmivora* by garden snails facilititates their germination (179).

P. citrophthora grows best at 24°-28°C and *P. parasitica* at 28°-32°C. Abundant production of sporangia of *P. citrophthora* occurs at 20° while that of *P. parasitica* at 30°C. Maximum recovery of *P. citrophthora* from rootlets in naturally infested soil is obtained at 15°-20°C and that of *P. parasitica* at 15°-30°C (135-137). Thus, in citrus areas where temperatures are low *P. citrophthora* is more dangerous than *P. parasitica*. A temperature of 24°C after infection favours development of *P. citrophthora* in roots.

The number of propagules rapidly increases immediately after irrigation in irrigated orchards or nurseries. Lutz and Menge (124) have reported that in long interval furrow irrigation the initial population of 17 propagules/g soil increased to 77 propagules/g soil two days after irrigation. According to them very low numbers of oospores of *P. parasitica* are found in the soil throughout the year. Low levels of dark coloured multi-papillate sporangia were also consistently found by them in soil. Agostini, *et al.* (4) have stated that conditions which allow abundant immature fibrous root development, a highly susceptible rootstock, and favourable soil temperature and moisture promote development of root rot and, consequently, high propagule density of *P. parasitica* in soil.

In addition to oospores and chlamydospores in the soil as the main survival structures for initiating primary infection. *P. parasitica* also survives on fallen fruits, twigs, leaves, and in cracks on the standing trees. The main source for the spread of these pathogens in Nagpur mandarin orchards in central India is reported to be infested nurseries. More than 20% nursery plants die due to Phytophthora diseases and almost all nurseries are infested (149).

When chlamydospores and oospores germinate in the presence of nutrients, optimal soil moisture and aeration they form sporangia which liberate zoospores. Zoospore release is optimal in saturated soils. Nutrient depletion stimulates sporangia production (175, 215, 216). Diurnal temperature changes in soil may serve to synchronize the release of zoospores. According to Duniway (60) the motile zoospores swim to short distances (cm) or are carried by soil water to long distances (meters). Dispersal of sporangia and zoospores is by wind, raindrop splashes, irrigation water, and even insects. Zoospores are attracted to roots by root exudates (101, 220, 221), especially when roots are damaged but living (107, 108). On the root surface they encyst, germinate and cause infection by penetrating the cortex. Citrus tissue is more susceptible during the periods of the year when trees are actively growing than during the months when trees are dormant (132).

Heavy or fine textured soils where drainage is impeded, high soil moisture, pH of 5.4-7.5 and a temperature of 24°C are conducive to disease development. Soil moisture is the most important plant-soil environmental factor that affects development of Phytophthora root rot of citrus. The host is pre-disposed to infection when roots are stressed or damaged in saturated or dry soil. Long periods of soil saturation are required for _P. parasitica_ to be an effective root decay agent (217). According to Stolzy _et al._ (200) the length of saturation time is more important than the frequency of saturation. The frequency and duration of irrigation can also influence the activity of the fungus and pre-disposition of roots to rot (74). In sandy loam soils the greatest destruction of feeder roots occurs in irrigation furrows where saturated conditions favour zoospore production and movement. Roots outside the furrow often remain healthy (92). If soils are saturated during irrigation, zoospores are released and can infect roots to form more sporangia. When soils do not dry sufficiently between irrigations, sporangia survive until the next irrigation and again release zoospores. Soils with drainage restricted by hardpans or clay layers or those with shallow water table that temporarily rises into the root zone provide ideal conditions for fibrous root rot and build up of Phytophthora propagules.

Availability of oxygen in soil atmosphere is closely related to soil moisture because pore space for air is reduced in saturated soil. When roots are subjected to low oxygen conditions they are damaged by reduced forms of minerals and by toxic metabolites of microorganisms on the root surface. Root regeneration is restricted, new roots are not formed and root exudation increases under flooded conditions. Thus, in presence of _Phytophthora_, reduced oxygen level in soil causes greater root decay (200).

Low grafting and nearness of bud union to soil line increase chances of infection from soil-borne inoculum. Populations of the pathogens around roots in soil are increased where, in addition to the favourable environments, there are abundant immature fibrous roots and a highly susceptible rootstock thus promoting development of root rot. Around resistant rootstock there is low density of the fungi in soil. Most rootstock are at least moderately tolerant to _Phytophthora_ (209) but their susceptibility varies (92, 166). Although mechanisms of resistance are not clearly understood, it is presumed that coumarin phytoalexins in infected roots play some role (1).

Microbial antagonism in the suppression of _Phytophthora_ spp. in soil, rhizosphere or in the infection court is reported (124). Processes of parasitism, predation, and competition may all be operating alone or together in reducing inoculum. The reason for rapid loss of _Phytophthora parasitica_ mycelium in soil is attributed to lysis induced by intense bacterial colonization of hyphae. Specific strains of _Streptomyces_ and fluorescent pseudomonads occur in soil that cause hyphal lysis of _P. citrophthora_ (22).

Vesicular-arbuscular mycorrhizae have been implicated in microbial antagonism of _P. parasitica_ in citrus but increased tolerance of mycorrhizal plants to root rot is probably due to improved host nutrition, mainly phosphorus (53, 91, 93, 94).

Management

A number of preventive measures can be recommended to reduce primary infection. The site for citrus orchards should be well drained land. Resistant rootstock such as Khatta and trifoliate orange (*Poncirus trifoliata*) may be used in areas where gummosis and virus diseases of citrus are serious problems (92, 166). The bud union should be about 30-45 cm above the base and at the time of planting care should be taken to keep the bud union well above the soil line. The irrigation system should be planned in such a way that water from below one tree does not flow to the other trees. Every year the trunk should be painted with Bordeaux paste up to a height of about 70 cm. If the soil is known to contain *Phytophthora* the walls of new pits should be dusted with a mixture of zinc sulphate, copper sulphate and lime (5 : 1 : 4). Cleanliness of the orchard is very important. All infected fruits, leaves, twigs, etc. that have fallen should be collected and burnt.

During the summer and rains the orchard should be regularly sprayed with copper fungicides such as Bordeaux mixture (4 : 4 : 50 or 5 : 5 : 50) or copper oxychloride formulations such as Blitox-50. Soil drenching with 1000 ppm terrazole is reported to totally inhibit *P. palmivora* up to 2.5 cm depth of soil (183). In more recent chemical treatments, systemic fungicides have been used (52, 68, 207, 204) for soil drench and, sometimes, for trunk spray. Working with root rot and crown rot (*P. citrophthora* and *P. parasitica*), Matheron and Matejka (134) reported that metalaxyl, fosetyl-Al (Phosethyl-Al), and sodium tetrathiocarbonate (Enzone, which releases carbon disulphide in soil) reduced production of sporangia in soil by 90%. There were no lesions when infested soil was treated with 10 microgram/ml metalaxyl. A single application of metalaxyl or fosetyl-Al can provide protection to citrus from colonization by *P. citrophthora* or *P. parasitica* for 2-3 months (131). In Turkey, fosetyl-Al is reported to protect trees against gummosis for at leat one year (66). Enzone is more effective than metalaxyl in eradicating *Phytophthora* from host debris in soil. According to Sastry and Hegde (183) metalaxyl could penetrate and cause inhibition of inoculum up to a depth of 1.25 cm of soil. According to Naqvi (149) the density of fungal propagules in soil treated with Ridomil drench or Ridomil spray as well as drench is reduced by 69.1-72.8%. The population of the pathogens in relation to feeder roots density is also decreased. Drench and spray treatment with fosetyl-Al (Aliette) or Ridomil significantly increased the feeder root density. For economic control applications of fungicides should coincide with periods favourable for pathogen activity and disease development. Extremes in temperature (35°C at the max. end) are not favourable for the pathogen and at such periods fungicide application is not required. Application of metalaxyl or fosetyl-Al is beneficial when threshold level of the fungus reaches 10-15 propagules per cubic centimeter of soil (*cf.* 4). In general, the chemical treatments are required to be done before rains start because the pathogen multiplies rapidly during rains.

Although chemical control of citrus feeder root rot in the field with fosetyl-Al and metalaxyl is effective, it also is expensive. Over the years, emphasis has shifted to the production of nursery trees free of *Phytophthora* spp. by preventive phytosanitary methods such as soil fumigation, treated irrigation water and sound hygiene. Such healthy nursery trees grow consistently better than infected nursery trees. However, nursery plants that are initially certified disease-free and planted in virgin soil eventually become infected by irrigation water sources which are frequently contaminated by *Phytophthora* spp. and nematodes. The contaminated irrigation water supplements existing *Phytophthora* and nematode populations in the soil, making chemical control more difficult. Use of bleaching powder (chlorine) in irrigation water for nursery beds can reduce *Phytophthora* spp. on the planting stock. Application of chlorine on a field scale was considered costly. Electrolytic method of chlorine generation makes the water treatment by chlorine cheap and makes it possible for field application. Chlorine kills the propagules of *Phytophthora parasitica*, *P. citrophthora* and *Fusarium* spp. Dipping roots of grafts for 6-10 min in water at 35°C or in 0.02% suspension of captan and soil fumigation with Vapam or Mylone had been routine practices in many citrus growing countries to reduce losses from gummosis and root rot.

An integrated chemical treatment for Phytophthora root rot and the citrus nematode, *Tylenchulus semipenetrans* which causes citrus decline, has been proposed by Le Roux *et al.* (120). They used metalaxyl or fosetyl-Al as fungicides and aldicarb (Temik) as nematicide. In 24 months treatments were given at 3 months interval with 5% metalaxyl granules as soil drench at the rate of 2 g a.i. per sq. metre of leaf canopy and 15% aldicarb at 4.5 g a.i. per sq. metre of leaf canopy. The treatments resulted in increased trunk diameter and canopy volume. After 32-44 months of commencement of treatment fruit yield was significantly increased.

Eradication of infection from standing trees is possible only if the disease is detected in the early stages. If infection is on thin branches they may be cut and burnt. In the early stages of root rot, affected roots may be removed and the soil drenched with a fungicide. On thick branches and trunks the infected portion may be removed by a sharp knife and the wound cleaned with 0.1% mercuric chloride or 1% potassium permanganate solution followed by application of Bordeaux paste on the wound.

Generally, the above mentioned chemical treatments reduce the chances of brown rot of fruit. Copper fungicides or captan applied prior to beginning of rains are usually quite effective. Pre-harvest application of systemic fungicides metalaxyl or fosetyl-Al to the canopy provides effective control of brown rot (43). Post-harvest disinfection of fruits with chlorine or sodium orthophenylphenate, recommended for canker affected fruits, can also be helpful. Hot water treatment of grapefruit (48°C, 3 min), lemon (52°C, 5-10 min) and orange (53°C, 5 min) has been reported (10). Hot water treatment of certain citrus fruits has limitations (108). Although lemons could routinely tolerate immersion in water at 46.1°-48.9°C for 4 min or longer without injury, release of rind oils leading to oleocellosis could occur if lemons were cold and turgid at the time of treatment (111). The immersion should, therefore, be delayed by 1-4 days after harvest to allow the rind to lose turgor. Without this conditioning the fruits can be injured even at 37.8°C. Using soap in post-treatment rinse of fruits entraps released oils and terpenoids to further reduce the chances of rind injury.

■ ANTHRACNOSE, DIE-BACK OR WITHER TIP

Anthracnose is one of the causes of citrus decline (6, 68, 166). The disease is especially serious on orange, grapefruit and lemon trees. It affects all mature, weakened or injured aboveground plant parts, including leaves, twigs and fruits. Anthracnose may occur on trees of any size, in the nursery or in the orchard, but it rarely develops on vigorously growing trees. It is common on trees that are weakened or injured due to inadequate fertilization, lack of water, low temperature, insect attack, etc.

Symptoms

Wither-tip and die-back: The main symptoms of this condition are falling of terminal leaves on the shoots and drying of the shoots (69). The drying starts from the tip and progresses backward. The dry portion is straw or ash-coloured. A large number of black dot-like structures representing acervuli of the fungus appear on the ash-coloured portion. These symptoms are more pronounced during the summer and rains than during the winter.

Anthracnose: When the symptoms due to necrosis appear on leaves, fruits and twigs, they are called anthracnose. Reddish brown spots appear on the leaves which are disfigured. The central portion of these spots turns gray or ash-coloured and acervuli can develop on this portion. On the shoots, in addition to die-back, tissue necrosis may occur at any point.

Similar grey or black spots appear on the inflorescence stalk and flowers which shed. In fruit infections, reddish brown, circular and sunken spots are seen. They may be tiny specks or dark brown or black areas of 5-10 mm diameter. Usually they are only skin deep but in over-mature fruits they affect the flesh also. The spots become dry and hard and sometimes are dotted with small black acervuli that exude pinkish masses of spores in humid weather. These spots may appear on any part of the fruit surface but usually they start from the stem-end. Over-ripe fruits are particularly susceptible to anthracnose. When spores of the fungus are washed down from the twigs onto the fruits they germinate there and cause anthracnose russetting. The russetting appears as a large blotch or as a tear stain. Diseased leaves, flowers, apical parts of shoots and fruits drop down prematurely. In severe infections there may be considerable defoliation. In grapefruit (*Citrus paradisi*) a drop of fruits before ripening has been found during the rainy season. Usually such fruits do not show any spotting. But isolations from the stem-end tissue yield the fungus associated with anthracnose (184). Infected tissues show reduced amount of different sugars. Infection affects the amino acid content of the tissue also. Some amino acids are decreased. This varies with the host species.

Post-bloom fruit drop: The post-bloom fruit drop is caused by a strain of the same fungus that causes typical die-back and anthracnose (67). Peach to orange-coloured necrotic spots are formed on flower petals. Under favourable conditions entire flowers and clusters are invaded resulting in blossom blight (211). Abundant acervuli are produced on the surface of blighted flowers. After flower infection fruitlets drop, but the buttons composed of peduncle, floral disc, calyx and nectaries remain (210, 213). These symptoms are same as those described above under anthracnose and for premature fruit drop (188).

The causal organism

Fawcett (69) had recorded *Colletotrichum gloeosporioides*, *Gloeosporium limetticolum* and *Gloeosporium foliicolum* as pathogenic and causing anthracnose of citrus. *C. gloeosporioides*, described under mango anthracnose is most common. There may be several races with varying levels of pathogenicity within each species of the anthracnose fungus.

Disease cycle

The anthracnose fungus is favoured by high temperature and humid or moist weather. A relative humidity of 95% is optimum for growth of the fungus. The conidia are released only when the acervuli are wet and are generally spread by splashing and blowing rain or by coming in contact with insects, other animals, tools, etc. (3, 209). They germinate only in the presence of free water. The germ tubes produce appressoria and penetration pegs enter the host directly. In the beginning the hyphae may grow rapidly, intercellularly and intracellularly, but cause little or no visible discolouration or other symptoms of disturbance. Then, quite suddenly, especially when fruits begin to ripen, the fungus becomes more aggressive and symptoms appear.

Some workers believe that die-back symptoms are not caused directly by fungi. Nutritional deficiencies and attack of viruses and citrus greening bacterium first cause the die-back and later the fungus grows saprophytically on the dead or weakened parts. *C. gloeosporioides* produces non-specific toxins also which are supposed to induce die-back (185).

Management

The disease has a close relationship with poor growth of the trees. Priority should be given to measures that help in vigorous tree growth. Presence of hard pan below the root zone, accumulation of salts near the roots, and general low fertility level of the soil should be taken care of before planting and during growth of the trees. The diseased twigs should be pruned before the warm rainy season starts. Fallen leaves, twigs, and fruits should also be collected and burnt.

Spray of copper fungicides during the summer and rainy season had been recommended in the past for the control of citrus anthracnose. After pruning the cut ends of the branches should be protected by a fungicidal paste. Fungicides recommended against the anthracnose disease include Bordeaux mixture, copper oxychloride, benomyl, zineb, maneb, mancozeb, chlorothalonil, captafol and folpet. Benomyl and captafol alone or in various combinations have proved the most effective fungicides for the control of post-bloom fruit drop. One to four applications during the bloom period are generally recommended. The number, rather than timing, of applications during the bloom period are important. When blossom blight incidence is low only one application of benomyl or captafol at mid-bloom or two applications, one at early bloom and one at mid-bloom control the blossom blight. In a severe incidence, weekly or 10-day schedules provide high degree of control of blossom blight and button formation (210). In acid lime (*C. aurantifolia*) three sprays of 1% Bordeaux mixture starting from the first week of July, at 21 days intervals, have been found most effective in India. Carbendazim (Bavistin) at 0.1% also gives effective control of die-back.

ARMILLARIA ROOT ROT

Armillaria root rot is of worldwide distribution and affects hundreds of species of fruit trees including citrus, apple, pear and grapevine both in the temperate and tropical regions. It has been described under different names such as 'shoe string root rot', 'mushroom root rot', 'crown rot' and 'oak root fungus disease' (5). The pathogen, *Armillaria mellea*, is a common fungus in forest soils. The disease is common in orchards or vineyards planted in recently cleared forest land. The losses are inconspicuous, appearing as slow decline and death of occasional trees, with greater number of trees dying from the disease during periods of moisture stress and after leaf fall (5).

Symptoms

The above ground symptoms on the tree are similar to those caused by other root diseases. These include loss of tree vigour, reduced growth, smaller, yellowish leaves, die-back of twigs and branches, and gradual or sudden death of the tree, especially when the crown or collar is affected. The disease starts from root infection. The fungus continues to grow underground on the roots but aboveground symptoms appear late. Sometimes, no symptoms are seen but the tree suddenly wilts. Occasionally, only the top of the tree dries or branches on one side dry and rest of the foliage remains healthy. In the orchard, if leaves on the tree are turning yellow and are few and the tree shows poor growth Armillaria root rot can be suspected (55). Initially, scattered trees show symptoms but later they may be found in circular patches.

Specific symptoms can be seen at the foot of the tree or on the roots. The bark shows decayed areas on the collar and the roots. Infected roots are swollen and as the fungus grows they are destroyed. Below the bark, radiating, white, velvety masses of fungus mycelium are seen. This mycelial web soon destroys the bark and wood and grows out on the surface or in surrounding soil as hard, black or deep pink ropes (rhizomorphs). The shape of these growing hyphal strands is somewhat similar to shoe laces. At the apical portion of this growth the fungus forms mushrooms or carpophores. These fruiting bodies of the fungus can be seen on the lower trunk or around the trunk on the soil. Soon these fruiting structures break and fall down on the ground. By this time most of the roots are destroyed. As a result of infection, the trees either suddenly wilt or signs of slow decline appear. This depends on the number of roots destroyed by the fungus.

The causal organism

Armillaria mellea (Vahl ex. Fr.) P. Kaarst (syn. *Armillariella mellea*) belongs to the mushroom group of fungi (Agaricales). The mycelium is white, velvety, and radiating in a fan-like manner. The individual hyphae are thin and incapable of penetrating the host tissue. They aggregate into thick strands of hyphae (rhizomorphs) which are 1-3 mm thick and only these bundles of hyphae can cause infection of the host. They consist of a compact outer layer (sheath) of black mycelium and a core of white or colourless mycelium. Often the rhizomorphs can form branched network on the roots, under the bark, or in severely decayed wood. Some strands spread out in to the soil surrounding the roots. These rhizomorphs enter the roots through medullary rays. At the base of dead or drying trees the apical portion of the rhizomorphs develops the mushroom or the b-sidiocarp (the fruiting body of the fungus). These could be few or few to many, honey coloured, speckled and 9 or more cm tall with a cap (pileus) of 5-15 cm diameter. The lower surface of the pileus contains numerous gills in which basidiospores are formed. The basidia are elongate clavate, 34-47 × 5-9 μm in size and bear four sterigmata which are up to 6 μm long. The basidiospores, borne on sterigmata, are short, ovoid to ellipsoid, and measure 7-12 × 5-7.5 μm. These basidiospores cannot infect living tissues of the host. They fall on dead wood, stumps, or injured surface and develop the mycelium and rhizomorphs. The latter grow and cause infection of roots.

Disease cycle

Disease severity in Armillaria root rot is associated with relative amount of residual woody debris, especially roots, from trees present when the land is converted to new plantations. In orchards where the disease has established, the fungus mainly survives through rhizomorphs in soil or as mycelium and rhizomorphs in infected roots and trunk in which the survival for a minimum of 6 years is reported (80). Basidiospores have little direct role in survival. The principal method of tree to tree spread of the fungus is through rhizomorphs or direct root contact. Rhizomorphs grow from roots of infected trees or from decaying roots or stumps through the soil to roots of adjacent healthy trees. The rhizomorphs must remain connected to a food base such as an infected root or stump in order to grow and to infect other roots. Nutrients are transported from the food base to the rhizomorph tip but not the reverse (174). However, from experimental evidence Morrison (144) suggested that rhizomorphs absorb nutrients from the soil. Soil rich in organic matter supplies more nutrients. Thus, soil is the principal source of nutrients for rhizomorphs. In countries like Zambia and Zimbabwe, where rhizomorphs are not formed (203) due to toxic materials in the soil, spread of *Armillaria* is only by root contact. Pieces of rhizomorphs can also be dispersed by various cultural operations in the orchard. The pathogen produces antibiotics that protect it from antagonistic fungi. But if the fungus is weakened antibiotic production is less and the mycelium can be destroyed by antagonists like *Trichoderma viride* (230).

The pathogen is active in warm, moist soil and is inhibited by cool and dry soil (176). The optimum temperature for growth of the fungus is 21°-25°C but it causes maximum infection of host roots at 10°-18°C when the growth of roots is slow (16). The optimum for growth of citrus roots is 17°-31°C, similarly for peach and apricot it is 10°-17°C while the optimum for root damage in these hosts is 15°-25°C. This suggests that the severity of the disease depends on adverse effect of temperature or other factors on the host. Host defoliation due to attack of gypsy moth is reported to predispose the trees to infection. Rhizomorphs continue to elongate while covered with a water film, but without the water film, more oxygen reaches the outer cells of the rhizomorph, pigmentation occurs and growth is checked.

Management

Eradication of *Armillaria* from orchards is a difficult and time consuming operation. Losses can be reduced by removing the substrates such as stumps and decayed roots and avoiding or delaying new plantations for several years on land that has been recently cleared from forest trees, especially oak trees. The badly affected trees should be dug out along with all possible root vestiges. A trench around the place of the removed tree should be dug and the area enclosed by the trench should be fumigated with carbon disulphide for several times at 6 months interval until all vestiges of roots have decayed (16). Trenching around lightly infected trees and applying fumigants in the trench also prevent spread of rhizomorphs to healthy roots. Dig and fumigate has been the most commonly adopted practice in citrus orchards in USA.

Soil fumigation with carbon disulphide had been shown to kill *Armillaria* as early as 1914 (160). By 1951, it was noticed that while high dosages of the fumigant killed the fungus in 2-3 days, lower or moderate dosages killed it only after 30-59 days in field soil and not at all in sterilized soil. *Trichoderma viride*, the antagonist was thought to attack *Armillaria* only when the latter was weakened by fumigation (cf. 45). Methyl bromide has similar effect as carbon disulphide. Compared to the pathogen, the antagonist has a high tolerance to the fumigant. Heating the soil to 33°C for 7 days with aerated steam also weakened the pathogen (146, 147).

Temperature of 36°C for 7 days or 43°C for 2 hours kills the pathogen. Since living citrus and peach roots are not injured by such temperatures, heat treatment of planted trees is a potential measure for protecting valuable specimen trees.

■ POST-HARVEST DECAY OF CITRUS FRUITS

Post-harvest rot of citrus fruits, particularly mandarins and oranges, may result from infections of the fruit on the tree such as *Colletotrichum gloeosporioides* causing anthracnose (24), *Phytophthora* spp. causing gummosis root rot and brown rot of fruits, *Phomopsis citri* causing melanose, *Diplodia natelensis* and *Lasiodiplodia* (*Botryodiplodia*) *theobromae* causing die-back. In these cases the pathogens may occur as microscopic surface contaminants of the fruit skin, or in soil and debris on the skin or stem tissues. The infection remains latent and symptomless and the fruit rot develops later when physiology of the fruit permits growth of the pathogen. The post-harvest decay can also result from infection of fruits during harvesting, transport and other handling operations due to spores of fungi (*Penicillium, Aspergillus, Rhizopus,* and *Alternaria*) being present in the atmosphere or on the orchard soil, in the stores and in the containers. The rot of fruits after harvest by one or more of these fungi is responsible for 10-15 % or even up to 30 % loss during transit, storage and in the market. There are many reviews of post-harvest fruit pathology (62, 63, 64, 159, 162).

Blue, Green, and Black Mold Rots

Various species of *Penicillium* cause the blue mold rot (*P. italicum*), and green mold rot (*P. expansum* and *P. digitatum*). These are the most common and the most destructive post-harvest diseases affecting all kinds of citrus fruits throughout the world in addition to apples, pears, quinces and grapes. Although some infection of citrus fruits can occur in the orchard (Penicillium mold of oranges), these molds are essentially post-harvest diseases and account for up to 90 % of the total loss due to various types of rot in transit, storage and in the market.

Penicillium rot at first appears as soft, watery, slightly discoloured spots of varying size on any part of the fruit. The spots are at first rather shallow but quickly become deeper and at room temperature most of the fruit or the whole fruit decays in just a few days. Soon after the decay starts a white mold begins to grow on the skin surface near the centre of the spot. Later, the growth starts producing spores. The sporulating growth has a blue, bluish green, or olive green colour and is usually surrounded by a narrow or wide band of white mycelium with a band of water-

Fig. 1.   Penicillium rot of mandarin orange.

soaked tissue ahead of the mycelium. The surface growth of the fungus develops on spots of any size as long as the air is moist and warm. In cool, dry air the surface growth is rare even when the fruits are totally decayed. Under storage conditions, small tufts of spore bearing hyphal branches appear on the surface of the spots. The quantity of spores produced is enormous and these spores get in to the atmosphere, landing on healthy fruits and spreading the rot. Decaying fruits have a musty odour and in very dry conditions may shrink and mummify while under moist conditions, when secondary fungi and yeasts also enter the fruit, it is reduced to a wet, soft mass.

Penicillium is the imperfect (conidial) stage of the Ascomycetous fungus *Talaromyces* (Eurotiaceae in Plectomycetes of Ascomycotina). Only the conidial stage is common in storage rots. The hyphae are branched and septate. The conidiophores arise in clusters (tufts) forming definite fascicles (coremia). They are smooth, 600-700 μm long in *P. expansum*, often intertwined and asymmetrically branched. The final branches bear conidia in tangled chains. These chains may be up to 200 μm long. The conidia are elliptical and measure 3-3.5 μm in diameter. Their wall is smooth and in mass gives the characteristic colour of the mold.

Penicillium spp. are saprophytes and continue growing and producing spores on decaying organic matter. The spores are present in the atmosphere. These fungi enter the fruit tissue through breaks in the skin (bruises and wounds) and through the lenticels. However, fruit-to-fruit spread during transit can occur without skin injury.

Generally, for infection, the spores must get dead tissue which the bruises and wounds on the fruit skin provide. Once lodged on the wounds the fungus feeds on the dead cells, produces macerating enzymes which kill more cells on which the mycelium advances. The water-soaked band ahead of the growing mycelium represents the area of enzymic action. Although most of the rot is seen during storage and at the market and there is spread during contact between fruits (11), the occurrence of these molds is greater when (i) the fruits are harvested and handled during wet, humid weather than in cool and dry weather, (ii) when fruits are delayed in going in to the storage, (iii) when fruits are injured during handling, (iv) when they are stored for long, and (v) when they are held at warm temperatures after removal from storage. The rot is favoured by high storage temperatures but the fungi continue growing even at temperatures near freezing. The green mold fungus (*P. digitatum*) can cause rot even at 0°-5°C. Some species produce ethylene which diffuses in to the containers and hastens ripening of fruits pre-disposing them to secondary infection and rot. Some species produce mycotoxins such as patulin which cause damage to intestines, kidney and liver. They can affect the nervous system and may also cause cancer.

The black mold rot of citrus fruits caused by *Aspergillus niger* also produces similar symptoms but the surface growth looks black. This fungus produces chains of conidia on small stalks on a vesicle formed at the tip of the conidiophores. In acid lime (*Citrus aurantifolia*) this species is reported to cause 10-15 % loss during transit and storage.

Stem-end Rots

In this type of decay, the rot generally starts from the stem-end of the fruit. The rind becomes soft and lighter or brown colored. If the fruit is cut open, the rot is seen growing in to the core and margins of juice sacs. Later, the entire fruit becomes soft and internal tissues totally rot.

Several fungi are reported to cause stem-end rot in citrus. Most important is *Phomopsis citri* (teleomorph *Diaporthe citri*) which incites the disease known as melanose in the orchards. This fungus causes small, round, black spots on young leaves. The spots have yellow margins. Later, the centre of the spots is raised and becomes rough and brown. Similar lesions are formed on twigs and fruits also. The other fungi which have been reported to cause stem-end rot are *Diplodia natalensis* (36) and *Botryodiplodia* (*Lasiodiplodia*) *theobromae* (199). These two species are described under post-harvest diseases of mango. In *Phomopsis citri* the conidial stage is common on green parts of the tree. Perithecial stage

(*Diaporthe citri*) develops on twigs lying on the ground. Conidia are produced in pycnidia which are scattered or clustered, at first immersed, becoming erumpent, black, conical to lenticular, ostiolate and up to 600 μm in size. Conidiophores are phialidic, hyaline, simple and cylindrical to obclavate. Conidia are hyaline, 1-celled, fusiform to ellipsoidal and 6-10 × 2-3 μm in size. Another type of spores (stylospores) may also be produced. These are hyaline, filiform, 1-celled, curved and often strongly hooked, and measure 20-30 × 0.5-1.0 μm. In its perithecial stage the fungus produces cylindrical to clavate asci containing 8 ascospores which are bi-celled, constricted at the septum, and germinate by 1-2 germ tubes.

In citrus, infection by *Lasiodiplodia* and *Phomopsis* via the floral parts is reported. These fungi colonize the injured peduncle and pedicel of citrus fruits, remain quiescent (latent) under the sepals or in the button (calyx + disc), and enter the fruit only after abscission occurs. Thus, chemically induced delay in abscission reduces stem-end rot. Post-harvest treatment of fruits with 2,4-D prevents abscission of the button. Uninjured fruit is infected by *L. theobromae* only through the exposed surface of the pedicel and pedicel scar. The fungus cannot enter the intact rind even if the fruit is fully ripe.

Alternaria Black Rots

Different species of *Alternaria* cause rot of a variety of fruits before or after harvest. Among citrus fruits, mandarins (*C. reticulata*) and sweet oranges (*C. sinensis*) are particularly susceptible to these rots. Two types of symptom development are seen. In one, spots are seen on the rind of the fruit and rot proceeds to different depths in the fruit. In the other, the fruit outwardly looks healthy without blemishes on the skin but inside the core and/or juice sacs show a black rot.

Fawcett (69) had mentioned the Alternaria rot of oranges and lemons caused by *Alternaria citri* Ellis and Pierce. Subsequently, Keily (106) described a brown spot disease of mandarins caused by *A. citri* in Australia. In India, Alternaria rot of sweet oranges and lemons caused by *A. citri* was reported in 1967 (2). A black core rot of mandarins in India was attributed to *A. tenuis* Auct (189). Logrieco *et al.* (121) have reported a similar rot of mandarins caused by *A. alternata* (Fr.:Fr.) Keissel from Italy.

In the infection of *A. citri*, brown spots appear on the rind, especially at the stem-end. These lesions increase in size to 2-3 cm. Dark blue or black growth of the fungus develops on the spots. The decay grows in to the fruit and destroys the tissues up to the core. In the core rot of mandarins caused by *A. tenuis* (189), initially the fruit looks healthy from outside. The fungal growth is seen only in the core where the empty space is filled with dark bluish growth of the fungus. Sometimes, there is

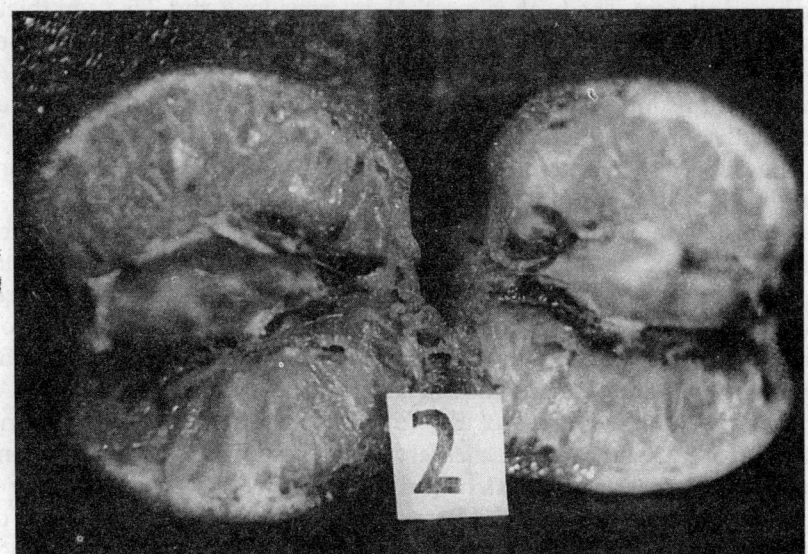

Fig. 2. Alternaria core rot of orange.

browning of the skin at the stylar end. When the fruit is opened, the rot appears to grow from the stylar-end or stem-end towards the core. The decay gradually spreads and destroys the juice sacs. After prolonged storage, when the internal tissues are destroyed, the decay reaches the rind and produces brown spots. In the rot of mandarins caused by *A. alternata* the symptoms are more or less similar to as described above but the rot is more in and around the juice sacs than in the core. In the first stage of the disease, fruits do not show any symptom on the outside but later the surface turns dark starting from the stem-end. In the advanced stage of the disease, the fruits generally fall down from the tree. In this case two strains of the fungus are involved. One produces a dark coloured growth and the other produces a grey coloured growth. In the dark strain there is heavy sporulation. The species is very similar to *A. citri* and produces several mycotoxins (121).

Alternaria belongs to Dematiaceae of Hyphomycetes in Deuteromycotina. In *A. citri* conidiophores are simple or branched, straight or flexuous, septate, light or olivaceous brown, up to 3-4 μm thick and with a terminal or 1-2 lateral scars. Conidia are solitary or in simple or branched chains of 2-7, straight or slightly curved, commonly obclavate or ovoid, pale- to mid- or sometimes dark-brown or olivaceous brown, with up to 9 transverse and numerous longitudinal or oblique septa. They are constricted at the septa. The conidia measure 8-60 (42) μm long including the beak when the latter is present and 6-24 (17) μm thick at the broadest part. The beak is mostly 8 microns long and 2.5-4 μm thick, hyaline or pale brown. In *A. tenuis* the conidia are similar but measure 14-45 (34.5) × 10-15 (13.4) μm with a beak length of 1-12 (3.6) μm (189). *A. alternata* forms long chains of conidia with a longer beak. It also produces chlamydospore in culture.

These species have a large host range and can survive through resistant conidia and as mycelium on plant debris. *A. tenuis* can survive on a suitable substrate as mycelium for 12 years. Conidia can also survive for many years. Spores are dispersed by wind and rains and reach the fruits on the trees. Where symptoms first appear as spots on the fruit surface, the infection is through wounds during handling operations. In core rot, latent infection of fruits occurs on the tree. In core rot of mandarins the optimum temperature range for invasion of fruits is 25°-30°C (190). At 20°C there is little growth of the fungus in the fruit and there is no infection at 5°C and 36°C. *A. alternata* grows best at 28°C. The pathogens cause early infection of the stem end in the growing season and enter a latent or quiescent state in the styler end soon after the initial stages of infection are completed. After harvest fruits gradually lose their resistance as they ripen. Generally, the problem of Alternaria rots is with ripe or over-ripe fruits and when there are wounds on the skin or when the tissue is weakened.

Sour Rot

Sour rot caused by *Geotrichum candidum* Link: Pers (31) is reported on many fruits such as citrus, apples, plum and litchi and on many vegetables including tomato and beans. It is a major post-harvest disease throughout the world. In India, the disease was first described in 1978 (186).

Infected areas on the fruit appear water-soaked and soft and the skin is easily punctured in the affected area. The decay spreads very rapidly, at first mainly inside the fruit, and eventually invades the whole fruit. Later, the skin frequently cracks over the affected area revealing a white, cheesy or scum-like fungal growth. A compact, cream-coloured fungal growth develops on the surface also. The entire inside of the fruit becomes a sour-smelling, decayed watery mass. Fruit flies attracted to the decaying tissue carry the fungus to other fruits.

Conidia of the fungus are aseptate, hyaline, with a thick and smooth wall. They measure (on citrus) 6.6-12.8 × 3.4-5.3 μm (126). The fungus is widely present in the soil and in decaying fruits and vegetables. Infection of healthy fruits occurs during or after harvest through stem scars, skin cracks, cuts and punctures of various types and by contact with decaying fruits (105). Mature green fruits are resistant to infection but can be artificially infected if submerged in water for 24-36 hours before storage (44). Pericarp of the fruit acts as a barrier against infection (122) hence only cuts and punctures or natural scars deep enough to allow entry into the endocarp favour infection. The disease develops rapidly on tree-ripe fruits (12, 61) especially if harvested when the rind is highly turgid and the fruits are kept in moisture-holding plastic bags or packages. Fruit ripeness, nutrition, pH, high water content in the rind, temperatures of 25°-30° C, and high atmospheric humidity favour infection and rapid decay. Low oxygen and high carbon dioxide environment stimulates growth of the fungus. Synergism of *G. candidum* and *P. digitatum* in infected citrus fruits is also reported (143).

CONTROL OF POST-HARVEST DECAY OF CITRUS FRUITS

The approaches to management of citrus decay can be divided into two parts: pre-harvest and post-harvest. The pre-harvest treatments involve management of such tree diseases as anthracnose, gummosis and brown rot which also cause post-harvest rot, and tree and orchard sanitation. In post-harvest treatments of the fruits, chemical, physical and biological control methods are involved. Integration of different approaches ensures better management of fruit rots.

1. Cultural and Sanitary Precautions

The preliminary precautions necessary for maintaining tree vigour involve proper selection of the site for planting trees, proper routine fertilization or manuring, and proper water management. These steps avoid tree stress and provide the right type of tissue strength to the fruits. The first step in sanitary precautions should be proper selection of planting stock. It should be free from infection or inoculum of any fungus, bacterial and virus or viroid disease, particularly anthracnose, root rot, and tristeza. Infected parts of the tree should be pruned off and all types of plant debris from the orchard floor should be removed and destroyed. The cut ends of the branches should be protected by a fungicidal paste and routine sprays of fungicides for control of tree diseases should follow the pruning operation.

2. Picking and Harvesting

Wet and cloudy weather is not suitable for harvest of fruits. Maximum care should be taken to avoid cuts and bruises on the fruits during packing and transporting them from the orchard. Some training of the workers in this regard is essential. The boxes or baskets for transport should be clean, preferably sanitized by some anti-microbial solution. In some diseases, a fungicidal spray just at or a week before harvest is recommended (26). At the time of harvest fruits should not be over-ripe. All damaged, over-ripe, and soft fruits should be segregated in the orchard itself. Such fruits can be disposed off in the orchard or packed separately. The time gap between picking and storage should be minimum. A temperature of 10°C in store for short or long duration storage is considered ideal for avoiding most of the storage rots. In advanced economies, storage and transport under low oxygen (5%) or increased carbon dioxide levels (5-20%) have been used to suppress respiration of both the host and the pathogen thereby suppressing post-harvest rots. The results are further improved by the addition of 10% carbon monoxide.

3. Post-harvest Treatments of Fruits

This aspect of post-harvest fruit and vegetable pathology has been extensively investigated in many countries. Physical treatments (gamma radiation, heat removal or chilling, low temperature storage), chemical treatments (dips or sprays with fungicides and antibiotics) and biological protection (use of antagonists as fruit dip or spray) have been used.

Physical treatments affect both the pathogen and the host. Low temperature halts growth of the pathogen and also slows down the ripening of fruits. However, post-treatment susceptibility to fresh infections also occurs following heating, refrigeration and irradiation. Various types of radiation such as cathode rays, ultra violet, X-ray, and gamma ray have been tried and found to be highly fungicidal. Although all kinds of UV radiation (wave lengths above 280 NM) damage plant DNA and physiological activities, radiation below 280 NM at low dosages induce resistance in citrus, apples and peaches against post-harvest storage rots, and improve the shelf life of fruits (57, 122, 123). Gamma ray treatment (13) has been found better than others because of better penetrability. Gamma radiation stops growth of the pathogen by impairing mitotic cell division. It has been used against Penicillium rot in citrus. Irradiation is a costly process and its facility is not available everywhere. It is recommended only in serious disease problems.

It is always desirable to store fruits at the lowest temperature that does not harm them. For citrus fruits temperatures around 10°C are satisfactory. Many molds continue to grow even at lower temperatures but the decay is considerably slowed down till disposal in the market or until consumption in the homes. Disease development could be stopped by storage at temperatures above the maximum for growth of the pathogen. This is possible when the pathogen has a relatively low maximum temperature for growth such as *Monilinia* on pome and stone fruits. The citrus mold fungi have relatively high temperature maxima for growth. Storage above these temperatures may spoil the fruits. Brief exposure of fruits to heat through hot water is, therefore, recommended.

The temperature-time requirement for hot water treatment of citrus fruits (10) is given below:

Grapefruit   *Phytophthora*   48°C, 3 min dip
Lemon   *P. expansum* and *Phytophthora* sp.   52°C, 5-10 min dip
Oranges   *Diplodia, Phomopsis, Phytophthora*   53°C, 5 min dip

Hot water treatment of oranges causes poor degreening. Limitations in hot water treatment of lemons were mentioned with brown rot control measures.

4. Chemical Fruit Treatments

Field sprays of fungicides, especially before harvest, not only manage the associated diseases of leaves, twigs, stem and blossoms but also check incipient infections of developing fruits before harvest, which later cause decay of ripening or ripe fruits (*Diplodia, Phomopsis, Colletotrichum, Botryodiplodia*, etc.). These decay fungi can also be checked to some extent by post-harvest chemical treatments of fruits.

With pathogens which attack fruits during or after harvest, the decay can be controlled by use of chemicals to prevent infection and suppress development of the pathogens on the surface of the fruit. Most commonly used chemicals and commercial fungicides have been borax, diphenyl, sodium orthophenyl phenate, dichloran (DCNA), thiabendazole (28, 134), benomyl, thiophenate methyl, benzimidazole, captan, iprodione, vinclozolin, imazalil (27) and triforine. Some fungicides such as dichloran and biphenyl and acetaldehyde vapours, ammonia-emitting or nitrogen trichloride-forming chemicals are used as in-package fungistats impregnated in paper sheets during storage and transport.

The effectiveness of a fungicide in post-harvest treatment depends on the depth of inoculation, growth rate of the pathogen, susceptibility of the fruit to infection, temperature and humidity, and depth to which the inhibitory concentration of the fungicide can penetrate. *P. digitatum* on citrus fruits grows at a rather slower rate than *Rhizopus* on some other fruits. An effective fungicide will satisfactorily control decay by *Penicillium* if applied within 24-36 hours of infection at room temperatures. The fungicides that can penetrate the rind to some depth or are locally systemic are better than fungicides which have only surface action. Development of resistance to systemic fungicides and even some protectant fungicides is common among molds and may cause failure of the treatment. Therefore, precautions must be taken to include, additional, preferably broad spectrum, fungicides in the control programmes.

Borax was among the first chemical treatments employed to control post-harvest decay of citrus fruits. Penicillium rot and Diplodia or Phomopsis-rot of oranges is controlled by a 5 min dip in 6-8% alkaline suspension of borax. The effectiveness is improved by heating the solution to 43.5°C. Applicability of borax treatment is limited by its low solubility and disposal of toxic water after fruit wash since fruits must be washed to remove borax residue. Subsequently, a widely used combination was 4% borax + 2 % boric acid at 43.5°C. Later, borax was replaced by sodium ortho-phenylphenate as a more effective chemical against Penicillium and stem-end rots of citrus fruits.

Chlorine has been used in water for washing the fruits and destroying superficially present fungal propagules. A 10 min dip in 2% bleaching powder solution is reported to keep the citrus fruits free from green mold decay for a long time. However, chlorine is not effective against established infections. It is generally recommended only for purifying the water used for washing fruits, containers, etc. in large establishments to reduce inoculum load.

Singh and Khanna (191) had reported that very low concentrations of copper sulphate are toxic to *Alternaria tenuis* causing black core rot of mandarins. They had also reported that zinc sulphate, ferrous sulphate, and boric acid also suppress spore germination. Kumar and Grover (116) had recom....ended fruit dip in Bordeaux mixture against Alternaria rot of oranges. The disadvantage with copper based fungicides is that they leave stains on the fruit skin.

Among the heterocyclic nitrogen compounds or dicarboximides, Captan considerably delays the blue and green mold rots of citrus fruits if used at 0.2% (10 min dip). Fruit dip for oranges in 5 ppm Phaltan (folpet) is effective against *A. citri* (116). The wettable powders are generally not recommended for post-harvest fruit treatment because they leave visible residue on the fruits.

Biphenyl (diphenyl) is a major fungicide for treatments of citrus fruits in many ways. The crystalline compound sublimes in the packed containers, fumigating the fruits during the entire period of transport and storage. It has a strong fungistatic effect against *P. digitatum, P. italicum, Diplodia natalensis, Phomopsis citri*, etc. The growth and sporulation of *Penicillium* on fruit surface is inhibited and the spread of decay to adjacent fruits is prevented in the containers. However, biphenyl does not affect bacteria, yeasts, Phycomycetous fungi and the resistant strains of *Penicillium* and *Diplodia*. It also does not control decay caused by *Geotrichum candidum* (sour rot), *A. citri* (black rot) and *C. gloeosporioides* (anthracnose rot). The most common use of biphenyl has been in the preparation of fungicide-impregnated paper sheets for packing, one sheet at the floor of packing case and another at the top.

Isolates of *Penicillium* sp. and *Diplodia natalensis* may develop stable resistance to biphenyl by constant exposure to this compound and failure of the treatment has been reported (98, 198). Biphenyl-resistant strains of

Penicillium may also develop by exposure to the chemically related orthophenylphenol and orthophenylphenol-resistance can develop by prolonged exposure to biphenyl. Pre-storage treatment of lemons with orthophenylphenate is now discontinued in USA.

Although biphenyl has been highly successful in controlling major fruit decays of citrus and is still used worldwide, there have been certain reservations about its use from the very beginning. One is the odour it leaves on treated fruits which the consumer does not like. This odour is temporary and dissipates in a few days. The other adverse effect is that it accelerates physiological decline of the fruit button (calyx + receptacle). This predisposes lemon fruits to attack of *A. citri*. In spite of these drawbacks use of biphenyl has continued throughout the world. The residue remaining on the fruit is safe from a toxicological viewpoint and the intensity of odour has no relationship with residue on the fruit (64).

Orthophenylphenols and sodium orthophenyl phenate came in to use for better volatility than biphenyl, reducing the unpleasant odour and for the same type of decay control as biphenyl. Citrus fruits are submerged or flooded with 0.5% alkaline (pH 11.5-11.8) solutions of sodium ortho-phenylphenate or the fungicide is incorporated in wax coating of fruits. In solution treatments the fruits are washed to prevent phytotoxicity.

To combat strains resistant to biphenyl and orthophenylphenol, the systemic fungicides of benzimidazole group such as thiabendazole (TBZ), benomyl (Benlate), carbendazim or MBC (Bavistin), thiophanates, and some of the sterol-biosynthesis inhibiting (SBI) triazole fungicides such as triforine were introduced for post-harvest fruit treatment. TBZ was first used in 1967 against *P. digitatum* and benomyl in 1969 against *P. italicum*. Thiabendazole (0.075-0.1%) was used as fruit dip or as fruit wax preparation against Penicillium decay of oranges (28). Residue of thiabendazole on the surface of citrus fruits sprayed with 0.1% suspension prevents sporulation of *P. digitatum* on decaying fruits (64). Generally, the same concentrations of benomyl or carbendazim are also used. Benomyl has been used as pre-harvest grove spray and found effective against Penicillium rot. It persists in orange fruits (26).

In 1980s, imazalil (Fungaflor), an imidazole SBI fungicide, was found effective against *P. digitatum* on citrus fruits as 15 sec dip in 1000 μg/L water. It prevents production of spores on infected fruit which could contaminate and rot the healthy fruits in packing cartons. The fungicide is systemic in fruits and penetrates the rind to a depth of 2 mm and prevents *P. digitatum* from invading fruits through injuries occurring after harvest (27, 29). The penetration is better while the fruit is still wet after treatment. Imazalil is also added to fruit waxes and applied to fruits as non-recovery sprays with good results. However, it has been shown that imazalil controls green mold (*P. digitatum*) better when used in water solution than when used in waxes which keep a major portion of the fungicide bound and prevent its penetration (25, 27, 29). Smilanick, *et al.* (194, 196) have reported that warm water suspension of imazalil is more effective. Antisporulant activity of the fungicide at 500 $\mu g/ml$ in water solution at 37.8°C is superior to 4200 $\mu g/ml$ in wax. Fenpropimorph, a morpholine SBI fungicides, is also effective against citrus sour rot and green and blue molds.

With the use of systemic fungicides, resistance to the above named fungicides in the fruit rot pathogens has also been reported. Bus, *et al.* (30) conducted a study of isolates of *P. digitatum* and *P. italicum* from mandarin (*C. reticulata*), orange (*C. sinensis*), lemon (*C. lemon*) and grapefruit (*C. paradisi*) from different geographic regions and found 37% isolates resistant to TBZ (10 mg/L), 34% resistant to benomyl (10 mg/L) and 17% resistant to imazalil (0.2 mg/L). Ninety per cent of isolates resistant to TBZ were resistant to benomyl also and 13% of these were resistant to imazalil. Eckert, *et al.* (65) also have reported resistance to imazalil in *P. digitatum*. Strains of *P. italicum* resistant to benzimidazoles are capable of surviving with sensitive strains even in the absence fungicide selection pressure (222). Khilare and Gangawane (109) have used medicinal plant extracts against *P. digitatum* (green mold of sweet orange). While extracts of margosa alone provided significant control, combination of plant extracts with thiophanate also significantly improved the efficacy of the latter against thiophanate-resistant strains.

Sodium carbonate (soda ash), though less toxic than borax to conidia of *P. digitatum* and *P. italicum*, is also used at 3-5% concentration for fruit treatment. It leaves sufficient alkaline residues in potential infection sites of the fruit surface to prevent establishment of the pathogens. Soda ash (sodium carbonate) treatments are equal to or superior to imazalil treatments. Soda ash reduces the green mold rot by more than 90% when applied to lemon 48 hours after inoculation (195). The most effective control is obtained by 1-2 min dip of fruits in hot (40.6° or 43.3°C) solution of 4-6% sodium carbonate even when the treatment is given 24 hours after inoculation of the fruit with *P. digitatum* (191). The use of soda ash for post-harvest fruit treatment against fungal decay is economical and non-hazardous. This treatment at high temperature (48°C) may pre-dispose lemon fruits to decay without visible symptoms.

Use of plant growth regulators in treatments aimed at host-pathogen interaction to control post-harvest decay of citrus and other fruits has been reported. Spores of *A. citri* are present under the buttons (calyx + receptacle at the stem-end) of most lemon fruits at the time of harvest.

Senescence of this part of the fruit precedes the onset of Alternaria stem-end rot during storage (*cf.* 64). It has been demonstrated that 2,4-D and 2,4,5-T at concentrations of 100-1000 ppm in a wax emulsion applied to lemons after harvest reduce button deterioration, rate of correlation, and water loss by the fruit during storage. Since these chemicals are only weakly fungistatic for *A. citri*, the control of Alternaria rot is attributed to physiologic action on the fruit tissue. The albido of the rind of lemons possesses considerable resistance to *A. citri* but during ripening this resistance is decreased. Pre-storage treatment of fruits with 2,4-D delays this loss of resistance by the fruit rind. Half min dip in 25 ppm solution of 2,4-D with 2% wax solution is reported to give protection to citrus fruits against stem-end rot (*Diplodia, Botryodiplodia, Phomopsis, Alternaria*) and also blue, green and black molds. Combination of 2,4-D and Bavistin used as 1 min dip is an effective safeguard against storage decay of sweet orange. Gibberellic acid also delays maturity and senescence of rind of oranges on trees sprays with this chemical before harvest. In China, the antitranspirant compound called 'gao-zhi-mo' was used as dip for effective control (upto 82%) of storage decay of sweet oranges (97).

5. Biological Control of Fruit Rots

Biological controls have been developed that are effective against major post-harvest pathogens of citrus (and other fruits and vegetables). Fungal and bacterial antagonists used as alternatives to fungicides for control of fruit rots of citrus, apples, peaches, pears, and cherries include *Bacillus subtilis, Enterobacter cloacae, Pseudomonas cepacia, Pseudomonas syringae, Trichoderma* spp., *Acremonium brevae* and several species of yeasts (237). Their commercial application is limited by lack of proper formulations, except for *Bacillus subtilis*. *B. subtilis* as a biocontrol agent (fruit dip in cell suspension) can inhibit 10 citrus fruit pathogens by antibiotic production (192). These include *Lasiodiplodia theobromae*. *Trichoderma viride* is effective against Penicillium rots of citrus. Application of *Pseudomonas cepacia* (as fruit dip) to lemons after harvest gives 80% control of green mold without any visible injury to the fruit. *Pseudomonas fluorescens*, although ineffective against *P. digitatum in vitro*, reduces decay of lemons by 70% (193). The yeast, *Debaromyces hansenii* (now renamed *Candida guilliermondii*), when applied before infection can protect citrus fruits against green and blue molds and sour rot for 21 days at 11°C. The yeast is more effective against green mold than the other two rots (37, 141). The efficacy of the treatment is enhanced if the yeast suspension is prepared in 2% calcium chloride solution. The yeasts control fruit rot pathogen by competition for nutrition and, to some extent, by pathogen cell wall degradation (135). Glucanase activity as the main mechanism is reported.

■ CITRUS CANKER

Citrus canker or Asiatic citrus canker (also known as true canker or A-form canker) is a widespread disease in all the citrus growing areas of the world. It is reported to have originated from China but in the herbaria of the Royal Botanic Gardens Kew, England canker lesions have been detected in *Citrus medica* specimen collected from India as early as 1827-1831 and in *Citrus aurantifolia* specimen collected from Indonesia in 1842-1844 (*cf.* 70, 83). Thus, the origin of the disease is supposed to be in the tropical areas of Asia, such as South China, Indonesia and India where *Citrus* species are presumed to have originated. The pathogen was distributed through planting material and spread to Europe and to USA (in 1910) and to other citrus growing areas of the world. Citrus canker is found in Africa, Asia, Australia, Oceania and North and South America. It is a major disease of citrus in India, China, Japan and Java. It is economically important because fruit lesions downgrade the appearance of fruits and when severe, cause premature fruit drop. Heavy foliage infection causes severe defoliation, leaving only bare twigs. Severe infections of newly planted stock may cause delay in growth and can also be fatal.

Although the disease was once reported to have been eradicated from Australia, New Zealand, South Africa and the United States, its reappearance was reported during the 1980s in some parts of Australia, Mexico and Florida. In the state of Florida in the USA where it was first recognized as a new disease in 1913, it became so severe that mass eradication of diseased trees and nursery stock, often the entire orchard, had to be undertaken to eradicate the pathogen from the state in about 10 yrs (100). It was claimed that citrus canker had been completely eliminated from USA by 1949. However, it has reappeared in Florida since 1984 and the same eradication measures had to be started. The form of citrus canker which reappeared in Mexico and Florida seemed to be different from that identified in Asia (the A-form canker).

Symptoms

The disease occurs on leaves, twigs, thorns, older branches and fruits as necrotic brown spots with rough surface. Fawcett (68) had mentioned canker incidence on exposed roots. In the tropical climate of south India, Reddy and Naidu (168) reported canker lesions on roots of 5 yrs old Kagzi lime (*C. aurantifolia*) seedling plants upto a depth of 70 cm and in a 20 cm radius. This root infection caused decline of the plants.

Leaf lesions first appear as small, round, watery, translucent spots. They are raised and become yellowish brown. They first develop on the lower surface of the leaf and then on both the surfaces. As the disease

Fig. 3. Citrus canker on leaves.

advances the surface of the spots becomes white or greyish and finally ruptures in the centre giving a rough, corky, or canker-like appearance. The spots increase in size (1 mm to 1 cm in dia) and may coalesce to form elongated lesions on fruits and twigs. The rough lesions are surrounded by a yellowish-brown to green raised margin and watery yellow halo. Spots occurring on petioles and midrib cause premature defoliation. On larger branches the cankers are irregular, more rough and more prominent. Cankers on fruits are similar to those on leaves except that the yellow halo is not visible and a crater-like depression in the centre is more prominent. The injury to fruits is only skin deep and no visible effect on pulp or juice is noticed. Cankers on twigs cause them to break. The leaves during their early stage of formation and fruits of about 2-4 cm diameter are most susceptible to infection by the bacterium (88). As the fruits increase in size they become resistant but water soaking and lesion formation continues to occur as long as the fruit is expanding.

Histopathology
Citrus canker lesions are characterized by over-development of parenchymatous tissues, each consisting of a large number of hypertrophic cells and a limited number of hyperplastic cells. In the early stages of invasion by the bacterium, the spongy cells near the site of infection show increased size as well as increase in the amount of cytoplasm, followed by rapid enlargement. The hypertrophic cells occupy the intercellular spaces. As the cells further increase in size the callus tissue expands, lifting the epidermis above the leaf surface and finally causing

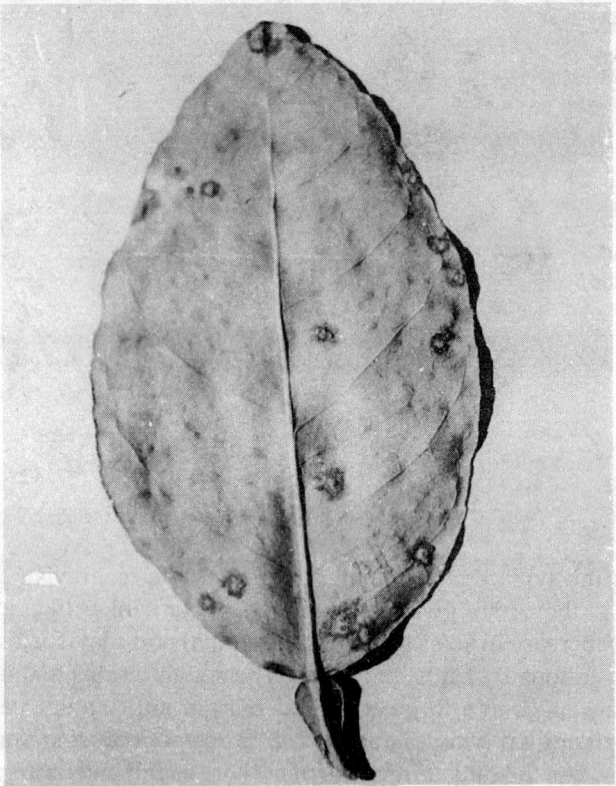

Fig. 4. Citrus canker. Close up of pustules.

disrution of the epidermis, exposing the internal callus tissue. Hyperplasia usually occurs in a few cells adjacent to the healthy tissue. The hyperplastic cells develop into hypertrophic cells without continuous cell division (83).

The Causal Organism

The bacterium causing citrus canker was known as *Xanthomonas campestris* pv. *citri* (Hasse) Dye [*X. citri* (Hasse) Dowson]. Gabriel *et al.* (76) had proposed reinstatement of the name *Xanthomonas citri*. However, Young *et al.* (239) had cautioned against the proposal. In the reclassification of *Xanthomonas*, Vauterin *et al.* (227) finally gave the name *Xanthomonas axonopodis* pv. *citri* (Hasse) Vauterin *et al.* A bacterial leaf spot of citrus is caused by *X. campestris* pv. *citrumelo* which was earlier considered a strain of *X. citri*.

The cells of the bacterium are rod-shaped and measure 0.5-0.75 × 1.5-2.0 μm. It forms chains and capsules but no spores and the cells are motile by a single polar flagellum (monotrichous). It is Gram-negative and aerobic. Colonies on beef agar are circular, straw yellow to amber yellow, slightly raised and glistening. The yellow pigment is the characteristic xanthomonadin. Oxidase test is weak or negative. Litmus milk turns blue and milk is peptonized without coagulation. Asparagine is not used as a sole source of nitrogen and carbon. The cells are positive for hydrolysis of starch, aesculin, casein, liquefaction of gelatin and pectate gel, and production of tyrosinase and reducing substances from sucrose and hydrogen sulphide. The bacterium is negative for arginine dihydrolase, nitrate reduction, production of 2-ketoglucanate, acetoin, urease and amino acid dehydrolases and for methyl red test. Growth requires methionine or cysteine and is inhibited by serine. Growth is inhibited by 0.02% trephenyl tetrazolium chloride and by 4% sodium chloride but not by 3% sodium chloride. Xylose, glucose, fructose, sucrose, galactose, mannose, maltose, lactose, trehalose, glycerol, dextrin, starch, malonate citrate, succinate and malate are utilized as the sole source of carbon. L-arabinose, rhamnose, raffinose, malicin, sorbitol, inositol, dulcitol, inulin, gluconate, oxalate, acetate and tartrate are not utilized. Optimal growth temperature is 28°C, minimal 6°-7°C, and maximal 36°-38°C. The doubling time is 79 min.

Fig. 5. Citrus canker on fruits.
Courtesy: Dr. H.B. Singh

Pathotypes within *X. c.* pv. *citri* (*X. axonopodis* pv. *citri*) have been identified by host range, geographic origin, bacteriophage sensitivities, plasmid content, and serology (*cf.* 83, 99). Pathotype A (the A-form canker) has the widest host range and a global distribution. Pathotypes B, C, and D are restricted to lemon (*Citrus lemon*) and lime (*Citrus aurantifolia*) in South America and Mexico. Three phages of the bacterium have been isolated and characterized. The phage susceptibility can be used for rapid identification of the citrus canker pathogen. Strains of citrus canker bacterium carry indigenous plasmids (42). The function of these plasmids is not yet clear. However, plasmids differ in size among different strains (A, B, C). Differentiation based on plasmids (99) is consistent with that observed on the basis of serology and of phage susceptibility.

Disease cycle

Studies have shown that the bacterium of citrus canker has a short life in soil or in fallen leaves. The short longevity in natural soil is attributed to microbial interactions, especially the predatory effect of protozoa (83). This is generally true when the temperature is warm enough to allow the soil microorganisms to compete with the bacterium. Graham *et al.* (90) have reported populations of the bacterium in soil in citrus nurseries in Maryland (USA) and in Argentina, both temperate climate regions. When the source of inoculum is removed (removal of affected leaves and twigs) the survival in soil is considerably reduced. In presence of active host tissue for support such as roots it can survive deep in soil (172). In tissues of fallen diseased leaves and twigs also the bacterium dies quickly. The survival is not for more than 3 weeks if the lesion-bearing leaves and twigs are wetted on the soil surface or are buried at a depth of 3-6 cm. If the plant debris is maintained under dry conditions the survival is increased to 2-3 months.

The extracellular polysachharides (EPS) which form a matrix in which bacterial cells are embedded play an important role in survival of the bacterium. If the EPS dries and is not disturbed the survival of the bacterial cell can be considerably prolonged because EPS forms a protective coat that prevents desiccation of the cells. When the EPS matrix is diluted with water to the level of lower than a million cells/ml the cells are rapidly killed. This lethal dilution effect is found in most xanthomonads and some other bacteria. It also implies that during dispersal by raindrop splashes the bacterial cells must carry with them sufficient EPS to remain viable (83).

Although the bacterium has been detected on certain weeds, this source does not support prolonged suvival. Epiphytic populations on citrus leaves away from lesions also quickly decline. However, since the host is a perennial plant the cankers on it can support parasitic survival of the pathogen indefinitely once the plant gets infected (49, 82, 83). In temperate climate countries where the citrus trees undergo dormancy during winter, survival in holdover cankers on the trees is most important. In the subtropical and tropical countries such as in India attacked twigs bearing old lesions on the standing trees are the main source of perennation of the pathogen.

According to Timmer *et al*. (208) when water is placed on lesions of detached leaves 10 to 100 thousand bacteria/ml from each lesion are released immediately. Exudation continues at a high level for 24 hr. Cumulative release is 100 thousand to 1000 thousand cells per lesion. Fewer bacteria are released and exuded more slowly from old lesions than from young lesions. The concentration of bacterial cells is largely dependent on the age of the lesions. In fresh lesions, bacterial density often reaches 10-100 million/drop (83). Individual canker lesions ranging 3-9 mm in diameter contain 1-10 million bacteria per lesion and lesions formed in spring flush continue to contain this level through summer and the rains (212). In winter the numbers decline.

In dispersal of the bacteria by rain, either the water running on the host surface or splashes of raindrops disperse the bacterium (49). In citrus nurseries with citrus canker, dissemination of the bacterium is primarily by splash dispersal. Rains driven by wind velocity in excess of 8 m/sec aid in dispersal (88). The bacterial numbers per ml of canker washings and their dispersal by rain-driven water splashes are, to some extent, effective as source of inoculum only if sufficient quantity of EPS accompanies the bacterial cells otherwise they may be desiccated before reaching the site for infection.

The role of strong winds has great importance in the epidemic of citrus canker. Strong winds cause many injuries to leaves and twigs. The nature of wounds varies from easily visible large wounds to small, invisible ones, such as small scratches or removal of the cuticle edges extending over the stomata. However, a single cell of the bacterium is enough to cause infection and the size of wounds does not matter. Storms help in long distance dispersal of raindrop-borne bacterial cells. In addition to driving rains, insects such as the citrus leaf minors (*Phyllocnistis citrella* and *Thosconyrsa citri*) also help in dissemination of the bacterium. Venkataswarlu and Ramapandu (228) have reported that the percentage of leaves affected by canker in presence of injury by leaf minors was 26 to 48 while in absence of the leaf minor injury it was 3 to 10. Man himself is the chief agent of dissemination and introduction of the pathogen into new localities through the transfer of infected nursery stock.

The result of numerous cells reaching various sites on host surface is the development of numerous secondary foci which later coalesce to form larger patches. The bacterium enters the host through natural openings (stomata) and through wounds such as those caused by insects, movements of thorns, etc. Susceptibility of leaves to infection through wounds is maximum when the ratio of leaf length at a given time to the length of the fully expanded leaf is 0.8 : 0.9 or more (*cf.* 83). Wounds sustained early in spring or late autumn take longer to heal and, therefore, expose the injured tissue to infection for a longer period of time. The greater the number and size of stomata per unit area, the greater the susceptibility of the organ. But the stomatal invasion by the bacterium is governed by developmental stage of the organ. In young organs such as leaves, stems, and fruits, the front cavity of the stomata has a wide opening because the thin cuticular layer of the epidermis is not enough to elongate the edges. As organs approach maturity and the tissues become harder, the cuticular layer of the epidermis becomes thicker so that the edges develop over the stoma, leaving a narrow opening between them. The slit is so narrow that surface tension prevents entry of rainwater carrying the pathogen into the opening of the mature stoma. Thus, availability of young stoma determines the susceptibility of leaves, stems and fruits. In very young leaves, just after emergence, the stoma are immature with no opening and, therefore, only slight infection occurs.

After entry, the bacteria multiply rapidly in the intercellular spaces, dissolve the middle lamella, and establish in the cortical region. The bacterial cells which enter the intercellular spaces adhere to the host cell walls through and interaction between EPS and citrus agglutinins (204, 205). The citrus agglutinins contain 96% proteins and 4% carbohydrates. They are active with EPS of various xanthomonads at pH lower than 6.0. The EPS induces localized water congestion enhancing the growth of the bacterium in the intercellular spaces.

In association with EPS, the bacteria show ethylene biosynthesis for several hours after inoculation (84, 85). Continuance of ethylene production is followed by leakage of electrolytes (86) and amino acids from the cells indicating damage to cell membrane. A large amount of ethylene is also produced after canker symptoms develop. At this stage ethylene production originates in the hypertrophic host cells within the canker lesions as well as the cells in the peripheral zones, which appear to be under the influence of auxin and sometimes form yellow halo. The high level of ethylene produced at this stage induces the formation of abscission layer at the base of the leaf petiole, resulting in defoliation. Presence of antimicrobial compounds (phytoalexins) in citrus cells has been reported. However, these compounds are present inside the cells and are not leaked into the intercellular spaces to act against the bacterium. Citrusnin-A is one such compound (231).

The disease is favoured by mild temperatures and wet weather. Temperatures between 20° and 30° C with good evenly distributed rains are most suitable. Presence of free moisture on the host surface for at least 20 min is essential for successful infection. The size, density, and age of stomata determine susceptibility and resistance of a citrus cultivar. Under unbalanced conditions of excess nitrogen, citrus trees produce more shoots allowing an increased number of large and tender leaves which bear larger lesions and cankers.

Management of the disease

The only effective method of control of citrus canker is complete destruction of the affected trees by burning (56). Though drastic and costly, this method has proved its efficacy in USA, Australia, South Africa, New Zealand, and Brazil. The new eradication programme followed in Florida, after reappearance of canker involves 1) burning of plants in nursery where an infected plant is found, 2) destroying all trees with canker symptoms within orchards and defoliation of surrounding trees, and 3) using fruits from diseased or exposed trees for only processing (89). Similar rigid eradication programmes had been implemented in Australia where the first and subsequent outbreaks of the disease had been eliminated by this method. In spite of total destruction of infected and suspected trees and apparent elimination of the citrus canker from the country, the reappearance or reintroduction of the pathogen in the same country has pointed out the importance and possible failure of quarantine regulations. Two documented examples are of Florida (USA) and Australia. The introduction of Asiatic citrus canker in to Florida during the 1910s was traced to infected trifoliate orange seedlings imported from Japan for use as rootstock (*cf.* 89). The reappearance in around 1984 in Florida was again traced to entry of canker affected citrus. Detection of infected material had been made during 1973-1983 at the ports of entry. In Australia also, the first outbreak of canker in 1912 was attributed to citrus trees imported from Japan and China along with fruits. Mass eradication of trees and rigid quarantine had eliminated the disease but subsequent outbreaks in 1981, 1984, and 1991 are suspected to have originated from illegal importation of citrus into isolated home gardens in one part of the country and subsequent spread to other parts (23). Both these countries had followed the method of mass destruction of trees and imposition of rigid quarantine regulations not only at the international level but also within the country. In India and in other areas where the disease is well established in most orchards, eradication of trees as a control measure is not feasible.

Since wind-driven rains and water soaking of tissues are essential for dissemination and ingress of bacteria and for epidemic development of citrus canker, wind breaks are essential. In absence of this precaution chemical and other methods of disease management remain inadequate (89). Other recommendations made to check the disease are: (i) use of disease free nursery stock for planting in new orchards, (ii) spraying the plants before planting in new orchards with a copper fungicide, and (iii) in old orchards pruning of the affected twigs and spraying with copper fungicides at periodical intervals, especially during the rainy season. The fallen canker affected twigs and leaves should be collected and burnt. Since inoculum present on fallen leaves is reduced when the leaves are buried deep in soil, periodical ploughing of the orchard floor is helpful. The plant vigour should always be maintained by suitable fertilizers and irrigation. Proper care should be taken to check the attack of leaf miners.

Rangaswamy *et al*. (169) had demonstrated that in orchards the disease could be controlled by antibiotic sprays. Streptomycin sulphate or crude agricultural preparations of streptomycin at 100 to 1000 ppm concentration sprayed at 15-day interval effectively checked the disease on 48-yr old lime trees while Bordeaux mixture was ineffective. Phytomycin (2500 ppm) was also effective. Four sprays of streptomycin-100 at 500 ppm are most effective in controlling the disease (114). In general, use of Bordeaux mixture (4 g copper sulphate with 4 g lime in 1 lit water) and Agrimycin or streptomycin is recommended in most countries. In China spraying citrus trees with copper ammonium during the summer and autumn is reported to reduce canker incidence by 86% and 90%, respectively. This control method is environment friendly, easy and cheap. Graham and Gottwald (89) have recommended chemical dip of fruits for shipment to prevent spread of the disease through infected or contaminated fruits. The treatments include 2-min dip in chlorine (200 μg/ml at pH 7) or 1-min dip in 2.0% sodium-*o*-phenylphenate.

Although chemical control has been claimed effective it is not very successful on all occasions. During rainy conditions, some bacterial cells may achieve direct access to the front cavity of stomata or wounds without being exposed to the chemical left on the leaf surface. This direct ingress of even very low number of bacterial cells may make the chemical sprays less effective (83). An effective bactericide against the citrus canker bacterium should not only be effective on the host surface but also reach into the substomatal cavity.

Studies on biological control of citrus canker are in a preliminary stage. A strain of *Pseudomonas syringae* is reported to show anagonism to the citrus canker bacterium and also prevents enlargement of canker lesions as well as subsequent defoliation of infected leaves (161). The antagonist probably stimulates phytoalexin (citrusnins) synthesis in the tissues. Strains of *Pseudomonas fluorescens* also are strong antagonists of *X. citri*. Kalita *et al*. (102) isolated *Bacillus subtilis* and *Aspergillus terreus* from phylloplane of citrus and reported that a strain of *B. subtilis* when sprayed on leaves in high concentration (100 million cfu) reduced canker incidence by 61.9%. A strain of *A. terreus* also reduced disease incidence by 47.5%. *Erwinia herbicola*, a common phylloplane microflora, grows more rapidly than the canker bacterium both *in vitro* and *in vivo* and eventually causes quick decline of the pathogen population (87). However, this bacterium grows only in the area where hypertrophic cells are established, but never in the front boundaries at which the pathogen attacks healthy tissues inducing development of hypertrophic cells. In the state of Andhra Pradesh in India, S. Vaheeduddin had reported in 1959 that spray of neem (margosa) seed cake at the rate of 80 kg/acre is highly effective against citrus canker as well as leaf miner. About 25 kg of the cake is soaked in 100 lit of water and allowed to decompose for a week. It is then sprayed without filtration. Some of the cake falls on the ground and becomes manure. Several sprays are required to produce good results. In experiments, two sprays at 3-week intervals during August-September reduced the disease from 5.8% in unsprayed plots to 2.5% in sprayed plots. Bordeaux mixture was not so effective. This control of canker was probably through enhanced microbial activity on the leaf surface which acted as a biocontrol agent.

Different species of *Citrus* show different degrees of susceptibility to the disease. These differences result from pathogen strain-host species interaction, ability of the host tissue to release phytoalexins in the intercellular spaces, behaviour of the stomata, level of density and size of stomata, presence or absence of thorns that cause injury to leaves and young twigs, etc. Reddy (171) had screened 144 varieties of *Citrus* spp. and related genera and has recognized 13 as immune to the disease.

■ CITRUS GREENING DISEASE

Citrus greening disease is common in Tanzania, South Africa, the Arabian Peninsula, India, Philippines, and many other countries (17, 110, 130, 142, 152, 153). In India, the disease is particularly serious in the northern states (152) and is considered more dangerous than tristeza (41) because of its widespread occurrence. In most trees greening is found along with tristeza. There is a synergistic relationship between tristeza virus and the bacterium of greening and the two are jointly responsible for citrus dieback and quick decline (130).

Symptoms

Citrus greening is characterized by yellows type of symptoms which are highly variable. Leaf chlorosis is the main symptom. It resembles the symptoms of zinc deficiency. Since chlorosis can be caused by nutritional disorders many scientists have claimed control of greening with micronutrient sprays which only temporarily mask the symptoms of greening. In the yellow tissue of the leaf lamina scattered green islands are seen. The leaf veins are also yellow. A characteristic feature of greening is that the yellow areas are surrounded on one side by the midrib and on the other side by lateral veins. The yellowing expands towards the margins. The size of leaves is also reduced. The leaves are thicker than normal and usually remain erect. The internodes of branches are shortened giving a bushy appearance to the branch. Such branches produce an excess of buds and later show die-back. The diseased trees look stunted, flower earlier than the healthy trees and produce smaller fruits. There is a considerable reduction in the number of roots.

The Causal Organism

Citrus greening was considered a virus disease for a long time (140, 150-153). Later, it was considered to be caused by mycoplasma-like organism and some workers claimed its isolation and culture. However, in 1970, D. Lefleche and J.M. Bove' had reported that the disease was caused by a phloem-inhabiting fastidious bacterium with double-membrane cell wall, distinct from MLO cells (51). Later the bacterium was identified as *Liberobacter asiaticum* for Asian citrus greening and *L. africanum* for African citrus greening (Jogousix, Bove' and Garnier. *Mol. Cell Probes* 10: 43. 1996).

The cells of phloem restricted fastidious bacteria are Gram-negative, non-motile, non-pleomorphic, rigid rods measuring 1.0-2.0 × 0.2-0.5 (1.3 × 0.3) μm. These cells are present in mature sieve elements, irregularly distributed among vascular bundles. They are sensitive to different antibiotics including penicillin and tetracyclines. Due to the presence of the cell wall they are more sensitive to penicillin, which interferes with cell-wall synthesis, than to tetracycline.

The citrus greening bacterium is transmitted through vegetative propagation and by two species of citrus psylla : *Diaphorina citri* and *Trioza erytreae* (17, 33). Electron microscopic studies have confirmed the role of *T. erytreae* in transmission of the organism (142). Prevalence of these vectors determines the regional prevalence of greening. After acquiring the bacteria, the vectors remain infective for their whole life. A single individual is enough to transmit the disease. After being acquired by the vector the bacteria have an incubation period of 8-10 days in the vector body before the latter becomes infective. The vector can acquire the bacterium in its larval stage also but cannot transmit it to healthy trees (130). It can do so only after becoming an adult. Transmission of the bacterium from sweet orange trees to periwinkle through dodder has also been reported (77).

Management

An integrated management system involving tree sanitation, production of disease-free nursery stock and eradication of insect vectors (110) is essential for control of the greening disease and also virus and viroid diseases of citrus that are transmitted by planting stock and insects. All badly affected and uneconomical trees should be cut down and destroyed. New plants raised from indexed stock or from nucellar seedlings (234) should be planted. In areas where greening affected trees and the vectors are present such healthy trees also become infected in a few years through the vectors. Eradication of citrus psylla by regular sprays of 0.02% of such insecticides as diazinon, endrin, or parathion reduces the spread of the disease. Kapoor and Cheema (103) and Cheema *et al.* (39) have reported 100% recovery of trees by spraying a mixture of Bavistin (carbendazim) and Ledermycin (500 ppm each) six times at 10 days intervals. Ledermycin alone is not effective while Bavistin alone reduces the disease incidence.

■ CITRUS TRISTEZA VIRUS

Tristeza, meaning sadness, caused by citrus tristeza virus (CTV), has been one of the most destructive diseases of citrus worldwide. The earliest report of the disease is from South Africa, about 1910, when it was considered as a quick decline of citrus scions propagated on sour orange rootstock (235). It was subsequently reported in Java in 1928, Argentina about 1931, Brazil in 1937 and USA in 1939. Later, similar diseases were reported from New Zealand, Australia, West Africa, Sri Lanka and Hawai. Now CTV is known to occur in most of the citrus growing areas of the world. It occurs throughout India and in neighbouring countries. Presence of the disease in India was suspected as early as 1953. It is presumed that the disease was present in the country even much earlier. In India tristeza is commonly associated with the greening disease (151). Synergism between the two has been reported (14, 130, 152). It has been speculated that tristeza originated in the Orient and was distributed worldwide by the movement of citrus budwood and plants. South Africa seems to be the first country to have imported the disease from the Orient (46). According to Wallace (229) nearly half the citrus trees in the world had been destroyed by tristeza up to the year 1959. Comprehensive reviews of the disease and the causal agent are available (7-9, 78, 118).

Fig. 6. Decline of citrus tree.
Courtesy: Dr. H.B. Singh

Symptoms

Tristeza is primarily a disease of sweet orange or other varieties grown on sour orange rootstock (235) and of grapefruit, lime and calamondin. The virus exists as many strains having different biological activities broadly grouped as mild, seedling yellows, decline on sour orange, stem pitting on grapefruit and stem pitting on sweet orange or no symptoms. Except for the mild or no symptom strains, the other categories of biological activity may occur alone or in any combination in a given CTV isolate.

The mild strains produce no noticeable effect on most commercial and commonly grown citrus varieties. In *Citrus aurantifolia* (Kagzi lime) mild strains cause only slight stem pitting, little or no vein clearing and flecking of leaf veins and veinlets. This host is an indicator plant for Indian CTV strains (32). In seedling yellows there is severe chlorosis and dwarfing of seedlings of sour orange (*C. aurantium*), lemon (*C. limon*) and grapefruit (*C. paradisi*). In decline of plants budded on sour orange rootstock, the plants are dwarfed, often chlorotic and show decline. The quick decline of sweet orange (*C. sinensis*), grapefruit or tangerine scions budded on sour orange rootstock can occur within 3-6 weeks. First the leaves turn yellow or golden colour and then they wilt and fall down, leaving only fruits hanging on the dead tree. An overgrowth just above the bud union can often be seen. If a piece of the bark is removed from the bud union, needle-like pegs originating from the bark (phloem) can be seen. Not all decline inducing strains of CTV cause a quick decline. Many cause a decline over a period of several years, causing the trees to stop growing and become less productive.

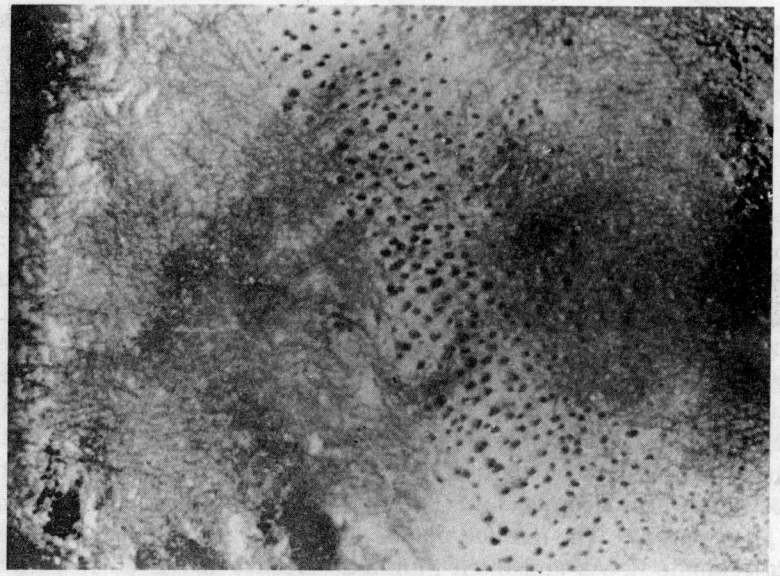

Fig. 7. Tristeza. Stem pitting.

Stem pitting is a common feature in grapefruit and sweet orange trees infected by CTV. When the bark is peeled off from the twigs pits can be seen on the wood. Plants also show stunting and chlorosis. In grapefruit, stunting can be very prominent producing large longitudinal ridges or depressions running up and down the trunk giving the trunk a rope-like appearance. The trees develop a bushy top (mushroom-like appearance) with occasional branches growing away from rest of the tree. The stem pitting occurs on the scion regardless of rootstock. Fruit size and productivity is considerably reduced. In sweet orange trees same abnormalities occur. Twigs tend to be brittle and break easily. Strains causing stem pitting on sweet orange may cause stem pitting on rootstock also. A given CTV isolate causing stem pitting on sweet orange may or may not cause stem pitting on grapefruit and *vice versa*.

The CTV strains having the above mentioned biological activity are serologically related and provide cross protection. The mild strains have been used to provide cross protection against severe strains. In the recent past, however, new and destructive strains of CTV have appeared in parts of the world which are insensitive to cross protection and have resulted in decline epidemics.

The Causal Agent

The cause of the decline of citrus scion on sour orange rootstock was unknown for many years and was thought to be graft incompatibility between scion and stock (214) or a nutrition problem. Fawcett and Wallace (72), for the first time, recognized it as a virus disease transmitted by aphid vectors. CTV is a phloem-limited closterovirus (7-9). Quick decline symptoms are caused by necrosis of the phloem. The virus particles are flexuous filaments measuring 11×2000 nm in size. The CTV virions contain a single-stranded, plus-sense RNA with an estimated size of 5.4-6.5 million or about 20 kilobases (7). The only gene product which has been identified is the coat protein which accounts for 3% of the total coding capacity of CTV genome (119). The genome can code for atleast 19 protein products ranging from 6 to 401 kDa (104). Two coat proteins have been identified. The larger coat protein (CP1) has a molecular weight of 23000 to 28000 daltons and the smaller coat protein (CP2) has a molecular weight of about 21000 daltons (117, 119). Sedimentation coefficient of full particle of CTV is 105 to 131 S. The genetic system of CTV is complex. The infected plants contain the genomic RNA, at least 9 subgenomic RNAs and often multiple defective RNAs (104). In many cases the plants are doubly or multiply infected with genetically distinct CTV isolates or strains and there are indications of high recombination potential between different virus-specific RNAs. This complicates the control of CTV with cross-protection.

Closteroviruses are characterized by the occurrence of inclusion bodies that are confined to the phloem and associated tissues and appear as large aggregates in arrays that are often cross banded. In CTV such bodies consisting of virus particles and related proteins occur in infected tissues (18). These inclusion bodies are found in phloem and occasionally in the ground meristem of newly forming stems. These may be seen with light microscopes after Azure A staining (20). The number of inclusion bodies is possibly related to strain severity and to virus titer (19). Within the plant the virus moves only through the phloem and fails to enter the xylem (35). Thus, the concentration of virus particles is highest in the phloem. In vectors the concentration of virus particles is low.

Transmission

CTV is transmitted by budding, grafting and by many aphid vectors (7). The aphid species transmit the virus in a semi-persistent manner (177). The virus is both stylet-borne and circulative. It is normally not sap transmissible. The first vector reported for CTV was the oriental citrus aphid, *Aphis citricidus*, now known as *Toxoptera citricida*. Later, other aphid species were reported as vectors. These include *Aphis gossypii*, *A. citricola*, *Toxoptera aurantii*, *A. craccivora*, *A. spiraecola*, and *Myzus persicae* but these are less efficient vectors. *T. citricida* (brown aphid) is the most efficient vector with 20% transmission by single aphid transfers. Geographic spread of tristeza with the spread of *T. citricida* in the Central American and contiguous countries has been reported (118). According to some recent studies efficiency of *A. gossypii* appears to have improved. There seems to be a lag period of about 30 years after introduction of CTV in a new area before A. *gossypii* becomes a relatively efficient vector. Perhaps, there is a helper factor which becomes widespread during the lag period and which enables the aphid to gain efficiency (*cf.* 118). Helper factors have been reported for other semi-persistently transmitted viruses. CTV has been transmitted from citrus to citrus by dodder (*Cuscuta subinclusa* and *C. reflexa*) also (173).

Costa and Grant (47) had reported that a 24-hour acquisition feeding was most effective for virus transmission by the vector although the virus can be acquired in a few seconds and transmitted in a few seconds of feeding. Aphids do not remain infective after feeding on healthy plants for 24 hour or longer. They lose the virus within 24 hour if not allowed to feed on healthy plants.

Management

The management of tristeza involves regulation, cultural management and biological control measures including mild strain cross protection, genetic engineering for virus resistance and breeding for virus resistance in commercially acceptable scions and rootstocks.

Management Measures for Citrus Tristeza Virus (114)

Prevalence Management measures available

CTV absent	Quarantine; budwood certification and/or clean stock programmes
CTV present	
Low incidence	Eradication/suppression programmes; budwood certification; clean stock
High incidence	Tolerant rootstocks; varietal tolerance to stem pitting; mild strain cross protection; genetically engineered resistance; budwood certification

Regulatory measures

Movement of CTV-infected budwood or plants is the means by which tristeza has spread around the world. In areas where the disease is not present, the first line of defence against its entry is quarantine at international and national level. Any citrus planting material should be free from the virus before it is brought to a clean area. Clean stock and certification programmes for citrus serve a useful purpose in providing propagative material and nursery stock which are true to type and free from viruses and graft transmissible diseases (e.g. citrus greening). However, under special situations where importation of doubtful material is unavoidable the material should be made disease-free by heat therapy and/or shoot tip grafting followed by indexing for certification. Heat therapy can eliminate some pathogens, including CTV, but not the viroids (CEVd). Apical shoot tip culture or grafting can eliminate viroids as well as other pathogens. Methods combining pre-heat treatment followed by shoot tip culture/grafting in practiced in Japan (113) for elimination of known graft-transmissible pathogens of citrus.

Such clean-stock and certification programmes are very effective for viruses which do not have a vector in a clean area. They are less useful against disease which are vector transmitted such as tristeza. In India, tristeza is present in most orchards and the vector is also present. Thus, even if clean grafts are used for planting they soon become infected through the vectors. Therefore, eradicative and vector control steps may be warranted even if disease-free stock has been planted.

Cultural measures

These include use of CTV-tolerant rootstock, switching between resistant or tolerant cultivars such as between grapefruit and sweet orange and proper water management. CTV is inactivated in seedlings or budwood by exposure to 35°-43° C for 87-107 days (95).

True seeds of citrus do not carry the tristeza virus and these can yield virus-free plants. But such plants are not true to type and have no commercial value. This problem was solved by obtaining seedlings from nucellus embryo which functions as a seed but has no virus (231). Such seedlings can be used either for direct planting or for raising resistant rootstock. Although, scientifically an encouraging programme, this approach also has limitations. First, the method takes 10-15 years to produce usable disease-free budwood for commercial propagation. Second, the existence of large contiguous orchards planted with a single variety promotes a rapid spread of viruses and virus-like agents. If the brown aphid is present in the area and if even a few infected plants are present in the locality, the plants raised from nucellar seedlings become infected in a couple of years. Providing chemical protection to the healthy plants also does not work because the insecticides do not instantly kill the vector which can transmit the virus in a very short time. In many countries, where the brown aphid is not common, rootstock raised by this method has been successfully used to reduce tristeza incidence. Waterlogging predisposes the plants to tristeza by killing or damaging roots.

Chemical protection

There is no direct chemical treatment against the virus. It has been suggested that insect control by chemicals over a large area may reduce spread of CTV by reducing aphid populations.

Biological Control

Disease resistance: Many citrus cultivars are tolerant of CTV infection meaning that CTV replicates in the host but no symptoms are expressed in infected plants (46). Some citrus relatives such as *Poncirus trifoliata* (trifoliate orange) are immune to CTV infection meaning that CTV does not replicate in the host (9). Trifoliate orange can be hybridized to citrus and many hybrid cultivars have been obtained which are immune to CTV replication. These cultivars are widely used as CTV-tolerant rootstocks (79). Somatic hybridization between protoplasts of sexually incompatible citrus relatives has also been successfully used to produce hybrids which can be used as tolerant rootstocks (96). However, such plants are not immune to CTV.

Mild strain cross protection: Cross protection against CTV usually deals with either decline on sour orange or protection against stem pitting. This has been a proven control strategy for citrus tristeza (48) and many other plant virus diseases (81). Mild strains are selected and inoculated on trees to provide protection against severe strains. In India, cross protection to CTV has been demonstrated in acid lime. The mild strains used must be mild in all citrus cultivars or susceptible hosts. Such viruses must be stable. There are reports that mild strain cross protection may fail to prevent superinfection with certain severe strains of CTV (164). Within the orchard, spread of severe strains is much higher than the mild strains (165).

Possibility of genetically engineered protection

Genetic engineering has the potential for CTV management. There are several possible approaches (118). First, satellite RNAs of cucumber mosaic virus and tobacco ringspot virus have been cloned and these cRNAs are inserted into vectors which enable transformation of plant tissue. This approach has afforded some degree of virus resistance in the transgenic plants. Second, antisense RNAs of cucumber mosaic virus, tobacco mosaic virus and potato virus X when expressed in transgenic plants have shown a small degree of virus resistance. Third, the incorporation of the viral coat protein gene and expression of the virus coat protein in transgenic plants provides a high degree of resistance in the transgenic plants. The plants transformed with the viral coat protein genes result in plants showing the benefits of cross-protection without the virus being present hence the disadvantages encountered in mild strain cross-protection are eliminated. This approach also provides a more broad based protection against diverse strains of CTV.

■ XYLOPOROSIS OR CACHEXIA

This virus disease of citrus was first noticed in the Philippines in 1934. It is now a destructive disease in California and Florida (USA), Brazil and Argentina (140). On the basis of its occurrence at international level, it is considered next to tristeza. There are unconfirmed reports of its occurrence in India also. However, since the susceptible rootstock (sweet lime and Orlando tangelo) are not used in India, chances of its occurrence are low.

Even in susceptible hosts the symptoms appear late, usually when the trees are 3-4 years old. Initial visual, but non-specific, symptoms are sick and weak looking appearance of the trees and stunting of leaves and the tree. In comparison to healthy trees the diseased trees fruit early and rather heavily. The fruits have more sugars and acids. Later, the trees cease fruiting.

When bark at the bud union is removed, pits measuring 1-4 mm in diameter when circular or 1-10 mm long when triangular are seen on the wood of rootstock and scion. Matching these pits there is raised, peg-like tissue on the inner side of the bark. In susceptible hosts such as Orlando tangelo, these pits are full of a gum-like material but in others this gum is not common. In advanced stages of the disease, the bark splits at the bud

Fig. 8. Xyloporosis. Pits on wood under the bark.

union. The basal portion of the stem loses strength, easily bends and appears bending on one side. Due to blockage of food movement, the bud union appears swollen.

The virus is transmitted during budding. It can be managed if care is taken to select rootstock and scion only from sources that are totally virus-free. Precautions recommended against tristeza can also give satisfactory control of this virus.

■ PSOROSIS OR SCALY BARK

This virus disease is known under different names such as psorosis A, psorosis B, concave gum, blind pocket, crinkly leaf and infectious variegation. The disease is serious in USA and many other countries. According to McClean (140) in majority of citrus growing areas of the world the number of trees ceasing to fruit due to this disease is higher

Fig. 9. Psorosis. Scales on the trunk.

than due to any other disease. The virus causing the disease has no relationship with the rootstock. Psorosis was reported in India in 1959 and is widespread in citrus orchards in the country (41).

The scaly bark is caused by strains of different viruses and symptoms are variable according to the strain involved. The common symptom is a bright yellow colour of the leaf veins. This is more conspicuous on new leaves. Leaf flecking also occurs all over the lamina. In the infection of psorosis A, characteristic scales develop on the bark of the trunk and thick branches. They may be absent in some cases. The bark scale is generally some shade of brown. The affected portion is spherical which first appears as pimples and progresses to girdle the entire stem or the branch. The outer 2 mm thick bark separates as a scale from the underlying living tissue. In advanced stages of the disease the entire bark and some part of the wood looks reddish brown. Sometimes gummosis also occurs. Decline of trees is very rapid when wood of the trunk or branches is also affected. In psorosis B gummosis occurs before scaling of the bark. The scaly bark progresses upward. Bright yellowing of veins and translucent spots are present on old leaves also. Such spots may develop on fruits also. In blind pocket psorosis, deep parallel furrows with straight wall develop longitudinally on the tree trunk. Sometimes the bark develops scales and there may be gummosis. In concave gum psorosis these furrows are longer and broader and leaf veins are shining like oak leaves. Usually, the furrows are covered by the bark but it may sometimes crack and gum exudation may occur. In leaf crinkle psorosis leaves look crumpled.

The viruses causing different types of psorosis are transmitted through use of infected budwood (50). Some of them are seed-borne also (40). For control of this disease it has been suggested that scion material (budwood) should be taken only from 20-25 years old trees to ensure that they are healthy. Trees which are not fruiting satisfactorily should be cut down. Planting stock can be given thermal treatment by exposure to 35° - 43° C in a room for 87-107 days (95).

- **CITRUS EXOCORTIS VIROID**

Exocortis is worldwide in distribution and affects trifoliate orange, citrange, Rangpur and other mandarins, sweet lime, some lemons and citrons. Orange, lemon, grapefruit and other citrus trees grafted on exocortis sensitive rootstocks show slight to great reduction in growth and yields are reduced by as much as 40%. Existence of exocortis in India was reported in 1968 (154, 163).

Infected susceptible plants show vertical splits in the bark and narrow, vertical, thin strips of partially loosened outer bark that gives the bark a cracked and scaly appearance. Since many of the exocortis-susceptible plants, such as trifoliate orange, are used primarily as rootstocks for other citrus trees, and because the scions make poor growth on such rootstocks, the enlarged, scaly rootstocks have given the disease the name "scaly butt". Infected exocortis-susceptible plants may also show yellow blotches on young infected stems, and some citrons show leaf and stem epinasty and cracking and darkening of leaf veins and petioles. All infected plants usually appear stunted to a smaller or greater extent and have lower yields (184).

The citrus exocortis viroid (CEVd) is similar to, but not identical with, potato spindle tuber viroid. It consists of 371 nucleotides arranged in a circular or linear form (54, 129). Host range of the viroid is largely restricted to the citrus family (Rutaceae) although solanaceous plants such as potato, tomato and petunia are also susceptible. There is no evidence of vector transmission. The viroid is readily transmitted from diseased to healthy trees by budding knives, pruning shears, or other cutting tools, by hand, and possibly by scratching and gnawing of animals. It is also transmitted by dodder and by sap to *Gynura*, *Petunia*, and other herbaceous plants. On contaminated knife blades CEVd retains its infectivity for atleast 8 days and, when partially purified, CEVd remains infective at room temperature for several months. The thermal inactivation point of extracted sap is about 80°C for 10 min but partially purified CEVd remains ineffective even after boiling for 20 min. The viroid also survives brief heating of contaminated blades in the flame of a propane torch (blade temperature about 260°C) and flaming blades dipped in alcohol or on contaminated blades treated with almost all common chemical sterilants except sodium hypochlorite solution.

The viroid can be isolated from all plant parts including roots and fruits. Actively growing regions of the tree contain the highest concentration of the viroid. The viroid apparently enters the phloem elements and spreads in them throughout the plant. It is associated with the host nuclei and internal membranes of cells and results in aberrations of the plasma membrane. Although the viroid apparently lacks the ability to serve as a messenger molecule or as an amino acid acceptor, it brings about several metabolic changes in infected plants. These changes include an increase in oxygen uptake and respiration and also in sugars and certain enzymes. Marked changes also occur in several amino acids. Marigold *(Tagetes patula)* has been reported as an experimental host of the viroid (75).

Exocortis can be managed only by propagating healthy nursery trees from certified foundation stock and use of sanitary budding, nursery and field practices. Apical shoot tip culture or grafting eliminates the viroids and many other pathogens (113). Tools should be disinfected between cuts in different plants by dipping in a 10-20 % solution of household bleach (sodium hypochlorite).

■ CITRUS ROOT NEMATODE

The citrus root nematode was first noticed in 1912 on orange trees in California (USA). In 1913 the nematode was described by N.A. Cobb who created the genus *Tylenchulus* with the type species *Tylenchulus semipenetrans* for the nematode. The nematode is present throughout the world wherever citrus is grown. In the United States it is placed among the economically most important phytonematodes because of the reduced fruit yields and poor quality of fruits. In India, where the nematode was first reported in 1961, upto 80% orchards were found infested in certain areas. The *Citrus* spp. found infected were *C. limon*, *C. sinensis*, *C. reticulata* and *C. aurantium*.

The nematode causes slow decline of citrus trees with die-back of twigs and debilitation of the trees. Normally, the trees are not killed but over the years they become nonproductive. Although, die back and decline could be caused by many other factors such as lack of moisture and nutrients, presence of hard pan in the subsoil, attack of bacteria (citrus greening) and virus (tristeza), the association of *Tylenchulus semipenetrans* as the major cause had been proved as early as 1923 (206).

Symptoms

Symptoms of slow decline are similar to those caused by poor nutrition. In early stages of root infection no above ground signs of the problem are visible. It is only after a few years that damage to roots show an adverse effect on the foliage. Affected trees show reduced terminal growth, chlorosis, shedding of terminal leaves, die-back of branches and considerable reduction in number and size of fruits. The fruit quality also deteriorates (15). Die-back symptoms first occur on the upper portion of the tree but later extend to the lower portion. Copper and zinc deficiency symptoms are more pronounced (15). Roots of such trees show brownish discolouration and unusual adherance of soil particles to the root surface. When roots are disturbed or gently pressed, the cortex readily separates from the axial portion. Ultimately, the roots decay resulting in reduced root volume. Up to 60% reduction in dry weight of root mass has been reported (224). The destruction of roots is not as much by the nematode invasion as by the invasion of secondary parasites such as *Fusarium solani* (225).

Morphology and Life Cycle

Tylenchulus semipenetrans is a sedentary semi-endoparasite of roots. Females are most commonly found on thick, stunted rootlets to which a layer of soil particles is clinging. These particles are held in place by a gelatinous mucus secreted by the female. The mucus and adhering soil particles protect the females and eggs hatched by them from their natural enemies. The young and egg-laying females can be seen in groups clinging to rootlets with their head and neck buried in root cortex.

Females are 0.35-0.40 mm long with a variable saccate body which is usually bent ventrally in vulvar region just anterior to the short, blunt tail. The excretory pore is unusually well developed and is located just anterior to the vulva. The single ovary has usually two flexures and reaches the esophageal region. The eggs average 33-67 microns in size. Usually only one egg is seen in the uterus at one time.

Larvae hatched from eggs under laboratory conditions in 1-14 days are of two types. One type is shorter and wider than the other (226). From the shorter and wider larvae, which do not feed, mature males develop within one week after three moults, one moult having occurred within the egg. Such males are about 26% of the total larval numbers. They measure 0.26-0.34 mm in different stages and are non-parasitic. The longer, more slender individuals fail to develop unless they feed on the host root. These become the females. After egg hatch, the second stage female larvae, require about 14 days to locate and feed on epidermal cells of host roots (223) until ready for moulting. During the period of searching for a feeding position 24-50 larvae congregate under fragments of cortical tissue, soil particles or organic matter. They are rarely seen on unprotected locations on the root. These second stage larvae measure 0.3-0.36 mm in length. This stage is most frequently found in soil samples and can live without host roots at a temperature of 15° C for more than a year but at 33° C only for four to five and half months.

The third and fourth stage female larvae are shorter with longer esophagus and a distinct vulval cleft. The length varies from 0.23 to 0.36 mm. Well developed spear measures 13.5 μm. The esophagus occupies almost half the body length. The vulva appears as a deep transverse slit. The excretory pore is very prominent and is located near the vulva. The tail is slightly convex-conoid dorsally, ending in a blunt terminus. The anus and rectum are not seen. The fourth stage female larvae and young females are seen about 21 days after inoculation. Within a week young females penetrate deeper in the cortex of the root, become sedentary and feed. The females excrete the gelatinous mucus in which eggs are deposited. Egg laying occurs 50 days after inoculation. The complete life-cycle from egg to egg requires 6-8 weeks at 25°C. Reproduction occurs without the help of males.

In the host root cortex, the feeding zone develops a nurse cell system consisting of uninucleate, not enlarged, discrete parenchyma cells (238). Synchytium is not formed. This type of nurse cells system is characteristic of this nematode. Feeding of the citrus nematode in cortical cells results in necroses. The injury does not extend to the stellar region of the root. Secondary invasion of the fungus pathogen *Fusarium solani* through injured tissue has been confirmed by many workers. Root decay in lemon by *F. solani* is increased when citrus nematode is also present (225). This could be an important factor in the lowering of tree vitality.

The spread of the nematode through soil is slow, the rate being approximately 15 cm per month when roots of adjacent trees are in contact. The nematode is, however, spread over long distances by movement of nematode infested soil on equipments, animals, and by irrigation water and even to longer distances by transfer of infected citrus nursery plants. The population of the citrus nematode is closely related to the stage of decline of the trees. Usually, the population is maximum on healthy looking trees with early stage of root infection and minimum on trees in advanced stage of decline. Seasonal and regional variations at the time of peak population of the nematode on roots are reported (155, 167). These depend on host cultivar, temperature and rainfall. In the northern states of India populations are at the lowest during January and February while in Maharashtra (south) maximum populations are reported in these months.The nematodes are found in large numbers 30-50 cm from the main trunk of the tree in the top 30 cm of soil. The highest numbers are in top 20 cm but some can be found upto a depth of 100 cm or even more (187). The number of nematodes on the root surface is very large, 108 nematodes found on a 4 mm piece of root (15). Under a single tree their number can reach the figure of 700 million at 15-60 cm depth.

Infection of roots is severe in sandy loam soil. In soils with 50% clay, reproduction of the nematode is extremely slow. The highest rate of reproduction and the greatest reduction of plant growth occurs in soils with 10-15% clay. Oxygen supply is important for activity of the nematode. A decrease in infectivity is associated with a corresponding decrease in motility and body contents of second stage females which are aged and starved at 27°C in soil and water (222). Survival of infective larvae is better in soil than in water. Optimum temperature for infection, growth, and reproduction of the nematode is 25°-30° C. There is much less infection at 16° C and at 35° C. At 20° C infectivity is greater than at 30° C but reproduction at 20° C is delayed by 4 weeks (155). For reproduction optimum temperatures are 28°-31°C. Above 31°C there is no reproduction. Temperatures suitable for citrus root growth are favourable for the activity of the nematode (155).

Host nutrition also plays a role in the severity of attack of citrus root nematode. Citrus seedlings growing in soil containing calcium carbonate, sodium and potassium levels insufficient for satisfactory growth of the host show maximum attack of the nematode (224). Leaf copper is reduced in the infected trees. Low levels of phosphorus also retard seedling growth by the nematode. The population of the nematode increases under low phosphorus conditions while high nitrogen suppresses it. Root phosphorus in excess of 0.3% retards the numbers of the nematode. Naidu *et al.* (148) have reported that root exudates of acid lime (*C. aurantifolia*) infected with tristeza (CTV) caused 44.8 and 70.2% mortality of larvae by exposure for 12 and 24 hours, respectively.

All species of *Citrus* and allied genera are attacked by *T. semipenetrans*. Other hosts are *Andropogon rhizomatus*, *Diospyros lotus* (Ebenaceae, date plum), *D. virginiana* (persimmon), *Makania batatifolia*, *Oleseuropea*, *Syringa vulgaris* (common lilac) and *Vitis vinifera* (grapes). The roots of *Citrus limon* (lemon), *C. aurantifolia* (lime), *C. medica* (citron) and *C. pennivesciculata* support very high populations of the nematode whereas *C. aurantium* supports the lowest population.

Management

Protection of rootstock is very important. Citrus nurseries should never be established on or near old citrus orchards. Nursery soil should be fumigated before planting for rootstock is done. Suspected rootstock should be denematized by hot water treatment at 45°C for 25 min. The seedling roots can also be treated by bare root dip method in nematicides such as zinophos (156, 157).

DBCP (Nemagon) had been extensively used for the control of citrus nematode. The nematicide was effective at a dosage of 15-45 lit a.i./ha and had reduced the population of the nematode by 98% with concurrent increase in fruit yield by 440-200% (34, 145). Fruit size in lemon increased by 11% with Nemagon treatment. This nematicide has been withdrawn from the market because of environmental pollution problem.

Dimethoate at 16 ml a.i. per tree and phorate (Thimet) at 15 g a.i. per tree followed by light irrigation are also reported to reduce nematode population by 82 and 90%, respectively, on sweet orange. Fensulfothion (Terracur P or Dasanit) at 24.7 kg a.i./ha, ethoprophos (Mocap) at 61.5 kg a.i./ha, and dichlofenthion at 27.8 lit a.i./ha give effective control of the nematode. Upto 90% reduction in nematode population and yield increase of up to 90% in the second year by application of ethoprophos at 40 kg a.i./ha are reported (145). Le Roux *et al.* (120) have recommended soil application of aldicarb (Temic) at 5 g a.i. per sq. metre of leaf canopy. Treatment was given at 3 months interval in 24 months. After 32-44 months significant reduction in root decay and increase in fruit yield was recorded.

Many reports suggest beneficial effect of organic amendment of the orchard soil. Steer manure, chicken manure, cotton waste, sugar beet pulp, lucerne hay and castor pomace (cake) are reported to reduce the population of citrus nematode (127, 128). Singh (187) obtained reduction in nematode population by applying castor cake at the rate of 10 kg/tree in trenches 30 cm wide and 30 cm deep on the periphery of the trees. Amendments with easily decomposable nitrogenous materials enhance microbial activity including microbivorus and fungivorus nematodes. The suppression of nematodes is correlated with high nitrate levels and decrease in soil pH. Mankau (127) had isolated nematophagus fungi *Arthrobotrys, Dactylella,* and *Dactylaria* from amended soils. Cultivation of marigold in citrus orchards (in the drip area) is reported to suppress nematode populations.

The use of resistant or tolerant rootstocks against the citrus nematodes is the most practical method of management. All species of *Citrus* and related genera are not equally good hosts of the nematode and a certain degree of resistance does exist in some species and hybrids. Trifoliate orange (*Poncirus trifoliata*) and Troyer citrange (*P. trifoliata x C. sinensis*) are resistant to the citrus nematode and their use as rootstock is one way of managing this nematode.

■ SPREADING DECLINE OF CITRUS

This disease is reported only from the Florida state of USA although the nematode species associated with it is universally present on other hosts including banana. The disease is a famous example of devastation caused by a phytonematode. It was first noticed in between 1926 and 1928 in a citrus orchard of Florida. The disease expanded to more areas by 1935-36. Final proof of *Radopholus similis* being the cause was given by Suit and Du Charme (201). The loss in fruit yield to the extent of 50-80% in grapefruit (*C. paradisi*) and 40-70% in orange are reported (59).

Symptoms

Spreading decline appears in citrus orchards as groups of stunted or unthrifty trees with sparse foliage, small fruits, and retarded terminal growth. Leaves have a tendency to wilt during hot and dry periods, but may respond to moisture and show temporary recovery. New branches do not come out or their development is retarded. After some years the trees have a large number of drying or dead branches. The number, size and quality of fruits declines.

Symptoms of decline spread steadily to more trees each year. Examination of the root system reveals that below 50 cm of soil young feeder roots are considerably reduced in number or may be absent, which causes the above ground symptoms of starvation usually a year after root infection. New roots that develop are also destroyed by the nematode. Typical lesions are present on roots that have not been totally destroyed.

Morphology and disease development

The nematode, *Radopholus similis* (burrowing nematode), is described under banana root and rhizome rot. All stages of the male and female can enter any part of the root and are found in root tissue as well as outside in the soil. The nematode spends its life and reproduces inside the cavities caused by destruction of cells in the root cortex where one life cycle is completed in 20 days. Although in banana *R. similis* does not enter the stele, in citrus it enters the stele through the endodermal passage cells. There, the nematodes accumulate in the phloem and cambium which they destroy in time and form nematode-filled cavities (58). In citrus roots a temperature of 25°-26° C is optimum for activity and development of the nematode (201). At this temperature one generation (egg to egg laying female) takes about 18-20 days.

The burrowing nematode can survive in sandy loam soil for 6 months in absence of the host roots. The life span depends on life span of the feeder roots which is shorter than the life time of the tree. In citrus orchards, the maximum numbers of the nematode are found between 0.3 and 1.8 meters but may be present as deep as 4 metres. The nematode is generally not found in the top 15 cm of soil.

Most of the spread from plant to plant is through root contact or near contact. The larvae migrate to a distance of several metres. However, long distance dispersal of the nematode and its introduction in an unaffected area occurs through transfer of infected planting stock. Once established in a new locality the nematodes can spread locally through root contact, soil, water and farm implements. Depending on the host the nematode can spread at the rate of 6 to 60 metres per year. Closeness of roots plays the major role in lateral spread. When roots of different trees are in close contact the lateral spread is fast. The average extent of spreading decline is 1.6 trees per year (202).

Management

Pull and treat method followed for control of spreading decline involves pulling out and destroying the infected trees and fumigating the soil. D-D was used for fumigation at the rate of 672 kg/ha. The land is either left fallow, keeping weeds out by herbicides, or such antagonistic crops as asparagus or marigold are cultivated for two years before replanting of citrus. To prevent introduction of the nematode in new areas through planting stock, bare rooted citrus plants are treated with hot water at 50° C for 10 min before planting.

■ NUTRITIONAL DISORDERS IN CITRUS

The normal growth of citrus trees requires nitrogen, phosphorus, potassium, calcium, magnesium, sulphur, iron, zinc, manganese, copper, boron and molybdenum as essential elements in their balanced nutrition. These elements are obtained by the tree from the soil. When one or more of these elements are in short supply, physiology of the tree is disturbed and aerial parts, especially leaves, show signs of hunger. Roots become weak and are ultimately dead. Loss of root functions weakens the absorption of minerals and water from the soil and nutritional deficiency symptoms appear on leaves and tender aerial parts.

The nutritional disorders show such symptoms as are commonly seen in the attack of fungal, bacterial, nematode and virus pathogens. Root rot, stunting, loss of green colour, and die-back are some such symptoms. Generally, these symptoms appear on grown-up or old trees when the demand for nutrients is high due to size of the tree. The nutritional deficiencies can be caused by any one or more of the following conditions: deficiency or loss of the particular nutrient in soil, leaching of the nutrient by irrigation and rains in light soils, non-availability of some elements such as manganese, zinc and iron due to excess of carbonates in the soil, loss of soil humus due to improper cultural practices, use of wrong combination of fertilizers and manures, defective rootstock and scion combination, poor irrigation, disturbed root and leaf functions due to attack of parasites and presence of hard pan below the root system. Before starting new orchards soil test for nutrients in soil and soil treatment against parasites are essential.

Nitrogen

Trees do not grow normally in nitrogen deficiency and remain stunted. New leaves remain small, narrow, and can easily detach. They are lighter green in colour. Old leaves are yellow. Such trees suffer from a certain amount of defoliation and look more open than normal trees. If the deficiency continues for a long time drying of some twigs may occur. Normally, the trees do not die but their productivity is decreased.

A soil having 250 kg/ha easily oxidized N can be considered N-deficient. Old leaves with less than 2% nitrogen (on dry weight basis) indicate N-deficiency in the tree (164). In these situations, nitrogen can be applied to soil around the tree as fertilizer or it can be sprayed as 2-5% urea. Foliar application needs several sprays during the year.

Phosphorus

Deficiency of phosphorus in the trees can be due to several reasons. Increased soil acidity hinders availability of phosphorus. This may happen if the soil contains sesquioxides of iron and aluminium. In heavy rainfall areas this nutrient is washed out from the soil. Signs of phosphorus deficiency are more or less similar to those of nitrogen deficiency. Apical growth of the tree and of the root is stopped. Twigs are short and thin and grow in a straight manner. Older leaves show loss of shining green colour and sometimes they show greenish bronze colour. Some of the leaf veins are destroyed and lamina surface shows spots at such points. Older leaves shed early. Young or new leaves normally show the hunger signs late because they draw the nutrients from older parts. Lateral buds dry and fall. Flowering and fruiting is reduced. The skin of the fruit is thick and rough and the fruit becomes soft before ripening.

In neutral or alkaline soils phosphorus deficiency can be corrected by application of soluble phosphorus. In acidic soils insoluble forms of phosphorus are also effective. Annual test of soil for phosphorus content is essential because continuous use of phosphorus increases its level in soil above the required limit which hinders availability of zinc and copper.

Potassium

There are no reports of harm to citrus trees as a result of potassium deficiency. In potassium deficiency, apical growth of the tree stops. Young twigs remain weak and break off before becoming hard. Loss of green colour, gummosis of branches, rolling, crinkling and spotting of leaves may be due to potassium deficiency.

If leaves contain 0.2% or less potash (on dry weight basis), the element is deficient in the tree. For good sized fruits and normal skin this level should be 0.4 % (71). If potassium deficiency is suspected this nutrient can be supplied through potassium sulphate or potassium chloride. Light soils usually need more potash. Excess of potash in soil hinders availability of manganese, zinc and magnesium.

Calcium

The early sign of calcium deficiency in the tree is yellowish colour of leaves. The midrib looks more yellow than the nearest tissue. Leaves at the tip of twigs start yellowing from the tip and margins. On the yellowish surface brown spots appear. Wither-tip and root rot are also signs of calcium deficiency. Calcium deficient fruits are prone to easy attack of soft rot bacteria and fungi.

In the type of soil commonly used for citrus cultivation the chances of calcium deficiency are not much. In acidic soils there is less calcium and plants take in more potassium. Application of' lime or gypsum at the rate of 5-9 kg/ tree is recommended for correcting acute calcium deficiency (38).

Magnesium

Deficiency of magnesium causes interveinal chlorosis of leaves. The loss of green colour spreads all over the lamina surface except the tip and basal portion which remain green. Sometimes veins are thickened and become rough. Continued deficiency of the nutrient leads to defoliation and wither-tip.

Highly acidic soils have low magnesium level. In alkaline soils magnesium is present but due to excess of potassium and calcium it becomes unavailable to the tree. Excess of phosphorus also reduces availability of magnesium. Signs of magnesium deficiency may appear in nitrogen-deficient trees. Magnesium is required by the plant for chlorophyll synthesis and seed setting. Therefore, symptoms of magnesium deficiency become more conspicuous when fruiting has started and is more common in trees with seeded fruits. Since magnesium is highly mobile in the plant it gets accumulated in growing twigs and young leaves. Thus, deficiency symptoms are most prominent in old leaves. Deficiency of magnesium can be corrected by application 7-9 kg magnesium sulphate per tree or by spray of 2 % magnesium sulphate solution.

Sulphur

Generally this element is not deficient in plants or in the soil. Different inorganic fertilizers used in cultivation supply the desired quantity of sulphur. Symptoms of sulphur deficiency are similar to those of nitrogen deficiency. New leaves show the signs of deficiency early.

Boron

Boron is an important element for formation of the middle lamella. Generally, its deficiency is not seen in lime, oranges and sweet orange. It is rarely deficient in alkaline soils. In the light sandy (17) soils with acidic reaction and in the hills its deficiency has been noted in fruit trees such as mango and apple. In apple, boron deficiency is the cause of browning and decay of the core. In alkaline soils boron may be in excess and may cause boron toxicity. Symptoms of boron deficiency are aggravated under conditions of continued soil moisture deficiency.

Trees suffering from boron deficiency show yellowing of leaves starting from the tip and margins. Veins get thickened and corky on the lower surface of the leaf and may rupture. In mandarin oranges, leaf surface at the base of veins turns brown. Destruction of the vascular tissue is a typical effect of boron deficiency. As a result, plants show drooping and wilting even when soil contains enough moisture. Defoliation and wither-tip follow this stage. In immature fruits, the inner white layer is thick. Due to rupture of tissue, gum exudation takes place which deposits as brown spots on the fruit. Such fruits fall down before ripening.

Excess of boron is toxic to plants. Before its application chemical analysis of soil and leaf tissue should be done to determine the quantity to be applied. Leaf tissue should have more than 25 ppm boron on dry weight basis. If less, boron can be applied as spray of boric acid (0.1%) or as soil application of borax (10-15 kg/ha).

Copper

Deficiency of copper in plants has been described under different names such as die-back and wither-tip, ammoniation, red rust, and exanthema. Cu deficiency occurs either due to lack of the nutrient in the tissues or due to disturbed balance between copper and nitrogen. Deficiency symptoms have been generally noticed when there is excess of nitrogen in soil. Roots of Cu-deficient plants fail to absorb zinc from the soil and zinc deficiency symptoms also appear.

The tender, long and pointed twigs of the tree twist to form a S-like structure. The leaves of these twigs are unusually long and deeper in colour. The midrib is slightly bent on the upper side. In acute deficiency of Cu, small leaves are formed on new twigs. They are mottled. Such leaves shed and the twig starts showing wither-tip. Old leaves are darker green and twisted. Blisters and cracks appear on the bark. Gum exudation occurs through these cracks. Brown spots may appear on skin of the fruit and they may also crack. The core of the fruit shows necrosis of tissue and gum deposit occurs in this area.

Normally, when copper fungicides are used to control gummosis and other fungal and bacterial diseases the deficiency of copper in the tree is not likely to occur. Otherwise, a mixture of 0.5% copper sulphate and 0.2 % lime should be sprayed. For soil treatment, 230 g of copper sulphate mixed with 115 g of lime can be applied per tree.

Zinc

Deficiency of zinc is described under different names in different countries such as frenching, leaf mottle and foliocellosis. Among the nutritional problems of plants, zinc is only second to nitrogen. In light soils in heavy rainfall areas zinc is washed out from the soil. In such areas zinc is deficient in the soil. In alkaline and heavy soils of high rainfall areas, zinc is not absorbed by the plants even if present in sufficient quantity in the soil. Attack of nematodes, shortage of humus, and unsuitable soil for the particular rootstock also cause zinc deficiency. The symptoms of Zn deficiency are so pronounced that they change or mask the symptoms of other nutrient deficiencies. Zinc and manganese deficiency symptoms usually appear together. Deficiency signs of iron, copper and magnesium can also be present simultaneously.

The symptoms of Zn deficiency appear early on new twigs and leaves. The twigs remain stunted. Leaves are small with interveinal chlorosis. These foliar abnormalities give the tree a bushy appearance with erect branches. Symptoms are more conspicuous on the side facing the sun. Die-back is also a symptom of zinc deficiency.

Zinc deficiency is easily corrected by foliar spray of 0.4-0.6% zinc sulphate solution 2-3 times a year. Addition of lime (half the quantity of zinc sulphate) to the mixture avoids chances of phytotoxicity. Generally, corrective measures for zinc and manganese deficiency are taken together.

Manganese

Manganese deficiency is found in all regions growing citrus. In alkaline soils manganese is generally deficient along with zinc. In acidic soils the amount of manganese may be low. High pH of the soil hinders uptake of manganese by roots.

Manganese deficiency symptoms are similar to those of zinc but the size of leaves is not decreased and mottling is more pronounced on shady side of the tree. Loss of green colour is not uniform all over the lamina and symptoms do not appear immediately after formation of twigs. Die-back is not acute. Manganese deficiency can be corrected by spray of 0.4- 0.6% manganese sulphate.

Iron

Deficiency of iron is not a problem in citrus trees. In alkaline soils uptake of iron is low. Leaves are lighter green or yellowish. Against the yellowish background network of dark green veins is very conspicuous. As the leaves grow, they become thin and translucent and sometimes become almost white. Only the tissue adjacent to midrib remains green. The chlorotic appearance of leaves is more pronounced during winter and spring.

Although sprays of 0.4-0.8% ferrous sulphate give some relief from iron deficiency, no effective method is known. Avoidance or correction of conditions that hinder uptake of iron is advised. Excess of lime or calcium, moisture, phosphorus, zinc, copper, manganese, and microbial activities reduce availability of iron to plants.

Molybdenum

White or yellowish spots appear on leaves in molybdenum deficiency. These spots may coalesce. After sometime they may be masked. On the underside of leaves below the spots gum deposit is seen. In acute shortage of this element leaves fall down. Die-back of twigs also occurs. Ammonium or sodium molybdate is sprayed to correct this deficiency.

From time to time, nutrient mixtures have been developed and recommended for use in orchards to avoid nutrient deficiencies. Such mixtures may prove harmful if one or more particular elements are not deficient and their addition through the mixture may harm the trees.

■ REFERENCES

1. Afek, U. and A. Sztejnberg. 1988. Accumulation of scoparine, a phytoalexin associated with resistance of citrus to *Phytophthora citrophthora*. *Phytopathology* 78: 1678.
2. Agarwal, G.P. and S.K. Hasija. 1967. Alternaria rot of citrus fruits. *Indian Phytopath.* 20: 259.
3. Agarwala, R.K. and R.N. Tandon. 1957. Studies on the anthracnose of lime in Uttar Pradesh. *Indian J. Agric. Sci.* 27: 205.
4. Agostini, J⁻. L.W. Timmer, W.S. Castle and D.J. Mitchell. 1991. Effect of citrus rootstocks on soil populations of *Phytophthora parasitica*. *Plant Dis.* 75: 532.
5. Agrios, G.N. 1988. *Plant Pathology*. 3rd ed. pp. 496-499; 255-259. Academic Press.
6. Aiyappa, K.M. and K.C. Srivastava. 1967. Citrus die-back disease in India. *ICAR Tech. Bull. (Agric.)* No. 14. pp. 77.
7. Bar-Joseph, M., S.M. Garnsey and D. Gonslaves. 1979. The closteroviruses: A distinct group of elongated plant viruses. *Adv. Virus Res.* 25: 93.
8. Bar-Joseph, M., C.N. Roistacher, S.M. Garnsey and D.J. Gumpf. 1981. A review on tristeza, an ongoing threat to citriculture. *Proc. Intern. Soc. Citric.* 1: 414.
9. Bar-Joseph, M., R. Marcus and R.F. Lee. 1989. The continuous challenge of citrus tristeza virus control. *Annu. Rev. Phytopathol.* 27: 291.
10. Barkai-Golan, R. and D.J. Phillips. 1991. Post-harvest heat treatment of fresh fruits and vegetables for decay control. *Plant Dis.* 75: 1085.
11. Barmore, C.R. and G.E. Brown. 1982. Spread of *Penicillium digitatum* and *P. italicum* during contact between citrus fruits. *Phytopathology* 72: 116.
12. Baudoin, A.B.A.M. and J.W. Eckert. 1982. Factors influencing the susceptibility of lemon to infection by *Geotrichum candidum*. *Phytopathology* 72: 1592.
13. Beraha, L. 1964. Influence of gamma radiation dose rate on decay of citrus, pears, peaches, and on *Penicillium italicum* and *Botrytis cinerea*. *Phytopathology* 54: 755.
14. Bhagabati, K.N. and T.K. Nariani. 1980. Interaction of greening and tristeza in Kagzi lime (*C. aurantifolia*) and their effect on growth and development of disease symptoms. *Indian Phytopath.* 33: 292.
15. Bindra, O.S. 1970. Nematode, pp. 56-63. In: *Citrus Decline in India: Causes and Control*. K.L. Chadha, N.S. Randhawa, O.S. Bindra, J.S. Chohan and C.L. Knorr (eds.). PAU, Ludhiana.
16. Bliss, D.E. 1951. The destruction of *Armillaria mellea* in citrus soils. *Phytopathology* 41: 665.

17. Bove', J.M. 1986. Greening in the Arabian Peninsula: Towards new techniques for its detection and control. *FAO Plant Prot. Bull.* **34(1)**: 7.

18. Brlansky, R. 1987. Inclusion bodies produced in *Citrus* spp. by citrus tristeza virus. *Phytophylactica* **19**: 211.

19. Brlansky, R.H. and R.F. Lee. 1990. Numbers of inclusion bodies produced by mild and severe strains of citrus tristeza virus in seven citrus hosts. *Plant Dis.* **74**: 297.

20. Brlansky, R.H., R.F. Lee and S.M. Garnsey. 1988. In situ immuno-fluorescence for the detection of citrus tristeza virus inclusion bodies. *Plant Dis.* **72**: 1039.

21. Broadbent, P. 1977. Phytophthora diseases of citrus: A Review. *Proc. Int. Soc. Citric.* **3**: 986.

22. Broadbent, P. and K.F. Baker. 1974. Association of bacteria with sporangia formation and breakdown of sporangia in *Phytophthora* spp. *Austral. J. Agric. Res.* **25**: 139.

23. Broadbent, P., P.C. Fahy, M.R. Gillings, J.K. Bradley and D. Barnes. 1992. Asiatic citrus canker detection in an orchard in northern Australia. *Plant Dis.* **76**: 824.

24. Brown, G.E. 1975. Factors affecting postharvest development of *Colletotrichum gloeosporioides* in orange fruit. *Phytopathology* **65**: 404.

25. Brown, G.E. 1984. Efficacy of citrus postharvest fungicides applied in water or resin solution water wax. *Plant Dis.* **68**: 415.

26. Brown, G.E. and L.G. Albrigo. 1972. Grove application of benomyl and its persistence in orange fruit. *Phytopathology* **62**: 1434.

27. Brown, G.E. and D.J. Dezmen. 1990. Uptake of imazalil by citrus after postharvest application and the effect of residue distribution on sporulation of *Penicillium digitatum*. *Plant Dis.* **74**: 927.

28. Brown, G.E., A.A. McCornack and J.J. Smoot. 1967. Thiabendazole as a postharvest fungicide for Florida citrus fruit. *Plant Dis. Rep.* **51**: 95.

29. Brown, G.E., S. Nagy and M. Maraujla. 1983. Residues from postharvest non-recovery spray application of imazalil to oranges and effect on green mold caused by *Penicillium digitatum*. *Plant Dis.* **67**: 854.

30. Bus, V.G., A.J. Bongers and L.A. Risse. 1991. Occurrence of *Penicillium digitatum* and *P. italicum* resistant to benomyl, thaibendazole and imazalil on citrus fruits from different geographic origins. *Plant Dis.* **75**: 1098.

31. Butler, E.E., R.K. Webster and J.W. Eckert. 1965. Taxonomy, pathogenicity and physiological properties of the fungus causing sour rot of citrus. *Phytopathology* **55**: 1262.

32. Capoor, S.P. 1961. Kagzi lime: an indicator plant of the citrus decline virus in India. *Indian Phytopath.* **14**: 109.

33. Capoor, S.P., D.G. Rao and S.M. Vishwanath. 1974. Greening disease of citrus in the Deccan Trap country and its relationship with the vector *Diaphorina citri*, pp. 43-49. In: *Proc. 6th Conf. Intern. Organiz. Citrus Virologist.* L.G. Weathers and M. Cohen (eds.). Univ. California, Berkeley.

34. Chhabra, H.K., O.S. Bindra and I. Singh. 1977. *Indian J. Plant Prot.* **5**: 160.

35. Chakraborty, N.K. and V.V. Chenulu. 1984. Movement and transmission of citrus tristeza virus in host tissue. *Indian Phytopath.* **37**: 174.

36. Chakravarti, D.K. and D.N. Srivastava. 1964. Stem-end rot of mango and orange fruits incited by *Diplodia natalensis*. *Curr. Sci.* **33**: 285.

37. Chalutz, E. and C.L. Wilson. 1990. Postharvest biocontrol of green and blue mold and sour rot of citrus by *Debaromyces hansenii*. *Plant Dis.* **74**: 134.

38. Chapman, H.D. 1959. Fertilization of citrus soils. *World Crops.* **11**: 251.

39. Cheema, S.S., S.P. Kaur and R.D. Bansal. 1985. Efficacy of various therapeutic agents against greening disease of citrus. *J. Res. PAU.* (Ludhiana, India) **22**: 479.

40. Childs, J.F.L. and R.E. Johnson. 1966. Preliminary report of seed transmission of psorosis virus. *Plant Dis. Rep.* **50**: 81.
41. Chohan, J.S. and L.C. Knorr. 1970. Diseases (of citrus), pp. 79-97. In: *Citrus Decline in India: Causes and Control.* K.L. Chadha *et al.*(eds.) PAU, Ludhiana, India.
42. Civerolo, E.L. 1985. Indigenous plasmids in *Xanthomonas campestris* pv. *citri. Phytopathology* **75**: 524.
43. Cohen, E. 1981. Metalaxyl for control of post-harvest brown rot of citrus fruit. *Plant Dis.* **65**: 672.
44. Cohen, E., C.W. Coggins Jr. and J.W. Eckert. 1991. Predisposition of citrus fruits to sour rot when submerged in water. *Plant Dis.* **75**: 166.
45. Cook, R.J. and K.F. Baker. 1983. *The Nature and Practice of Biological Control of Plant Pathogens*, pp. 89-99. APS Press.
46. Costa, A.S. 1957. Present status of the tristeza disease of citrus in South America. *FAO Plant Prot. Bull.* **4**: 97.
47. Costa, A.S. and T.J. Grant. 1951. Studies on transmission of the tristeza virus by the vector *Aphis citricidus. Phytopathology* **41**: 105.
48. Costa, A.S. and G.W. Muller. 1980. Tristeza control by cross protection: A U.S.-Brazil cooperative success. *Plant Dis.* **64**: 538.
49. Danos, E., R.D. Berger and R.E. Stall. 1984. Temporal and spatial spread of citrus canker in a grove. *Phytopathology* **74**: 904.
50. Dauthy, D. and J.M. Bove'. 1965. Experiments on mechanical transmission of citrus virus, pp. 250-253. In: *Proc. 3rd Conf. Intern. Organiz. Citrus Virol.* (ed.) W.C. Price.
51. Davis, M.J., R.F. Whitcomb and A.G. Gillaspie Jr. 1981. Fastidious bacteria of the vascular tissue and invertebrates (including so-called rickettsia-like bacteria), pp. 2171-2188. In: *The Prokaryotes*, Vol. II. M.P. Starr *et al.* (eds.). Springer-Verlag, Berlin.
52. Davis, R.M. 1982. Control of Phytophthora root and foot rot of citrus with systemic fungicides metalaxyl and phosethyl aluminum. *Plant Dis.* **66**: 218.
53. Davis, R.M. and J.A. Menge. 1980. Influence of *Glomus fasciculatus* and soil phosphorus on Phytophthora root rot of citrus. *Phytopathology* **70**: 447.
54. Diener, T.O. 1979. *Viroids and Viroid Diseases.* Wiley, New York.
55. Doepel, R.F. 1962. Armillaria root rot of fruit trees. *J. Agric. W. Austral.* **3**: 39.
56. Dopson, R.N. 1964. The eradication of citrus canker. *Plant Dis. Rep.* **48**: 30.
57. Droby, S., E. Chalutz *et al.* 1993. Factors affecting UV induced resistance in grapefruit against the green mold decay caused by *Penicillium digitatum. Plant Pathology* **42**: 418.
58. Du Charme, E.P. 1959. Morphogenesis and histopathology of lesions on citrus roots by *Radopholus similis. Phytopathology* **49**: 338.
59. Du Charme, E.P. 1968. Burrowing nematode decline of citrus: A review, pp. 20-37. In: *Tropical Nematology.* G.C. Smart and V.G. Perry (eds.). Univ. Florida Press.
60. Duniway, J.M. 1983. Role of physical factors in the development of Phytophthora diseases. In: *Phytophthora: Its Biology, Taxonomy, Ecology, and Pathology.* D.C. Erwin *et al.*(eds.). APS Press.
61. Eckert, J.W. 1959. Lemon sour rot. *Calif. Citrogr.* **45**: 30, 35.
62. Eckert, J.W. 1990. Recent developments in the chemical control of postharvest diseases. *Acta Hortic.* **269**: 477.
63. Eckert, J.W. and J.M. Ogawa. 1985. Chemical control of postharvest diseases: sub-tropical and tropical fruits. *Annu. Rev. Phytopathol.* **23**: 421.
64. Eckert, J.W. and N.F. Sommer. 1967. Control of diseases of fruits and vegetables by postharvest treatments. *Annu. Rev. Phytopathol.* **5**: 391.
65. Eckert, J.W., J.R. Sievert and M. Ratnayake. 1994. Reduction of imazalil effectiveness against citrus mold in California packinghouses by resistant biotypes of *Penicillium digitatum. Plant Dis.* **78**: 967.

66. Erkilic, A. and Y. Canihos. 1999. Determination of the effect of fosetyl-Al against citrus gummosis disease caused by *Phytophthora citrophthora*. *Turkish J. Agric. Forest.* **23**: 419.

67. Fagan, H.J. 1979. Post-bloom fruit drop of citrus, a new disease associated with a form of *Colletotrichum gloeosporioides. Ann. Appl. Biol.* **91**: 13.

68. Farih, A., J.A. Menge, P.H. Tsao and H.D. Ohr. 1981. Metalaxyl and fosetyl aluminum for control of Phytophthora gummosis and root rot of citrus. *Plant Dis.* **65**: 654.

69. Fawcett, H.S. 1936. *Citrus Diseases and Their Control.* McGraw-Hill, New York.

70. Fawcett, H.S. and A.E. Jenkins. 1932. Records of citrus canker from herbarium specimens of the genus *Citrus* in England and the United States. *Phytopathology* **22**: 820.

71. Fawcett, H.S. and L.J. Klotz. 1948. Diseases and their control, pp. 495-498. *Citrus Industry,* Vol. II. Univ. Calif. Press., Berkeley.

72. Fawcett, H.S. and J.M. Wallace. 1946. Evidence of the virus nature of citrus quick decline. *Calif. Citrogr.* **32**: 50, 88.

73. Feld, S.J., J.A. Menge and J.E. Pehrson. 1979. Brown rot of citrus. A review of the disease. *Citrograph* **64 (5)**: 101.

74. Feld, S.J., J.A. Menge and L.H. Stolzy. 1990. Influence of drip and furrow irrigation on Phytophthora root rot of citrus under field and greenhouse conditions. *Plant Dis.* **74**: 21.

75. Fonseca, M.E.N. and E.W. Kitijima. 1993. French marigold *(Tagetes patula)*: A new experimental host of citrus exocortis viroid. *Plant Dis.* **77**: 953.

76. Gabriel, D.W., M.T. Kingsley, J.E. Hunter and T. Gottwald. 1989. Reinstatement of *Xanthomonas citri* (ex Hesse) and *X. phaseoli* (ex Smith) to species and reclassification of all *Xanthomonas campestris* pv. *citri* strains. *Int. J. Syst. Bacteriol.* **39**: 14.

77. Garnier, M. and J.M. Bove'. 1983. Transmission of the organism associated with citrus greening disease from sweet orange to periwinkle by dodder. *Phytopathology* **73**: 1358.

78. Garnsey, S.M. and R.F. Lee. 1988. Tristeza, pp. 48-50. In: *Compendium of Citrus Diseases.* J.O. Whiteside, S.M. Garnsey and I.W. Timmer. (eds.). APS Press.

79. Garnsey, S.M., H.C. Barrett and D.J. Hutchinson. 1987. Identification of citrus tristeza virus resistance in citrus relatives and its potential applications. *Phytophylactica* **19**: 187.

80. Garrett, S.D. 1956. *Biology of Root Infecting Fungi.* Cambridge Univ. Press.

81. Gonslaves, D. and S.M. Garnsey. 1989. Cross protection techniques for control of plant virus diseases in the tropics. *Plant Dis.* **73**: 592.

82. Goto, M. 1972. Survival of *Xanthomonas citri* in the bark tissue of citrus trees. *Can. J. Bot.* **50**: 26.

83. Goto, M. 1992. Citrus canker, pp. 170- 208. In: *Plant Diseases of International Importance.* Vol. III. *Diseases of Fruit Crops.* J. Kumar, H.S. Chaube, U.S. Singh, and A.N. Mukhopadhyay (eds.). Prentice- Hall, New Jersey.

84. Goto, M. and H. Hyodo. 1985. Role of extracellular polysaccharides of *Xanthomonas campestris* pv. *citri* in the early stage of infection. *Ann. Phytopathol. Soc. Japan.* **51**: 22.

85. Goto, M., Y. Yaguchi and H. Hyodo. 1979. Ethylene production in citrus leaves infected with *Xanthomonas citri* and its relation to defoliation. *Physiol. Plant Pathol.* **16**: 343.

86. Goto, M., I. Takemura and K. Yamanaka. 1979. Leakage of electrolytes and amino acids from susceptible and resistant citrus leaf tissues by *Xanthomonas citri. Ann. Phytopathol. Soc. Japan* **45**: 625.

87. Goto, M., Y. Tadauchi and N. Okabe. 1979c. Interaction between *Xanthomonas citri* and *Erwinia herbicola* in vitro and in vivo. *Ann. Phytopathol. Soc. Japan* **45**: 618.

88. Gottwald, T.R., L.W. Timmer and R.G. McGuire. 1989. Analysis of disease progress of citrus canker in nurseries in Argentina. *Phytopathology* **79**: 1276.

89. Graham, J.B. and T.R. Gottwald. 1991. Research perspectives on eradication of citrus bacterial diseases in Florida. *Plant Dis.* **75**: 1193.

90. Graham, J.B., T.R. Gottwald, E.L. Civerolo and R.G. McGuire. 1989. Population dynamics and survival of *Xanthomonas campestris* pv. *citri* in soil in citrus nurseries in Maryland and Argentina. *Plant Dis.* **73**: 423.

91. Graham, J.H. 1986. Citrus mycorrhizae: Potential benefits and interactions with pathogens. *Hort Science* **21**: 1302.

92. Graham, J.H. 1990. Evaluation of tolerance of citrus rootstocks to Phytophthora root rot in chlamydospore-infested soil. *Plant Dis.* **74**: 743.

93. Graham, J.H. and L.W. Timmer. 1992. Phytophthora diseases of citrus, pp. 250-269. In: *Plant Diseases of International Importance.* Vol. III. *Diseases of Fruit Crops.* J. Kumar *et al.*(eds.). Prentice Hall.

94. Graham, G.H. and D.S. Egel. 1988. Phytophthora root rot development on mycorrhizal and phosphorus-fertilized nonmycorrhizal sweet orange seedlings. *Plant Dis.* **72**: 611.

95. Grant, T.J. 1958. Heat treatment for obtaining sources of virus-free citrus budwood. *Proc. Fla. Sta. Hort. Soc.* **71**: 51.

96. Grossner, J.W., F.G. Gmitter Jr. and J.L. Chandler. 1988. Intergeneric somatic hybrid plants from sexually incompatible woody species: *Citrus sinensis* and *Severinia disticha*. *Theor. Appl. Genet.* **75**: 397.

97. Han, J.-S. 1990. Use of antitranspirant epidermal coating for plant protection in China. *Plant Dis.* **74**: 263.

98. Harding, P.R. Jr. 1959. Biphenyl induced variations in citrus blue mold. *Plant Dis. Rep.* **43**: 649.

99. Hartung, J.S. 1992. Plasmid based hybridization probes for detection and identification of *Xanthomonas campestris* pv. *citri*. *Plant Dis.* **76**: 889.

100. Hewitt, H.B. and L. Chiarappa (eds.). 1977. *Plant Health and Quarantine in International Transfer of Genetic Resources*, Ch. 9. CRC Press, Florida.

101. Hickman, C.J. 1970. Biology of *Phytophthora* zoospores. *Phytopathology* **60**: 1128.

102. Kalita, P., L.C. Bora and K.N. Bhagabati. 1996. Phylloplane microorganisms of citrus and their role in management of citrus canker. *Indian Phytopath.* **49**: 234.

103. Kapur, S.P. and S.S. Cheema. 1983. Chemotherapeutic control of citrus greening disease. *Pesticides* **17**: 13.

104. Karasev, A.V., W.D. Dawson, M.E. Hill, S. M. Garnsey and A. Hadidi. 1998. Molecular biology of citrus tristeza virus: Implications for disease diagnosis and control. *Acta Hortic.* **472**: 333.

105. Kaul, J.L. and R.L. Sharma. 1978. Mode of entry of *Geotrichum candidum* causing sour rot of citrus fruits. *Indian Phytopath.* **31**: 77.

106. Keily, T.B. 1964. Brown spot of Emperor mandarins. *Agri. Gaz. N.S. Wales* **75**: 854.

107. Khew, K.L. and G.A. Zentmyer. 1973. Chemotactic response of zoospores of five species of *Phytophthora*. *Phytopathology* **63**: 1511.

108. Khew, K.L. and G.A. Zentmyer. 1974. Electrostatic response of zoospores of seven species of *Phytophthora*. *Phytopathology* **64**: 500.

109. Khilare, V.C. and L.V. Gangawane. 1997. Application of medicinal plant extracts in the management of thiophanate resistant *Penicillium digitatum* causing green mold of mosambi. *J. Mycol. Pl. Pathol.* **27**: 134.

110. Kiritani, K. and H.J. Su. 1999. Papaya ring spot, banana bunchy top and citrus greening in the Asia and Pacific region: Occurrence and control strategy. *Japan Agric. Res. Quarterly* **33**: 23.

111. Klotz, L.J. and T.A. De Wolfe. 1961. Limitations of the hot water immersion treatment for the control of Phytophthora brown rot of lemon. *Plant Dis. Rep.* **45**: 264.

112. Klotz, L.J. and T.A. De Wolfe. 1961. Brown rot contact infection of citrus fruits prior to hot water treatment. *Plant Dis. Rep.* **45**: 268.

113. Koizumi, M. *et al.* (eds.). 1998. Production systems of fruit tree nurseries to control graft transmissible disease. *J. Japan. Soc. Hortic. Sci.* **67**: 1093.

114. Krishna, A. and A.G. Nema. 1983. Evaluation of chemicals for the control of citrus canker. *Indian Phytopath.* **36**: 348.

115. Kuch, T.K. and K.L. Khew. 1982. Survival of *Phytophthora palmivora* in soil and after passing through alimentary canal of snails. *Plant Dis.* **66**: 897.

116. Kumar, S. and R.K. Grover. 1964. Evaluation of fungicides for the control of black rot of sweet orange. *Indian Phytopath. 17: 328.*

117. Lee, R.F. and L.A. Calvert. 1987. Polypeptide mapping of citrus tristeza virus strains. *Phytophylactica* **19**: 205.

118. Lee, R.F. and M.F. Rocha-Pena. 1992. Citrus Tristeza Virus, pp. 226-249. In: *Plant Diseases of International Importance.* Vol. III. *Diseases of Fruit Crops.* J. Kumar *et al.* (eds.). Prentice-Hall, New Jersey.

119. Lee, R.F., L.A. Calvert, J. Nagel and J.D. Hubbard. 1988. Citrus tristeza virus: characterization of coat proteins. *Phytopathology* **78**: 1221.

120. Le Roux, H.F., F.C. Wehner and J.M. Kotze. 1991. Combining fosetyl-Al trunk injection or metalaxyl soil drenching with soil application of aldicarb for control of citrus decline. *Plant Dis.* **75**: 123.

121. Logrieco, A., A. Visconti and A. Bottalico. 1990. Mandarin fruit rot caused by *Alternaria alternata* and associated mycotoxins. *Plant Dis.* **74**: 415.

122. Lu, J.Y., C. Stevens *et al.* 1991. The effect of ultraviolet radiation on shelf life and ripening of peaches and apples. *J. Food Qual.* **14**: 299.

123. Lu, J.Y., S.M. Luombo *et al.* 1993. Effect of low dose UV and gamma radiation on storage rot and physicochemical changes in peaches. *J. Food Qual.* **16**: 301.

124. Lutz, A.L. and J.A. Menge. 1991. Population fluctuations and the number and types of propagules of *Phytophthora parasitica* that occur in irrigated citrus grove. *Plant Dis.* **75**: 173.

125. Malajczuk, N. 1983. Microbial antagonism of *Phytophthora.* In: *Phytophthora: Its Biology, Taxonomy, Ecology and Pathology.* D.C. Erwin *et al.* (eds.). Am. Phytopath. Soc. Press.

126. Mall, S. and G.P. Mall. 1982. Morphology and pathogenicity of *Geotrichum candidum* causing sour rot . *Indian Phytopath.* **35**: 562.

127. Mankau, R. 1963. Effect of organic amendments on nematode populations. *Phytopathology* **53**: 881.

128. Mankau, R. and R.J. Minteer. 1962. Reduction of soil populations of the citrus nematode by the addition of organic material. *Plant Dis. Rep.* **46**: 375.

129. Maramorosch, K. and J.J. McKelvey (eds). 1985. *Subviral Pathogens of Plants and Animals: Viroids and Prions.* Academic Press.

130. Martinez, A.L., D.M. Nora and W.C. Price. 1971. Observations on greening in the Philippines, *Second Int. Symp. Plant Pathol.* New Delhi (India), p. 133.

131. Matheron, M.E. and J.C. Matejka. 1988. Persistence of systemic activity of fungicides applied to citrus trunks to control Phytophthora gummosis. *Plant Dis.* **72**: 170.

132. Matheron, M.E. and J.C. Matejka. 1989. Temporal changes in susceptibility of citrus phloem tissue to colonization by *Phytophthora citrophthora* and *P. parasitica. Plant Dis.* **73**: 408.

133. Matheron, M.E. and J.C. Matejka. 1990. Differential virulence of *Phytophthora parasitica* recovered from citrus and other plants to rough lemon and tomato. *Plant Dis.* **74**: 138.

134. Matheron, M.E. and J.C. Matejka. 1991. Effect of sodium tetrathiocarbonate, metalaxy and fosetyl-Al on development and control of Phytophthora root rot of citrus. *Plant Dis.* **75**: 264.

135. Matheron, M.E. and J.C. Matejka. 1992. Effect of temperature on sporulation and growth of *Phytophthora citrophthora* and *P. parasitica* and development of foot and root rot of citrus. *Plant Dis.* **76**: 1103.

136. Matheron, M.E. and J.C. Matejka. 1997. Distribution and seasonal population dynamics of *Phytophthora citrophthora* and *P. parasitica* in Arizona citrus orchards and effect of fungicides on tree health. *Plant Dis.* **81**: 1384.

137. Matheron, M.E. and M. Porchas. 1996. Colonization of citrus roots by *Phytophthora citrophthora* and *P. parasitica* in daily soil temperature fluctuations between favourable and inhibitory levels. *Plant Dis.* **80**: 1135.

138. McCornack, A.A. and G.E. Brown. 1967. Thiabendazole, an experimental fungicide for fresh citrus fruit. *Proc. Fla. Sta. Hortic. Soc.* **82**: 235.

139. McLaughlin, R.J., C.L. Wilson, S. Droby, T. Ben-Arie and E. Chalutz. 1992. Biological control of postharvest diseases of grape, peach and apple with the yeasts *Kloeckera apiculata* and *Candida guilliermondii*. *Plant Dis.* **76**: 470.

140. McLean, A.P.D. 1957. Virus infection in citrus trees. *FAO Plant Prot. Bull.* **4**: 88.

141. Mehrotra, N.K., N. Sharma, R. Ghosh and M. Nigam. 1996. Biological control of green and blue mold diseases of citrus by yeast. *Indian Phytopath.* **49**: 350.

142. Moll, J.N. and M.M. Martin. 1973. Electron microscope evidence that citrus psylla (*Trioza erytreae*) is a vector of greening disease in South Africa. *Phytopathology* **63**: 41.

143. Morris, S.C. 1982. Synergism of *Geotrichum candidum* and *Penicillium digitatum* in infected citrus fruit. *Phytopathology* **72**: 1336.

144. Morrison, D.J. 1982. Effect of soil organic matter on rhizomorph growth by *Armillaria mellea*. *Trans. Brit. Mycol. Soc.* **78**: 201.

145. Mukhopadhyaya, M.C. and M.R. Dalal 1971. Effect of two nematicides on *Tylenchulus semipenetrans* and on sweet lime yield. *Indian J. Nematol.* **1**: 95.

146. Munnecke, D.E., W. Wilber and E.F. Darley. 1976. Effect of heating or drying on *Armillaria mellea* or *Trichoderma viride* and the relation to survival of *A. mellea* in soil. *Phytopathology* **66**: 1363.

147. Munnecke, D.E., M.J. Kolbezen, W.D. Wilber and H.D. Ohr. 1981. Interactions involved in controlling *Armillaria mellea*. *Plant Dis.* **65**: 384.

148. Naidu, P.H., A. Mani, M.R. Reddy and G.S. Reddy. 1986. Influence of tristeza virus infected acid lime root exudates on *Tylenchulus semipenetrans*. *Indian Phytopath.* **39**: 299.

149. Naqvi, S.M. 1994. Efficacy of some fungicides in control of Phytophthora diseases of Nagpur mandarin in Central India. *Indian Phytopath.* **47**: 430.

150. Nariani, T.K. 1977. Greening disease of citrus in India, pp. 53-58. In: *Mycoplasma Diseases of Trees*. S.P. Raychaudhuri (ed.). Associated Publishing Co., New Delhi (India).

151. Nariani, T.K. and S.P. Raychaudhuri. 1968. Occurrence of tristeza and greening viruses in Bihar, West Bengal and Sikkim. *Indian Phytopath.* **21**: 343.

152. Nariani, T.K., S.P. Raychaudhuri and R.B. Bhalla. 1967. Greening virus of citrus in India. *Indian Phytopath.* **20**: 146.

153. Nariani, T.K., S.P. Raychaudhuri and R.B. Bhalla. 1967. Viruses associated with die-back disease of citrus in northern and central India, pp. 613-618. In: *Plant Disease: Problems*. S.P. Raychaudhuri (ed.). Indian Phytopathological Society, New Delhi.

154. Nariani, T.K., S.P. Raychaudhuri and B.C. Sharma. 1968. Exocortis in citrus in India. *Plant Dis. Rep.* **52**: 834.

155. O' Bannon, J.H. 1968. Observations on seasonal population changes of *Tylenchulus semipenetrans* and influence of temperature on egg hatch. *Nematologica* **14**: 12.

156. O'Bannon, J.H. and A.L. Taylor. 1967. Control of nematodes in citrus seedlings by chemical bare root dips. *Plant Dis. Rep.* **51**: 995.

157. O'Bannon, J.H. and A.T. Tomerlin. 1971. Control of nematodes on citrus seedlings by chemical dips. *Plant Dis. Rep.* **55**: 154.

158. O'Bannon, J.H., R.C. Leathers, and H.W. Reynolds. 1967. Interaction of *Tylenchulus semipenetrans* and *Fusarium* spp. on rough lemon (*Citrus limon*). *Phytopathology* **57**: 414.

159. Ogawa, J.M., R.M. Sonoda and H. English. 1992. Post-harvest diseases of tree fruits, pp. 405-422. In: *Plant Diseases of International Importance*. Vol. III. *Diseases of Fruit Crops*. J. Kumar *et al.* (eds.) Prentice-Hall.

160. Ohr, H.D. and D.E. Munnecke. 1974. Effect of ethyl bromide on antibiotic production by *Armillaria mellea*. *Trans. Brit. Mycol. Soc.* **62**: 65.

161. Ota, T. 1983. Interaction *in vitro* and *in vivo* between *Xanthomonas campestris* pv. *citri* and antagonistic *Pseudomonas* spp. *Ann. Phytopath. Soc. Japan* **49**: 308.

162. Pathak, V.N. 1997. Post-harvest fruit pathology: Present status and future possibilities. *Indian Phytopath.* **50**: 161.

163. Patil, B.P. and D.C. Warke. 1968. A note on the existence of exocortis virus in India. *Curr. Sci.* **37**: 469.

164. Powell, C.A., R.R. Pelosi and M. Cohen. 1992. Superinfection of orange trees containing mild isolates of citrus tristeza virus with severe Florida isolates of citrus tristeza virus. *Plant Dis.* **76**: 141.

165. Powell, C.A., R.R. Pelosi and R.C. Bullock. 1997. Natural field spread of mild and severe isolates of citrus tristeza virus in Florida. *Plant Dis.* **81**: 18.

166. Prasad, M.B.N.V. and V.N. Rao. 1983. Reaction of some citrus rootstock hybrids for tolerance to Phytophthora root rot. *Indian Phytopath.* **36**: 726.

167. Prasad, S.K. and M.C. Chawla. 1965. Observations on population fluctuation of citrus nematode *Tylenchulus semipenetrans*. *Indian J. Entomol.* **27**: 450.

168. Randhawa, N.S. 1970. Nutrition, pp. 34-43. In: *Citrus Decline in India: Causes and Control*. Chadha *et al.* (eds.) PAU, Ludhiana (India).

169. Rangaswami, G., R.R. Rao and A. Lakshamanan. 1959. Studies on the control of citrus canker with streptomycin. *Phytopathology* **49**: 221.

170. Reddy, G.S. 1968. *Citrus Diseases in India and their Control*. ICAR Tech. Bull. (Agric.) No. 19.

171. Reddy, M.R.S. 1997. Sources of resistance to bacterial canker in citrus. *J. Mycol. Plant Pathol.* **27**: 80.

172. Reddy, M.R.S. and P.H. Naidu. 1986. Bacterial cankers on roots of acid lime (*Citrus aurantifolia*)-a new report. *Indian Phytopath.* **39**: 588.

173. Reddy, M.R.S. and V.D. Murti. 1988. Transmission of citrus tristeza virus by dodder laurel from acid lime to acid lime. *Indian Phytopath.* **41**: 131.

174. Redfern, D.B. 1975. The influence of food base on rhizomorph growth and pathogenicity of *Armillaria mellea* isolates, pp. 69-73. In: *Biology and Control of Soil-borne Plant Pathogens*. G.W. Bruehl (ed.). APS Press.

175. Ribeiro, O.K. 1983. Physiology of asexual sporulation and spore germination in *Phytophthora*. In: *Phytophthora, Its Biology, Taxonomy, Ecology and Pathology*. D.C. Erwin (ed.). American Phytopath. Soc. Press.

176. Rishbeth, J. 1978. Effect of soil temperature and atmosphere on growth of *Armillaria* rhizomorphs. *Trans. Brit. Mycol. Soc.* **70**: 213.

177. Roistacher, C.N. and M. Bar-Joseph. 1987. Aphid transmission of tristeza virus: A review. *Phytophylactica* **19**: 163.

178. Sandler, H.A., L.W. Timmer, J.H. Graham and S.E. Zitko. 1989. Effect of fungicide application on populations of *Phytophthora parasitica* and on feeder root densities and fruit yields of citrus trees. *Plant Dis.* **73**: 902.

179. Santhakumari, P. and R.K. Hegde. 1991. Study of the germination of oospores of *Phytophthora palmivora*. *Indian Phytopath.* **44**: 345.

180. Sastry, M.L.N. and R.K. Hegde. 1987a. Pathogenic variation in *Phytophthora* species affecting plantation crops. *Indian Phytopath.* **40**: 365.

181. Sastry, M.L.N. and R.K. Hegde. 1987b. Distribution of mating types and their role in perpetuation of *Phytophthora palmivora* and *P. meadii*. *Indian Phytopath*. **40**: 370.

182. Sastry, M.L.N. and R.K. Hegde. 1988. Survival of *Phytophthora palmivora*. *Indian Phytopath*. **41**: 118.

183. Sastry, M.L.N. and R.K. Hegde. 1992. Soil percolation and efficacy of fungicides on the inoculum of *Phytophthora palmivora* MF 4, the incitant of black pepper wilt. *Indian Phytopath*. **45**: 71.

184. Semanick, J.S. 1980. Citrus exocortis viroid. *CMI Description of Plant Viruses*. No. 226.

185. Sharma, M.C. and B.C. Sharma. 1969. Toxic metabolite production by *Colletotrichum gloeosporioides* causing citrus die-back in India. *Indian Phytopath*. **22**: 67.

186. Sharma, R.L. and J.L. Kaul. 1978. Incidence of sour rot of citrus in Himachal Pradesh. *Indian Phytopath*. **31**: 214.

187. Singh, H. 1971. Studies on the control of citrus nematode. *M. Sc. Thesis* U.P. Agric. Univ. Pantnagar, India.

188. Singh, R.S. and R.P. Sinha. 1954. The fruit drop in grapefruit due to *Colletotrichum gloeosporioides*. Sci. & Cult. **20**: 41.

189. Singh, R.S. and R.N. Khanna. 1966. Black core rot of mandarin oranges caused by *Alternaria tenuis* Auct. *Plant Dis. Rep*. **50**: 127.

190. Singh, R.S. and R.N. Khanna. 1966. Physiological studies on *Alternaria tenuis* Auct., the fungus causing black core rot of mandarin oranges. *J. Indian Bot. Soc*. **45**: 277.

191. Singh, R.S. and R.N. Khanna. 1969. Effect of certain inorganic chemicals on growth and spore germination of *Alternaria tenuis*, the fungus causing black core rot of mandarin oranges in India. *Mycopath. et Mycol. Appl*. **37**: 89.

192. Singh, V. and B.J. Deverall. 1984. *Bacillus subtilis* as a control agent against fungal pathogens of citrus fruit. *Trans. Brit. Mycol. Soc*. **83**: 487.

193. Smilanick, J.L. and R. Denis-Arrue. 1992. Control of green mold of lemons with *Pseudomonas* species. *Plant Dis*. **76**: 481.

194. Smilanick, J.L., D.A. Margosan and D.J. Henson. 1995. Evaluation of heated solution of sulphur dioxide, ethanol and hydrogen peroxide to control postharvest green mold of lemons. *Plant Dis*. **79**: 742.

195. Smilanick, J.L., B.E. Mackey, R. Reese, J. Usall and D.A. Margosan. 1997. Influence of concentration of soda ash, temperature, and immersion period on the control of postharvest green mold of oranges. *Plant Dis*. 81 379.

196. Smilanick, J.L., I.F. Michael, M.F. Mansour, B.E. Mackey, D.A. Margosan *et al*. 1997. Improved control of green mold of citrus with imazalil in warm water compared with its use in wax. *Plant Dis*. **81**: 1279.

197. Smith, G.S., D.J. Hutchinson and C.T. Henderson. 1991. Comparative use of soil infested with chlamydospores to screen for relative susceptibility to Phytophthora root rot in citrus cultivars. *Plant Dis*. **74**: 402.

198. Smoot, J. J. and J.R. Winston. 1967. Biphenyl resistant citrus green mold reported in Florida. *Plant Dis. Rep*. **51**: 700.

199. Srivastava, M.P. and R.N. Tandon. 1969. Some storage diseases of orange. *Indian Phytopath*. **22**: 282.

200. Stolzy, L.H., J. Letey, L.J. Klotz and C.K. Labanaskas. 1965. Water and aeration as factors in root decay of *Citrus sinensis*. *Phytopathology* **55**: 270.

201. Suit, R.F. and E.P. Du Charme. 1953. The burrowing nematode and other plant parasitic nematodes in relation to spreading decline of citrus. *Plant Dis. Rep*. **37**: 379.

202. Suit, R.F. and H.W. Ford. 1950. Present status of spreading decline. *Proc. Fla. Sta. Hort. Soc*. **63**: 36.

203. Swift, M.J. 1968. Inhibition of rhizomorph development of *Armillaria mellea* in Rhodesia forest soils. *Tran. Brit. Mycol. Soc*. **51**: 241.

204. Takahashi, T. and N. Doke. 1984. A role of extracellular polysaccharides of *Xanthomonas campestris* pv. *citri* in bacterial adhesion to citrus tissue in a pre-infectious stage. *Ann. Phytopath. Soc. Japan* **50**: 565.

205. Takahashi, T. and N. Doke. 1985. Purification and partial characterization of agglutinins in citrus leaves against extracellular polysaccharides of *Xanthomonas campestric* pv. *citri*. *Physiol. Plant Pathol.* **27**: 1.

206. Thorne, G. 1961. *Principles of Nematology*. McGraw Hill.

207. Timmer, L.W. 1979. Preventive and systemic activity of experimental fungicides against *Phytophthora parasitica* on citrus. *Plant Dis. Rep.* **63**: 324.

208. Timmer, L.W. and W.S. Castle. 1985. Effectiveness of metalaxyl and fosetyl-Al against *Phytophthora parasitica* on sweet orange. *Plant Dis.* **69**: 741.

209. Timmer, L.W. and J.A. Menge. 1988. Phytophthora-induced diseases, pp. 22-24. In: *Compendium of Citrus Diseases*. J.O. Whiteside, S.M. Garnsey and L.W. Timmer (eds.). APS Press.

210. Timmer, L.W. and S.E. Zitko. 1992. Timing of fungicide applications for control of post-bloom fruit drop of citrus in Florida. *Plant Dis.* **76**: 620.

211. Timmer, L.W. and S.E. Zitko. 1993. Relationships of environmental factors and inoculum levels on the incidence of post-bloom fruit drop of citrus. *Plant Dis.* **77**: 501.

212. Timmer, L.W., T.R. Gottwald and S.E. Zitco. 1991. Bacterial exudation from lesions of Asiatic citrus canker and citrus bacterial spot. *Plant Dis.* **75**: 192.

213. Timmer, L.W., J.P. Agostini, S.E. Zitko and M. Zulfiqar. 1994. Post-bloom fruit drop, an increasingly prevalent disease in the Americas. *Plant Dis.* **78**: 329.

214. Toxopeus, H. J. 1937. Stock-scion incompatibility of citrus and its cause. *J. Pom. Hort. Sci.* **14**: 360.

215. Tsao, P.H. 1969. Studies on the saprophytic behaviour of *Phytophthora parasitica* in soil. *Proc. First Int. Citrus Symp.* **3**: 1221.

216. Tsao, P.H. 1971. Chlamydospore formation in sporangia-free liquid culture of *Phytophthora parasitica*. *Phytopathology* **61**: 1412.

217. Tsao, P.H. and M.J. Garber. 1960. Method of soil infestation, watering, and assessing the degree of root infection for greenhouse in situ ecological studies with citrus Phytophthoras. *Plant Dis. Rep.* **44**: 710.

218. Tsao, P.H. and J.L. Bricker. 1964. Soil fungistasis in relation to zoospore germination of *Phytophthora parasitica*. *Phytopathology* **54**: 910.

219. Tsao, P.H. and J.L. Bricker. 1968. Germination of chlamydospores of *Phytophthora parasitica* in soil. *Phytopathology* **58**: 1070.

220. Turner, P.D. 1963. Influence of root exudates of cacao and other plants on spore development of *Phytophthora palmivora*. *Phytopathology* **53**: 1337.

221. Turner, P.D. 1965. Behaviour of *Phytophthora palmivora* in soil. *Plant Dis. Rep.* **49**: 135.

222. Ullasa, B.A. *et al.* 1998. Competitive behaviour of benzimidazole-resistant strains of *Penicillium italicum* in citrus. *Indian Phytopath.* **51**: 72.

223. Van Gundy, S.D. 1958. The life history of the citrus nematode, *Tylenchulus semipenetrans*. *Nematologica* **3**: 283.

224. Van Gundy, S.D. and J.P. Martin. 1961. Influence of *Tylenchulus semipenetrans* on the growth and chemical composition of sweet orange seedlings in soils of various exchangeable cation ratios. *Phytopathology* **51**: 145.

225. Van Gundy, S.D. and P.H. Tsao. 1963. Growth reduction of citrus seedlings by *Fusarium solani* as influenced by citrus nematode and other soil factors. *Phytopathology* **53**: 488.

226. Van Gundy, S.D., A.F. Bird and H.R. Wallace. 1967. *Phytopathology* **57**: 559.

227. Vauterin, L., B. Hoste, K. Kersters and J. Swings. 1995. Reclassification of *Xanthomonas*. *Int. J. Syst. Bacteriol.* **45**: 472.

228. Venkataswarlu, Ch. and S. Ramapadu. 1992. Relationship between incidence of canker and leaf miner in acid lime and Sathgudi sweet orange. *Indian Phytopath.* **45**: 227.

229. Wallace, J.M. 1969. Tristeza disease investigations; an example of progress through cooperative international research, pp. 29-39. In: *Citrus Virus Diseases*. J.M. Wallace (ed.). University of California Press, Berkeley.

230. Wargo, P.M. and C.G. Shaw. 1985. Armillaria root rot: The puzzle is being solved. *Plant Dis.* **69**: 826.

231. Watanabe, K., M. Miyakado, N. Ohno, T. Ota and F. Nonaka. 1985. Citrusnin A: a new antibacterial substance from leaves of *Citrus natsudaidal*. *J. Pesticide Sci.* **10**: 137.

232. Waterhouse, G.M. 1963. *Key to the species of Phytophthora*. CMI Misc. Pap. No. 92. Kew, England.

233. Waterhouse, G.M. and J.M. Waterston. 1964. *Phytophthora citrophthora*. Description of pathogenic fungi and bacteria. No. 33. CMI, Kew, England.

234. Weathers, L.G. and E.C. Calavan. 1969. Nucellar embryo as means of freeing citrus clones of viruses. In: *Citrus virus diseases*. J.M.Wallace (ed.). University of California Press.

235. Webb, H.J. 1943. The tristeza disease of sour orange rootstock. *Proc. Am. Soc. Hort. Sci.* **43**: 160.

236. Whiteside, J.O. 1970. Factors contributing to the restricted occurrence of citrus brown rot in Florida. *Plant Dis. Rep.* **54**: 608.

237. Wilson, C.L. and E. Chalutz. 1989. Postharvest biocontrol of Penicillium rots of citrus with antagonistic yeasts and bacteria. *Scientia Horticulturae* **40**: 104.

238. Wyss, U. 1988. Pathogenesis and host-parasite specificity in nematodes, pp. 417-432. In: *Experimental and Conceptual Plant Pathology*. R.S. Singh, U.S. Singh, W.M. Hess and D.J. Webber (eds.). Oxford and IBH Publishing Co. (P) Ltd. New Delhi, India.

239. Young, J.M., J.F. Bradbury, L. Garden, T.I. Gvozdyak, D.E. Stead, Y. Takikawa and A.K. Vidaver. 1991. Comments on the reinstatement of *Xanthomonas citri* (ex. Hesse 1915) Gabriel *et al.*, 1989 and *X. phaseoli* (ex. Smith 1897) Gabriel *et al.*, 1989: Indication of the need for minimal standards for the genus *Xanthomonas*. *Int. J. Systematic Bacteriol.* **41**: 172.

Diseases of Pome and Stone Fruits

■ APPLE SCAB

Scab, one of the most serious diseases of apple throughout the world, was reported from Sweden as early as 1819. Subsequently, it was reported from Germany in 1833, United States in 1834, England in 1855, and Australia in 1862. It is now known to occur in different parts of Europe, in North and South America, South Africa, Australia and New Zealand, apart from India and Pakistan. In the Indian subcontinent apple scab was first reported in the Kashmir Valley and in Lahore (in Pakistan) in 1935. It has been destructive in the major apple growing states of Kashmir and Himachal Pradesh since 1973 when its first epidemic was recorded. The disease remained confined to the Kashmir Valley till 1977. In late 1977 it was seen in a number of apple orchards in Shimla and Kulu areas of Himachal Pradesh. During the next 5 years (1978-1983) the scab affected apple area in Himachal Pradesh increased from 150 to 40,000 hectares (3). In Kashmir over 60% of the apple orchard area had become scab affected by 1985. In addition to the Kashmir Valley and Himachal Pradesh, where its destructiveness is now well known, the disease is also reported from some orchards in the Kumaon Hills of U.P., Sikkim and Arunachal Pradesh in the north-east India, and also in Conoor and the Kodaikanal hills of Tamilnadu in south India.

Losses due to scab result from (i) premature defoliation of the tree, (ii) weakening of the tree, (iii) quantitative loss in fruit yield, (iv) low market price of affected fruits due to malformation and scab spots, and (v) increased cost of cultivation and disposal of fruits. In India, the first epidemic that occurred in 1973 in the Kashmir Valley ruined the apple crop worth Rs. 54 lakh (US $ 540,000) in a single season (93). In the Shimla hills, each year, 16-34% fruits are rendered worthless due to scab. In Himachal Pradesh, the loss amounted to Rs. 1.5 crores (US $ 1.5 million) in 1983.

The apple scab pathogen mainly attacks the genus *Malus*. All cultivated varieties of apple including the Delicious group are susceptible. Some varieties such as Cox's Orange Pippin and Jonathan, which are rated as highly susceptible to scab in many European countries, are either free from infection or show only mild infection in India. In addition, the pathogen is reported on *Cotoneaster bacillaris* (88) and *C. aitchinsonii* (201) in India.

A similar scab of pear caused by *Venturia pirina* Aderhold is also reported from Kashmir. It is an economically important disease of pear throughout the world. Conditions for infection and development of this disease are similar to those for apple scab. Major aspects of apple scab are covered in several reviews (93, 160).

Symptoms

Typical symptoms of scab appear on leaves and fruits and, in a severe attack, on 1-3 year old shoots. Petioles, pedicels, and sepals are also attacked. However, the symptoms on leaves and fruits are very distinct. The first symptoms are seen in spring on young leaves and flower buds. Light brown or olive-green, irregular spots develop on the lower surface of leaves as this surface is exposed to the outside atmosphere at this stage of tree growth. The spots may not be very conspicuous due to hairy underside of the leaf. Later, more pronounced spots are seen on the upper surface of leaves. The spots are darker with velvety, greyish dark surface and are more circular in outline. Soon these lesions become even more circular and metallic black in color. The tissue surrounding these spots is often thickened and sometimes bulged upward. Such spots persist on the leaf which develops resistance to fresh infections with advancing age (222). There is premature fall of leaves and floral buds. The leaves of susceptible apple varieties produce dark coloured spots with good sporulation of the fungus while on resistant varieties the spots are of a lighter colour with poor sporulation.

The scab spots on fruits are usually well-defined in shape and appearance against the shining and coloured background of the fruit skin. As the fruits mature in the orchard they gain resistance to infection (223, 270). The size and shape of fruit lesions varies with host variety and stage of development of the fruit at the time of infection. The spots are initially dull in appearance, brownish black and become almost black with passage of time. Early scab infection results in splitting of the skin of the fruit in the area occupied by the spot. Later, this forms corky layers with deep cracks. When infection occurs early near the stem end or on calyx the fruits are often deformed. Late infection of fully grown fruits on the tree does not cause cracking of the skin and formation of corky layer.

Fig. 10. Scab spots on apple leaves. Different levels of lesion severity.
Courtesy: Dr. G.K. Gupta

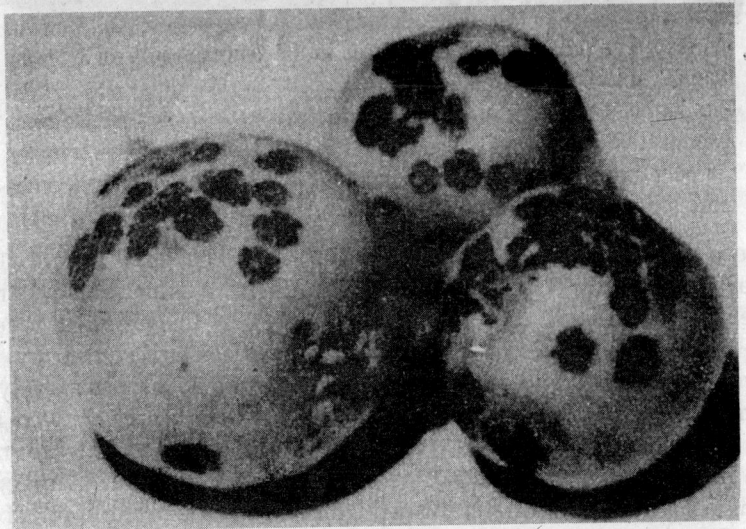

Fig. 11. Scab lesions on apple fruits.
Courtesy: Dr. G.K. Gupta

These spots are usually smaller. McIntosh is more susceptible than Jonathan and Golden Delicious (270). Scab may also develop on apparently clean fruits during storage. The storage scab probably results from late

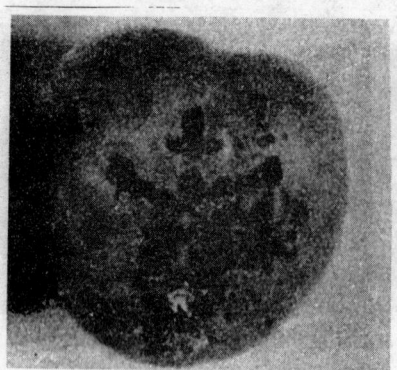

Fig. 12. Scabbed apple fruits snowing cracks in skin.
Courtesy: Dr. G.K. Gupta

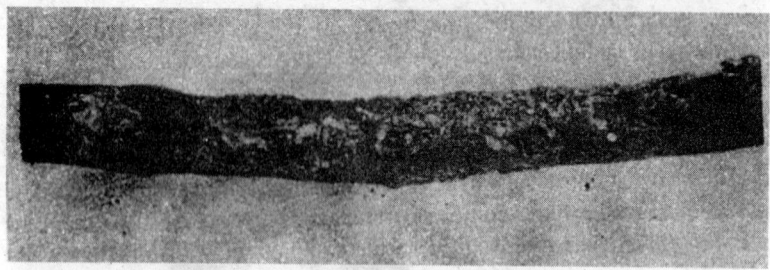

Fig. 13. Twig infection of apple scab.
Courtesy: Dr. G.K. Gupta

season, pre-harvest infections. According to Gupta and Verma (100) apparently healthy Golden Delicious fruits from scab infected trees developed 5-8 lesions after 90 days of storage at 4°C. During storage of scab bearing fruits the scab area increases, new lesions develop, and the entire fruit may rot. The storage rot developed after 45 days in a study by Sharma and Kaul (228) who reported that Royal Delicious showed less scab (3-5 lesions). Scab lesions on apple fruits are known to provide entry sites for many fruit rotting pathogens in transit and storage. Also, the storage of fruits results in diminished fruit quality due to shrinking of the fruit. Shrinking during storage is more severe in scabbed fruits (228).

In shoot infection of scab reported from many temperate climate countries as well as from India (94, 113) the lesions, on 1-3 year old twigs, are small, raised and cinnamon brown in color. The bark is peeled away at the point of emergence of the lesions and looks silvery grey in appearance. These lesions contain conidia during the active season.

- **THE CAUSAL ORGANISM**

Apple scab is caused by the fungus *Venturia inaequalis* (Cke.) Wint. (Ascomycotina-Loculoascomycetes, Pleosporales-Venturiaceae). The conidial stage is *Spilocaea pomi* Fr. (97) which was earlier described in India as *Fusicladium dendriticum* (Wallr.) Fckl. A related species, *Venturia asparata* Samuels and Sivanesan, with a *Fusicladium* conidial stage has been reported in New Zealand occurring side by side with *V. inaequalis* on dead fallen leaves of apple. The perfect stage (*Venturia*) is saprophytic and grows on dead leaves while the conidial stage (*Spilocaea*) is parasitic on living host tissues.

Venturia is the largest and most important genus of the family Venturiaceae. In *V. inaequalis* the septate mycelium is at first light in colour but later turns grey in culture and brownish in the host tissues. The cells of the hyphae are all uninucleate. In young leaves the mycelium develops as branched, radiating, parallel strands. In older tissues and in the fruits these strands are compact and thicker. In the living tissues the mycelium is generally located between the epidermal cells and the cuticle. With age it becomes compact and several layers in thickness. In dead leaves, the saprophytic mycelium grows throughout the mesophyll in which it forms a network of hyphae composed of dark brown, irregularly thick-walled cells which ultimately give rise to ascostromata.

The brown, continuous or rarely septate, conidiophores arise from the hyphal strands or from the more compact stroma. This stroma is 10 or more cells in thickness at the centre of the scab spot but at the margins it is only one cell thick. The length of conidiophores varies within the same lesion. The continuous conidiophore or each cell of the septate conidiophore is uninucleate. Each conidiophore successively cuts off a number of uninucleate conidia at the tip, the conidiophore elongating after each conidium is produced. The conidia are generally 1-celled but often they become 2-celled through septation. They are ovate to lanceolate, with a truncate base and somewhat pointed apex, and smoky brown in colour at maturity. They measure 12-22 × 6-9 μm on the host and 13-36 × 6-12 μm in culture. Variability in colony characters and conidial size in culture is also reported (283). On the host, size of conidia is influenced by cultivar, temperature and relative humidity or rainfall. Size of conidia produced in April is more as compared to those produced later in the growing season. Length of conidia is inversely correlated with temperature and positively correlated with rainfall. Conidial pustules on the fruits show the dark olive conidia surrounded by a fringe of silvery white, torn cuticle.

Late in the season in the dead tissues of fallen leaves, the thick-walled hyphae in the interior begin to form ascostromata (initials of ascocarps or perithecioid pseudothecia). A small coil of hyphae consisting of uninucleate cells initiates the process. As this develops, a coil of multinucleate cells representing the ascogonium is differentiated inside the young stroma and the trichogyne pushes through and protrudes from the stromatal wall. In the meantime an antheridium is formed from a hypha. The fungus is heterothallic and requires two different mating lines for further development of a mature ascocarp (97). In monoascosporic or monoconidial cultures the above developments will take place i.e., ascogonium with trichogyne and antheridia will be seen but the antheridium will not fertilize the ascogonium of the same thallus. No further development is possible unless the antheridium is from a hypha of an opposite mating strain. When two mating strains are present contact is soon established between the antheridium and the trichogyne. Through a pore dissolved in the trichogyne wall at the point of contact the antheridial nuclei pass into the ascogonium and pair with the ascogonial nuclei. The nuclear pairs pass into the ascogenous hyphae which now develop from the base of the ascocarp. Asci develop from croziers. In the meantime the stroma continues to grow and forms the pseudothecium (perithecium).

The pseudothecia in dead leaves mostly mature in early spring. Mature ascocarps are spherical, dark brown to black, and possess a short beak (neck) and distinct ostiole around which single-celled dark setae are present. These fruit bodies measure 90-150 μm in diameter. The perithecial wall is composed of brownish cells, 3-5 layers thick. The number of asci in each perithecium may be as high as 242 but the usual number varies between 50 and 100. The ascus is slightly spatulate or saccate in shape and measures 55-75 × 6-12 μm. It is thin-walled with a short stalk. The young ascus contains a single, large nucleus formed by the fusion of the two nuclei in the pair. Three successive nuclear divisions in the ascus result in the formation of 8 nuclei around which the 8 ascospores per ascus are formed.

The ascospores are oval to boat-shaped, hyaline, yellowish or olivaceous in colour and 2-celled. The upper cell is shorter and somewhat wider than the lower cell. The unequal size of the 2 cells gives the species its name, *inaequalis*. These spores measure 11-15 × 5-7 μm (average 13 × 5 μm). At first the young ascospores are spherical, uninucleate and unicellular. A nuclear division followed by a cell division occurs and each spore is then composed of two equal, uninucleate cells. The spore then elongates rapidly, the two cells growing unequally.

Variability in monoconidial cultures of *V. inaequalis* had been reported in 1934 (194). At least 5 strains of the fungus are known. These strains differ in their morphologic, physiologic, cultural, and pathogenic reaction (241). Two strains occur in the Kashmir Valley (96). The earlier known race which was first detected on the variety Ambri is a mild strain while the strain occurring now on Red Delicious is more virulent with heavy sporulation and resembles Race 1 of the pathogen. It is of worldwide occurrence. A host specific strain occurs on crab apple (*Malus baccata*). The strains occurring on *Cotoneaster* spp. are also other host specific strains.

Disease cycle

The apple scab fungus has two distinct stages in its life cycle: the saprophytic pseudothecial or over-wintering stage in dead fallen leaves, and the conidial, summer stage which is parasitic on leaves, flower buds, fruits and shoots on the tree during spring and summer. The saprophytic stage is considered the major source of survival of the pathogen during dormancy of the host in winter and provides the inoculum in the form of ascospores for primay infection of new leaves and flower buds in spring. This initiates the active parasitic stage by formation of lesions on leaves in which millions of conidia are produced. These conidia are responsible for repeated summer cycles of the life history of the pathogen and spread of the disease during the entire summer. Due to the short generation time of 8-10 days under ideal conditions of humidity and temperature several cycles (conidia-infection-conidia) occur within 3-4 months and epidemics develop. With leaf fall in early winter, the pathogen enters its saprophytic stage on dead leaves where ascocarps develop and mature towards the end of the winter just before activity of the tree starts in spring. Good snow coverage and alternate periods of wetness and dryness in winter and early spring favour development of pseudothecia in the leaf tissue. The pseudothecia are formed within 30 days of leaf fall occurring from September to November.

The presence of scab lesions on shoots of standing trees in the orchard may be another source of survival of the pathogen during winter (113). Becker *et al.* (22) have emphasized the role of conidia in perennation and primary infection. They detected viable conidia in dormant apple buds and on early developing apple tissues. The numbers varied from zero to 142 viable conidia from inner tissues of flower buds. When these conidia were inoculated on apple tissues at an early stage of bud growth there was an increase in sepal and fruit infection. The conidia were not detected on surface of buds or lesions on infected shoots. The shoot lesions as source of perennation and primary inoculum may be of importance in areas where fallen leaves decompose rapidly and early, eliminating the pseudothecial inoculum. In India, no experimental evidence is available to consider shoot lesions (94) as a source of primary inoculum (3). Conidia have a short life of about 14 days. It can be presumed that by the time the shoot lesions become active in spring to produce conidia there is already enough primary inoculum in the form of ascospores. Thus, these lesions may be playing only an additive role. The scab bearing shoots are important for long distance dispersal of the pathogen from country to country and, in India, from state to state, when they are used for vegetative propagation.

The maturation of ascospores (74, 161, 162), release of mature ascospores, infection of the host parts, and repetition of conidial cycles are highly influenced by weather conditions such as autumn and spring rains, temperature, and leaf wetness periods (114-116). Heavy snow and rains in late winter favour the development of ascocarps and the discharge of ascospores. Ascospore maturity occurs at a wide range of temperatures from 4°C to 24°C with optimum at 16°-20°C while temperatures above 24°C retard ascospore development. Optimum temperature for early growth of the ascocarp is 13°C. Dew and rains wet the leaves on the ground and cause swelling of pseudothecia and favour ejection of ascospores. Dew is not as effective as rains (114). Intermittent rains at short intervals or even heavy dew influencing the period of wetting and drying of leaves increases the rate of maturity and discharge of ascospores provided the temperature remains below 24°C. Continuous sunshine is lethal for ascospores. In a single scab infected over-wintered leaf up to 2 million ascospores are formed (3). Each pseudothecium starts producing 1475 ascospores/ml which rises to 69500/ml at petal fall stage of the tree (227).

The discharge of ascospores starts at a time when the buds on the trees are about to burst open in spring. The total period of ascospore discharge from fallen leaves varies from 4 to 10 weeks depending on weather and then the perithecia are exhausted. It may continue till May, 2 to 3 weeks after petal fall (3) depending on the prevalence of cool and mild rainy weather. Confirming these observations, Sharma and Gupta (227) have stated that mature ascospores are available in the overwintered leaves as early as silver tip stage of apple in March-April with peak ascospore maturity and production occurring at petal fall stage in April-May. The average ascospore emission period is 74 days. On an average 9 apple scab infection periods occur in the spring and early summer in Himachal Pradesh (India).

While the ascospore discharge is taking place the lesions on shoots also produce conidia adding to the inoculum load. In cloudy, damp, and windy weather ascospores may be disseminated upto a distance of several kilometres from the source of production. Leaves that fall in September mature pseudothecia at a much faster rate and discharge ascospores a month earlier than those falling in November and December. Thus, pseudothecia formed in leaves fallen early are not as important as the leaves falling late for source of primary inoculum.

The efficiency of ascospores, landing on leaves, to form lesions is reported to be 5-14% (10). The efficiency on cluster leaves of apple flower buds is 6-16%, 3-9%, and 0.4-0.6% at tight cluster, first pink, and full pink to bloom, respectively (214). Thus, susceptibility of leaves and fruits to scab generally decreases with the increasing age of the tissue (222, 223). Ascospores begin to germinate on a wet leaf surface in 2-4 hours and penetrate the cuticle by means of penetration pegs to form hyphal pads (stroma) between the cuticle and the epidermal wall. Radiating hyphae arise from these pads. Germination of ascospores and subsequent infection of young leaves and flower buds are optimum at 10°-22°C. At and above 28°C these processes are inhibited. Germination of ascospores and conidia on the host surface depends on the period for which free water or 90% or more relative humidity is present. This determines the period for which the leaf is susceptible on a given day. The leaf wetness period necessary for infection varies with temperature but minimum is 9 hours. It is longer at low temperatures and shorter at relatively higher temperatures. In pear scab (*V. pirina*) minimum wetness period for foliar infection varies from 10 hrs at 23.9°C to 25 hrs at 7.2°C. It is similar for apple scab infection requirements. These wetness durations determine the infection periods. Sunshine and rise in temperature dries the leaf surface preventing germination of spores and infection. Wetness for a longer period is necessary for the germination of spores at temperatures above or below the optimum. Although Mills (181) had indicated that one third less wetting period is needed for infection with conidia than with ascospores, Roosje (212) had reported that about the same wetting period is required for infection by both types of spores.

The primary infection results in the development of visible lesions within 7-8 d at 19°C, within 17 days at 9°C, and in 21 days at 5°C. The vegetative growth of the pathogen in the host can occur over a temperature range of 0°-30°C with the optimum at 19°-20°C. At higher temperatures there is more mycelial development and less sporulation. Usually, more abundant conidial production occurs at 8°-16°C than at 20°C or above (194). However, some strains of the fungus have the optimum at 16°C and others at 20°C.

Thus, under the ideal temperature and humidity conditions a fresh crop of conidia is produced in 7-8 days. A build up of conidial inoculum for secondary spread is favoured by the presence of susceptible varieties, low temperatures, and number of wet days (151). In the late summer months when temperatures may rise to 30°C or more the exposed conidia are killed. The conidia produced by primary infection are washed down by rains and carried to other green leaves and fruits to start secondary infection. Conidia also germinate best at 10°-20°C, the optimum being 20°C (97). The conidial germ tubes produce disc-like appressoria from which infection pegs arise. On susceptible leaves several appressoria, and thus a large number of infection pegs, are formed by a single germ tube.

The spread of the disease continues throughout the summer and early autumn. During spring the primary infection by ascospores and secondary spread by conidia are overlapping. Kumar and Gupta (152) have reported that fertilizer application increases the susceptibility of leaves to apple scab. Nitrogen contributes the maximum and potassium the least. However, combination of nitrogen and potash reduces disease incidence. In pear scab (*V. pirina*) susceptibility is enhanced by over fertilization with nitrogen which increases total N and Mn and decreases calcium content of the leaf. Some studies in Germany have indicated that manuring and pruning do not significantly affect incidence of apple scab.

Forecasting: To prevent epidemics of apple scab by timely and effective fungicidal sprays forecasting of the disease appearance is considered essential. Control measures are often ineffective if spraying is started after primary infection has taken place on a large scale and primary lesions with conidia have developed. The precise mode of survival on easily located sources (fallen leaves in the orchard) for prediction of primary inoculum and relationship between weather and infection have provided the basis for developing forecasting systems for specific areas. Forewarning of the incidence of apple scab can be done on the basis of an approximate quantity of primary inoculum (ascospores) present at the time when host leaves are susceptible and the frequency of infection periods in the season.

The quantity of primary inoculum and the likely time when ascospore showers will start can be roughly estimated by a periodical examination of the leaf material collected from apple orchard floor (79, 80). The examination of this material in the laboratory can reveal the stage of ascospore maturity and approximate quantity of ascospores likely to be discharged and at what time (79, 80, 89). Such examinations are particularly important in early spring when the host parts are just coming out to receive the inoculum. The most critical period, as regards the host, is the time the buds start swelling until about 2-3 weeks after petal fall. A second, though short, critical period occurs in autumn when cool moist weather prevails permitting severe late infections of fruits and leaves. More recently, it has been shown that for better precision in making the predictions such additional informations as the number of effective ascospores (spores that actually land on and infect a given area of leaf surface) and the relationship between morphological maturity of ascospores and physiological maturity of asci are also required. Lesion forming efficiency of ascospores varies with source of production, host variety and environmental conditions.

Infection period is most important in the prediction of initiation of the apple scab. It is based on leaf wetness periods and accompanying temperature prevalent in spring (181). For instance, if the mean temperature is 17.2°-24°C the incubation period is 9 days provided the leaves are continuously wet. Nine hours of leaf wetness will result in light infection, 12 h of leaf wetness will result in moderate infection and 18 hours of leaf wetness at the above temperatures will result in heavy infection. Such conditions frequently occur during spring and autumn. For measuring wetness, special recorders are used in the orchards. Hours of wetness period interrupted by more than 8 hours of dry period are not included in determining the infection periods. Thakur and Khosla (261) have investigated the relevance of Mills' scab infection criteria for improving efficiency in the monitoring of apple scab control programme in Himachal Pradesh (India). Observations on potted and orchard plants showed that the light infection period required 3 days more for development of scab symptoms at a leaf wetness duration of 9-13 hours and average temperature range of 8.5°-21.4°C whereas moderate and severe infection periods (> 15 hours leaf wetness) required 1 day more to exhibit symptoms. Rescheduling of fungicide applications on the occurrence of infection periods could save at least 3 sprays. They concluded that forecasting and monitored spray programme may be devised in accordance with prevailing weather conditions of the region.

Strict adherance to Mills's criteria for prediction based on leaf surface wetness has not been possible in many countries. MacHardy and Gadoury (163) have revised these criteria. In England, the leaf wetness duration has been substituted with the hours of 90% or higher relative humidity after rains. It gives an equally good prediction. In Germany, a mean day temperature aggregate of 105°F (over 15 days) after March 1 favourable for pseudothecial maturity is also taken into consideration along with leaf wetness period. In the Netherlands, warning of the possible disease development can be issued on the basis of (a) ascospores are ready to mature, (b) ascospore release is expected, (c) ascospore release has taken place, and (d) infection periods have occurred.

Gadoury *et al.* (81) have qualified the above observations with a report that there is disparity between morphological maturity of ascospores and physiological maturity of asci which can affect the timing of first fungicidal spray to control primary infection. According to them, at present the criteria used to judge the stage of development of asci include the delimitation of ascospores, ascospore shape and colour, and extrusion of the endoascus in discharged asci. This morphological rating of ascospore maturity may not be related with the physiological maturity of asci. The ability of asci to release ascospores (physiological maturity) lags behind morphological maturity of the ascospores by several days. The disparity is significant for the effects of spraying in early stages of primary infection season. Physiologically, the asci become ready for discharge of mature ascospore several days after bud break and spraying could be postponed beyond the date estimated by spore showers.

Predictive models and procédures have been developed in United States and Europe. These avoid the cumbersome process of microscopic examination of leaves for ascospore discharge. One such systems, Vintem TM (a PC based programme for farms), developed at the East Malling Horticultural Station in UK alerts about infection periods and then forecasts the scab severity. It calculates the infection efficiency (IE) of ascospores and conidia. The IE values (0-100%) alert the growers that weather conditions favour infection. Then it forecasts scab intensity specific to the particular orchard (48). Aylor (16) has reported a mathematical model of ascospore dispersal. The factors considered in formulating the model were seasonal variation of major factors affecting the risk of infection, susceptibility of leaves to infection, and lesion-causing efficiency of ascospores and seasonal variation of ascospore maturation and release. The model predicts probable numbers of infections occurring in an apple orchard due to ascospores originating in the orchard, compared with those coming from an abandoned, heavily diseased orchard several kilometres away.

Although spray schedules based on weather reduce the number of sprays per season, some studies in New Zealand have suggested that reduced fungicide use by timing sprays with weather information shows higher incidence of scab compared to standard calendar-based spray schedules (27). A chemical indicator, described in Russia, reveals scab spots 3-4 days before they become optically visible. The technique is very useful in adjusting the timing of the spray schedule and reducing the number of applications needed every year (*Entomophaga* 41: 461. 1996).

Management

Control of primary infection by ascospores (also conidia at some places) is the most important step in the management of apple scab. The following steps have been suggested to achieve this:

1) Elimination or reduction of perithecial production
Fallen leaves, bearing stroma for ascospore development, are the main source of survival of the pathogen. Attempts have been made to destroy this source in the orchards. In 1961, it had been demonstrated in Canada that perithecial formation in cultural media was inhibited by excess nitrogen. Practical application of this information was made in England where 5% urea was sprayed on infected trees after fruit harvest at the pre-leaf fall stage. This treatment caused about 97% reduction in ascospore

discharge from fallen leaves in the next spring A pre-bud spray of 2% urea in spring on over-wintered leaves not only reduced ascospore numbers but also prevented release of ascospores from perithecia in the remaining leaves (40, 43). Urea not only suppresses ascocarp development it also hastens decomposition of leaves thus destroying the perithecia. The population of such fungal antagonists as *Trichoderma* and *Cladosporium* spp. is increased (64). Beneficial effects of urea spray are reported in India also (92, 95). Spraying of trees with 5% urea in autumn prior to leaf fall and again 2% urea just before bud burst was recommended before starting fungicidal sprays. Complete suppression of ascospore discharge from over-wintered leaves occurs (282) when 3-5% urea is sprayed on the trees just before leaf fall (last week of October). Post-harvest spray of fungicides, such as 0.4% Benlate or 0.005% organic mercurials also reduces perithecia formation in fallen leaves. Topas-C (penconazole plus captan) and carbendazim spray also is as effective as urea spray. One per cent urea, dodine (Syllit) and fenarimol (Rubigan) were 90% effective. Spotts, *et al.* (248) have reported that an autumn application of dolomite lime to infected leaves on the orchard floor results in a decrease in the percentage of leaves with pseudothecia, in the number of pseudothecia per leaf, and the number of asci per pseudothecium of *V. inaequalis* and *V. pirina*. Application of 5.08 metric tons per hectare reduced the ascospores dose the following spring up to 88% for pear and 92% for apple (248). In an earlier study (179), autumn application of lime caused reduction in ascospore discharge in the following spring but urea was more effective than lime. Collection and burning of fallen leaves in winter, where feasible, is also recommended for elimination of source of primary inoculum. In Canada, thermal treatment of orchard floor by specially designed propane flamers (temperatures raised to 150°-200°C) has been demonstrated to reduce the ascospore maturation inside fallen leaves (*HortScience* **32**: 267. 1997).

2. Fungicidal sprays to prevent primary and secondary infection

A large number of fungicides have been found effective against the apple scab pathogen *in vitro* and *in vivo*. The list includes the sulphur and dithiocarbamate fungicides (flowable sulphur, zineb, mancozeb, cuman, polyram, etc.), the benzene fungicides chlorothalonil (Daconil) and dinocap (Karathane), the heterocyclic nitrogen fungicides captafol (Difolatan) and captan, and many other protectant fungicides such as dodine (Cyprex or Syllit), Dikar (a combination of mancozeb and dinocap), Glyodex (dodine plus glyodin), the mercurials such as phenyl mercury chloride, the systemic benzimidazole fungicides such as benomyl (Benlate), carbendazim (Bavistin, MBC), thiophanates such as Topsin-M and Cercobin-M, and the sterol biosynthesis inhibiting fungicides such as triforin (Saprol), myclóbutanil (Systhane, Eagle, Rally), difenoconazole (Score), fenarimol (Rubigan), and many others.

Among the protectant fungicides, captafol (Difolatan), mancozeb and captan had been extensively used for apple scab control in many countries. Difolatan has high resistance to weathering, persists for a much longer period than other protectant fungicides on the leaf surface and has very low phytotoxicity, thus can be used in high concentrations. In Canada and the United States a single spray of 0.3% Difolatan at the late dormant stage of the trees was found to have a persistent effect against apple scab (189). Since the massive dose is applied before bud burst, it is exclusively deposited on the bark of woody tissue. This deposit serves as a reservoir during subsequent several weeks and the fungicide is redistributed by rains to emerging foliage which is protected against early infection by the apple scab fungus. In the United States a single application of Difolatan at the green tip stage followed by 0.2% captan at petal fall stage and for cover sprays was commonly recommended. Mancozeb (0.25%) is included for two sprays after petal fall in a schedule recommended in India (86). Six sprays of 0.3% mancozeb were found as effective as carbendazim (200). Gupta (87) had reported that spring and summer sprays of Daconil, Cuman L, Euparen-M, dodine, carbendazim, Aureofungin, flowable sulphur and mancozeb provided almost complete control of both leaf and fruit infection. Difolatan (0.15%) gave 81% and 91% control of leaf and fruit scab, respectively. Residue of dodine applied for scab control at the rate of 3 or 6 kg a.i./ha could be detected for 20 days on fruits and for 30 days on leaves (225). The safe waiting period was determined as 14 days with a half life of 2.5-4.6 days on fruits and 4-4.9 days on leaves. Although more specific and effective sterol inhibiting fungicides have been introduced, the protectant fungicides are still included in calendar-based spray schedules against apple scab.

In on-season sprays the protectant fungicides dodine (Syllit 0.15%) and Difolatan (0.3%) provide better control of primary scab on leaves and fruits than the systemic fungicides (90). Delan, benzimidazoles, Daconil, mancozeb, zineb, and Cumal L are good for control of secondary infection. Six sprays of 0.05% Bavistin, Dithane (Indofil) M-45 (first spray at 0.4% and subsequent sprays at 0.15%), or Cuman-L (first spray at 0.4% and subsequent sprays at 0.3%) gave 93-100% control of leaf and fruit scab. A schedule of 5 sprays starting with 0.3% Difolatan and followed by 0.05% Bavistin, 0.25%, Dithane M-45, 0.3% Cuman L, and finally 0.3% Difolatan was highly effective. Thakur and Gupta (259) have reported the effect of pre-symptom sprays on conidial production and secondary spread. Mancozeb retains efficacy (98-100% disease protection in South Africa) in all wetting regimes but some fungicides such as triforine (Saprol) are effective when exposed to light mist only. Increased precipitation reduces their efficacy (224).

The systemic fungicide Benlate (benomyl) was found an excellent fungicide for apple scab control as early as 1968. In India, Bavistin (carbendazim) was used in place of Benlate. These fungicides have preventive as well as curative action. They are equally effective as post-harvest and pre-bloom sprays. The spray of Benlate or Topsin-M during the growing season prevents the formation of perithecia and ascospores in over-wintered leaves (87, 92) thus reducing the primary inoculum load.

Use of sterol biosynthesis inhibiting fungicides

Warner (275) had suggested addition of any sterol biosynthesis inhibiting fungicide to a broad spectrum protectant to prevent development of resistance in the fungus. Attempts have been made to reduce the number of fungicidal sprays (thus reducing the cost of chemical control) by deciding the most suitable timing for the first spray (81) and by choosing sterol biosynthesis inhibiting fungicides which have long duration protective, curative, and eradicative properties (98, 99). Dahmen and Staub (65) reported that the triazole fungicide difenoconazole (Score) is a highly effective fungicide meeting the above requirements and gives 90-100% control of scab. Wilcox *et al*. (289) evaluated several sterol-demethylation inhibitor fungicides for apple scab control. Fenarimol (Rubigan), flusilazol, or myclobutanil (Enzone) could control the disease by only 4 sprays given at tight cluster, pink bud, petal fall and 10 days after petal fall stages. Curative activity of myclobutanil, fenarimol, hexaconazole, bitertanol and flusilazole was demonstrated by Khosla and Gupta (147) who noted that the curative effect was best when the fungicides were applied 72 hrs after inoculation. Beyond this interval the activity decreased or was not reliable. They have recommended the use of bitertanol (Baycor) at 3.0 kg/ha for commercial sprays for control of apple scab in Himachal Pradesh. Others have also reported the antisporulant activity of myclobutanil (0.075%) and other SBI fungicides (260, 261). Ninety to 100% control of scab by application of the anilopyrimidine fungicide cyprodinil (100 mg a.i./lit) is also reported. This anilopyrimidine fungicide significantly reduces fungal infection stages that are formed after penetration of leaf. In apple scab it reduces the growth of subcuticular stroma (148). The choice of SBI fungicides and anilopyrimidines (cyprodinil) should be based on the prevailing temperature. The SBI fungicides are more effective at higher temperatures and anilopyrimidines at lower temperatures (64). Resistance development against cyprodinil is not reported even when used 43 times over 4 years (153).

Bitertanol (0.075%), hexaconazole (0.03%), myclobutanil (0.04%) and penconazole (0.05%) were tested by Sharma and Verma (229) for curative effect (after infection), and eradicant (post symptom) activity. Applied within 72 hrs of start of the infection period, all the fungicides completely suppressed appearance of scab lesions on leaves. Pre-symptom activity was noticed when the fungicides were applied 168 hrs after the infection period. Penconazole was most effective. When applied to lesion bearing leaves, conidia formation was significantly reduced which was more pronounced after 7 days. Myclobutanil was equally effective. Two consecutive sprays of these two fungicides resulted in 89-93% reduction in conidia production by the 14th d. Difenoconazole (Score) and penconazole (Topas) are used on a large scale against apple scab in Russia. Funt *et al.* (77) compared protectant spray programme with curative after-infection sprays, based on predicted infection periods for control of apple scab and grape black rot. Both provided equally good control but after-infection sprays were more economical. The new DMI fungicide fluquinconazole is reported to be more active than a pyrimidine fungicide (215) in *in vitro* tests. Vision is a new specialist polyvalent fungicide comprising two active ingredients in a concentrated suspension. It contains fluquinconazole (50 g/litre) and pyrimethanil (200 g/litre). It is particularly effective against scab and powdery mildew.

The sterol biosynthesis inhibitor fungicides have special importance for areas where incessant rainfall may continue for days because they can be applied with full efficacy even several days after infection periods. In a Brazilian study (33) post-infection applications of fenarimol, bitertanol and tebuconazole were compared. Fenarimol increased the level of control by increasing the dose (2.4, 3.6, 4.8, and 6.0 g/100 L) when applied 96, 120 and 144 hrs after the infection period. Tebuconazole was better than fenarimol when applied 96 hrs after the infection period. Application of fenarimol (6 g/100 L) 120 to 144 hrs after the infection period followed by another spray 7 days later improved scab control. There was good control when a mixture of fenarimol and dodine (6 g + 39 g/100 L) was applied followed by one application of fenarimol 7 days later.

A new generation of microbial fungicides (the strobilurin fungicides) was reported in late 1990s. The fungicide is derived from the natural antibiotic strobilurin A, produced by the symbiotic fungus *Strobilurus tenacellus*. These fungicides have a very broad and balanced spectrum of activity as a foliar fungicide. They dissipate rapidly from soil and surface water and are unlikely to cause undue hazard to non-target organisms. They have preventive and curative action against apple scab and apple powdery mildew. Several fungicides have been developed in this group, viz., kresoxim methyl, beta methoxyacrylates, CGA 279202, azoxystrobin, etc. (63, 165, 272). In addition to apple scab and apple powdery mildew these fungicides are highly effective against powdery mildews of field crop (*Sphaerotheca, Erysiphe*) powdery mildew of grapevines (*Uncinula necator*), downy mildew of grapes (*Plasmopara viticola*), powdery mildew of cucurbits, Cercospora diseases of groundnut and banana, rust of groundnut, and many other foliar diseases (165). These can work in combination with sterol biosynthesis inhibitor fungicides. Fifty four treatments over 6 years suggested that no resistance to these fungicides had developed at the sites sampled (153). Strobilurin compound CGA 279202 provides a long lasting protection when used in combination with DMI fungicides. At the low use rate of 3.75-5.0 g a.i./100 L it significantly controls scab.

Use of mixtures of protectant and systemic fungicides

The apple scab fungus has the tendency to develop resistance to the single site action systemic fungicides (112, 140). The possibility of development of adaptive mutants of *V. inaequalis* that could be resistant to sterol biosynthesis inhibitor fungicides has been reported by many scientists (140, 210, 249). This has necessitated the use of broad spectrum, multi-cite acting protectant fungicides in combination with the systemic fungicides. Mixtures of protectant and systemic fungicides are preferred over alternate application because they have better chances of preventing resistance development. Combinations of SBI fungicides and protectant fungicides have also been suggested. A combination of tebuconazole and captan has been found to provide excellent control of scab and with no residue problem (126). Combination of penconazole with captan (Topas-C) also is used as an effective, anti-resistance-developement combination.

Other approaches to chemical control

Plant oils have also been used against *Ventiria inaequalis* (190) but they are not as effective against scab as against powdery mildew.

Calendar-based spray schedules in India

In Himachal Pradesh (India) the chemical control of apple scab was suggested on the following lines (3, 91):
1. One spray of 5% urea on the trees before general leaf fall in October-November (post-harvest spray).
2. One spray of Difolatan (330 g/100 L water) or a benzimidazole such as Bavistin, MBC or Topsin-M (50-100 g/100 L water), at the silver tip to green tip stage in March-April.
3. One spray of Dithane (Indofil) M-45 at 200 g/ 100 L water or captan at 200 g/100 L water, or if not used in the preceding spray, a benzimidazole (30-50 g/100 L water) at the petal fall stage in April-May.

4. One to three sprays of Dithane M-45 (250 g/ 100 L water) at the post-blossom stage in late summer at 10-15 days interval.

In the Kashmir Valley, the calender based spray schedule recommended to farmers is as follows:

1. First spray of mancozeb (0.3%) or ziram at the silver tip to green tip stage (mid-March to mid-April).
2. Second spray with captafol or captan or dodine 10-12 days after first spray (pink bud stage).
3. Third spray with MBC or Bavistin or Topsin-M (50 g/100 L water) 12-15 days after second spray (petal fall stage).
4. At the fruitlet stage, 12-15 days after the third spray, the fourth spray of mancozeb or ziram or captan or captafol.
5. The above fungicides or dithiocarbamates can be again used for the fifth spray 15-20 days after the fourth spray.
6. During the fruit development stage, 25-39 days after the fifth spray, the sixth spray should be given with a benzimidazole fungicide.
7. If necessary, a pre-harvest spray (25-30 days before harvest) can be given with ziram, mancozeb, or dodine provided the prevailing temperature is not above 27°C.

Sharma (226) made a comparative study of 4 schedules to control apple scab and found them all effective in controlling leaf scab by 86-96% and fruit scab by 88-99%. These schedules consisted of the recommended protectant and systemic fungicides and a last spray of 5% urea at leaf fall stage. Better protection was given by schedules that started with a protectant fungicide at silver tip stage than those in which the first spray was of a systemic fungicide. On the basis of these observations, modifications have been made in the spray schedules mentioned above. Systemic fungicides are avoided in the sprays during silver tip to green tip stage. Captan is preferred over difolatan because of residue problem in the latter and sterol biosynthesis inhibiting fungicide is introduced in the spray at the fruit set stage.

Addition of one spray of a mildewcide at dormant or bud swell stage keeps powdery mildew of apple under check (239). Two pre-harvest sprays of 0.2% captan, 30 and 15 days before harvest, effectively check development and spread of storage scab up to 90 days in the store. Bitertanol (0.075%) is also effective but carbendazim is not so effective in the control of storage scab (228). When dodine is used in a pre-harvest spray the incidence of storage scab is reduced (228). Post-harvest dip of fruits in 0.075% bitertanol also checks storage scab upto 90 days in stores. Pre-harvest sprays of curative fungicides generally control storage scab under the conditions of low inoculum potential but in moderate and high potential the use of protectant fungicides, during the growing season is necessary. If a high possibility of late infections of fruits exists, the fruits should not be put under long term storage.

In the above given schedules it should be ensured that no fungicide is repeated. It is always better to pick up a fungicide from those listed only for one spray. In addition to the above steps precautions should be taken to restrict movement of diseased propagation material.

Biological control

Athelia bombacina and *Chaetomium globosum* are promising biocontrol agents against *Venturia inaequalis* (111). Used as foliar spray, preferably with calcium nitrate as foliar fertilizer, they reduce the scab intensity on leaves. Yohalem *et al.* (293) have reported significant reduction in scab area on leaves by weekly sprays of water extract of anaerobically fermented spent mushroom compost during green tip to petal fall stages.

Resistant cultivars

Sources of resistance to the apple scab fungus exist in wild species and forms of *Malus floribunda* (Vf resistance gene), *M. micromalus* (Vm resistance gene) and *M. pumila* (Vr resistance). During the last two decades about 30 resistant varieties of apple have been released, mostly in the United States and Canada, a few in UK and Europe. These varieties have a high degree of resistance which is mostly monogenic. But these varieties are not resistant to other diseases of apple. The race 6 of *V. inaequalis* is capable of overcoming the monogenic resistance conferred by Vf gene. Thus, genetic resistance can be used as a valid compliment to the use of fungicides rather than as an alternative to chemical control. The apple scab resistant cultivars introduced in India from the United States and Canada are Prima, Priscilla, Sir Prize, Macfree, Coop-12, Red Free, Nova Easygro, Libert, and Freedom. Coop-12 and Red free are being used in a hybridization programme in Himachal Pradesh. Four hybrids, evolved at Mashobra (HP) have shown high field resistance to scab. In comparison to susceptible Red Delicious they exhibit few lesions with poor sporulation under epiphytotic conditions (226). So far none of the resistant varieties have been accepted in commercial planting mainly because of their uncertain market compared to established varieties which are susceptible.

■ POWDERY MILDEW OF APPLE

The powdery mildew is present in apple growing regions of all the countries in the world. Originally, it was considered a disease of nursery stock but is now recognized as a serious threat to bearing trees also. Economic damage from powdery mildew in bearing orchards results from reduction in tree vigour, trunk growth, and blossom bud production, and from aborted blossoms, and fruit russett. Infection can reduce fruit size, weight, and market value. Severe infection can reduce the amount of bloom and almost eliminate the crop in the following season. In nurseries and young plantings, powdery mildew stunts the tree growth and causes poorly formed, misshapen trees. In north India, the disease damages the nursery plants more readily than the older plants during the months of April-June. In certain years it causes considerable damage to leaves and young shoots of bearing trees resulting in reduced yield of fruits. Pear (*Pyrus communis*), peach (*Prunus persica*), and quince (*Cydonia vulgaris*) are also susceptible to the same disease (128).

Symptoms

The disease appears soon after the buds develop into new leaves and shoots. Sometimes, the buds are so heavily infected in the previous season that they are killed before developing into leaves. Even if they grow, the new leaves and shoots are heavily mildewed and weakened. Early symptoms on such leaves consist of small patches of white or grey powdery mass on the under surface but as the disease progresses both surfaces of leaves and twigs become covered with the powdery mass. Affected leaves grow longer and narrower than normal leaves. The margin is curled. Later, the leaves turn brown from the tip downward. A severe attack of the disease results in partial defoliation of the tree. In nursery plants the disease prevents formation of wood in the stem. Petals of infected flowers are pale yellow or green and covered with mycelium. The flowers are shrivelled and may fail to set fruit. They are more susceptible to frost (246). Fruit buds suffer more damage than the vegetative buds. They may not bear any fruit at all. When fruits are attacked, they remain small and deformed and tend to develop a rough surface.

Physiological effects of infection include reduced photosynthesis, transpiration, and carbohydrate content of the host (72), increase in phenolic content (21), accumulation of sulphur at the site of young mildew colonies, and reduced calcium transport to infected leaves (288).

The causal organism

Powdery mildew is caused by *Podosphaera leucotricha* (Ellis and Ever.) Salm. The ectophytic mycelium forms saccate haustoria in the epidermal cells. Aerial conidiophores arise from this mycelium on leaves and shoots. Each conidiophore bears a chain of oval, hyaline conidia which measure 25-30 × 10-12 μm. They contain distinct fibrosin bodies. Perithecial stage of the fungus is rare in India. The cleistothecia are globose, black, partially embedded in the mycelial web and measure 75-96 μm in diameter. Two types of appendages are formed on the surface of these cleistothecia. Some are long and stiff and apically formed while others are basal, short, and tortuous and serve to anchor the fruit to the substrate. The apical appendages are 3-7 times as long as the diameter of the cleistothecium whereas the basal appendages are rarely well developed. Each cleistotheciumcontains a single ascus measuring 55-70 × 40-50 μm. Eight ascospores, 20-26 × 12-14 μm in size, are produced in each ascus (41, 128). The fungus is heterothallic (61).

Disease Cycle

Probably the cleistothecia of the fungus, produced on heavily infected leaves and shoots, do not play any major role in the disease cycle. The ascospores do not germinate readily (292). The disease cycle is mostly conidia-mycelium-conidia. At most places the fungus survives in the form of dormant mycelium or encapsulated haustoria in the dormant terminal and lateral shoot buds (28, 41) and in blossom buds produced and infected in the previous growing season. Survival is greatly affected by temperature (246, 247). Extreme cold in winter (-12°C) may kill the mycelium and buds may be freed from infection. When the buds open in the next season the surviving resting structures produce abundance of conidia which are wind-borne and serve as secondary inoculum. Fruit buds are the earliest to emerge and, therefore, provide the earliest source of secondary inoculum (39). Terminal shoot buds become active slightly later than fruit buds but constitute a more important source of inoculum because of their larger infected area and continued growth, and spore production on new growth well into the season (39). Lateral buds emerge still later and usually do not grow as long but become dominant if the terminal bud is killed. Usually, the healthy buds have come out slightly earlier than the infected buds. As a result when conidia have been formed and are released there is sufficient fresh young tissue for infection (292). The maximum number of conidia still attached to conidiophores are found in 7-12 days old colonies (47). A diurnal periodicity of spore release is reported with a peak of spore content in the atmosphere at mid-day or early afternoon. This is positively correlated with wind velocity, temperature and solar radiation and negatively correlated with relative humidity and leaf wetness (254).

Conidia can germinate at temperatures of 5°-30°C but optimum temperature is between 19°and 22°C (292). The ectophytic mycelium grows well at 20°C. Temperatures of 33°C and above are fatal to conidia. Conidia do not germinate below 88.5% relative humidity. High atmospheric humidity is essential for penetration of the leaf by germinating conidia (125). On germination the conidium produces a germ tub which penetrates the cuticle (41). Following penetration of the cuticle the hypha becomes thin and peg-like, penetrates the epidermal cell wall and forms a haustorium. The penetration of the cuticle is by enzymatic action (292). No appressorium formation has been reported. The infection process does not occur when the leaf surface is covered with a water film.

Management

The old chemical control methods included the use of sulphur dust or spray of lime sulphur. Lime sulphur was recommended for spray according to following schedules:
1) At green tip stage (when buds are green) 1:5 dilution
2) At the open cluster stage, 1:35 dilution
3) At blossoming or full pink stage, 1:60 dilution
4) At petal fall stage when about half the petals have fallen, 1:100 dilution.

More sprays could be given afterwards if necessary. Cupric hydroxide (125 g Cu/100 lit) reduces the disease but causes fruit russet. Addition of slaked lime (2.2 kg/100 lit) reduces fruit russet (27).

Fungicides in use as substitute for elemental sulphur are dinocap (Karathane), oxythioquinox, benomyl (Benlate), carbendazim (Bavistin), thiophanate methyl (Topsin-M) and the sterol biosynthesis inhibiting fungicides fenarimol (Rubigan), triadimefon (Bayleton), bitertanol (Bacor), triforine (Cella, Funginex or Saprol), bupirimate (Nimrod), myclobutanil (Systhane), etaconazole (Vangard), difenoconazole (Score), penconazole (Topas), flusilazole, triflumazole and fluquinconazole (42, 49, 238). Tetraconazole (1 lit/ha), hexaconazole (0.28 lit/ha), triadimenol (0.18 lit/ha) and bitertanol (0.4 lit/ha) are reported to give good control of mildew on some apple cultivars. The strobilurin fungicides (mentioned under apple scab) are also highly effective against the powdery mildew (165). The compound CGA 279202 of this group at 5-7.5 g/100 L provides high level of protective and curative action.

Combinations of SBI fungicides have also been used to enhance spectrum of disease control. Thus, a mixture of Bacor 25 WP (bitertanol) at 0.05% and Bayleton (triadimefon) at 0.025% gives good control of powdery mildew and scab both. Combination of the SBI fungicide fluquinconazole (Palisade 25 WP) and the anilopyrimidine pyrimethanil (Clarinet 200 SP or Vision) also gives control of both diseases (74).

In post-symptom activity, Bayleton reduces the number of normal conidia produced 10 days after treatment and etaconazole and sulphur give reductions 20 days after treatment (55). Four sprays of Bavistin (0.05%) or Morocide (0.1%) starting at bud swell stage and repeated at 15 days interval give control of the disease (236). Gupta and Gupta (101) observed that 0.05% triadimefon (Bayleton) gives the best control by persisting on the host surface for up to 15 days. They have also reported that Baycor, Rubigan and Saprol give a better reduction of conidial production than Karathane. Sharma and Gupta (237) reported the maximum reduction in germination of spores and germ tube growth by triforine, followed by tridemorph, bitertanol and carbendazim. The best disease control in the field was given by bitertanol followed by carbendazin and triforine. Antisporulant activity of bitertanol and carbendazim was noticed for 21 days. In a comparative study of Bavistin (carbendazim), Bacor (bitertanol) and Karathane (dinocap), four sprays of Karathane or Bacor were found superior to Bavistin (238). Sharma and Sharma (239) have recommended that the addition of one dormant or bud swell stage spray of Bavistin to the schedule recommended for apple scab control gives control of powdery mildew also.

The sterol biosynthesis inhibitors are the most advanced mildewcides in use. However, while resistance to benzimidazole fungicides in the fungus is reported, possibility of resistance to SBI fungicides has also been expressed (140). In addition to this problem, the most important period for application of fungicides is also important for economy and effectiveness of the treatments. Winter application of any fungicide is considered useless since the fungus in the buds is protected by thick scale leaves. In highly susceptible cultivars the most suitable time for starting fungicide sprays is the tight cluster or pre-pink bud stage and continued till mid-summer when terminal shoot growth ceases (292). Blossoms must be protected as early as pink bud stage to prevent fruit infection (66). Sometimes, fungicide sprays may be required in the late season to protect unseasonal flush of new growth. Triadimefon can be applied from bloom stage (29). Phytotoxicity of some fungicides during the growing season affecting pollen germination, or causing fruit russet is reported (44, 49).

The length of interval between two sprays is more important than the fungicide concentration. The control is enhanced when the interval is shortened rather than increasing the fungicide rate (46). An interval of 10 days between sprays during the period after bloom until terminal growth stops is recommended (46). Bupirimate and triadimefon applied at consecutive days at two weeks interval are reported to give a better mildew control than when applied at the same rate at one or two weeks intervals.

Plant oils have been more effective against powdery mildew than against scab of apple. Oils of sunflower, olive, maize, soybean, and rapeseed have provided more than 99% control of *P. leucotricha* under controlled conditions when applied to foliage one day before or after inoculation. Mechanically emulsified rape oil was comparable to dinocap (Karathane) and gave 99% control when applied 1-7 days after inoculation (190).

Pruning of dormant shoots infected with mildew in the previous season has been recommended as a means of reducing primary inoculum (39). Removal of these infected shoots on fungicide treated trees reduces secondary infection by nearly half (146). Pruning of dormant tips of all shoots longer than 15 cm also reduces the early disease incidence. However, pruning of infected shoots is not economical when there is heavy infection of the tree. In powdery mildew control it is also important to ensure that the pathogen is prevented from establishing in young trees one to three yr after planting.

The possibility of biological control of apple powdery mildew exists. *Ampelomyces quisqualis*, a hyperparasite of the powdery mildew fungi (described in grapevine powdery mildew) is reported to overwinter in mildewed apple buds and cleistothecia of the fungus and its growth is favoured by wet weather (292).

Differences in cultivar susceptibility to apple powdery mildew are widely recognized. The cultivar Jonathan is highly susceptible (67) while cultivars Red Delicious and Golden Delicious are rated only slightly susceptible or resistant at some locations. Cultivars of apple with commercial quality and resistance or immunity to scab, cedar apple rust, and resistance to powdery mildew are now becoming available in USA (292).

■ PEACH LEAF CURL

The disease is prevalent in peach orchards of Kumaon, Kulu and other parts of the sub-Himalayan range in India. It is a widespread disease in UK, the United States, Japan, China, parts of Australia, New Zealand and South Africa. The disease causes much defoliation and distortion of leaves thus reducing the life span of the tree.

Symptoms

The symptoms first appear in early spring. Soon after the leaves are well out of the bud some of them appear twisted, thickened, puckered, curled downwards and are often greatly distorted. In some leaves only a part of the lamina may be affected but more often the entire leaf is malformed. In the beginning the affected leaves are pale green or yellowish but finally they change to a reddish purple tint which makes them very conspicuous. These leaves are fleshy and thicker than the normal green leaves. The reddish velvety surface of the lamina soon becomes covered with a whitish bloom which represents the fungal fructifications. This growth is more common on the upper surface than on the lower surface. Young shoots attacked by the fungus are swollen and distorted. Even flowers and fruits are sometimes attacked. The affected leaves fall off early and in heavy infections the trees may suffer badly from premature defoliation in late spring which may lead to small fruits and fruit drop. The recurrent attacks of the disease from season to season drain off the vitality of the tree thus reducing its life span.

The Causal Organism

The disease is caused by the Ascomycetous fungus *Taphrina deformans* (Berk.) Tul. The fungus is characterized by the absence of a definite fruit body, the asci being produced in a naked layer on the host surface. The mycelium is intercellular, mostly subcuticular, but often becomes subepidermal or may be situated in deeper tissues. The hyphae are septate with binucleate or multinucleate cells. The binucleate cells are capable of becoming the asci. The host-parasite reaction results in increased growth of the cells on the upper surface of the leaf. A layer of hyphae develops beneath the cuticle of deformed leaves. This layer produces a large number of ascogenous cells which grow vertically rupturing the epidermis. They ultimately develop into naked asci which appear more like exposed terminals of hyphae than the usual clavate cells typical of asci. They measure 25-40 × 8-11 μm. Each ascus contains 8 or less globose, unicellular ascospores measuring 3-4 μm in diameter. Yeast-like budding of the ascospores within the ascus is common. The resulting spores (blastospores) are called sprout conidia and represent the asexual or conidial stage of the fungus. They can again multiply by budding. They are minute and enough in number to fill the entire ascus. The sprout conidia are oval in shape and measure 2.5-6 × 4.8 μm. The ascospores and conidia are discharged violently from the ascus in tiny balls and are carried away by wind or raindrop splashes. This usually occurs in early summer. These spores germinate by producing a very short germ tube of swollen vesicular nature. The ascospores and conidia are uninucleate and haploid. On the surface of the host these conidia either cause immediate infection or may continue budding producing more thin-walled conidia. At the time of germination of these conidia the single nucleus divides and the pair of nuclei passes into the germ tube, thus maintaining the binucleate condition of the hyphae.

Disease Cycle

The asci and sprout conidia can withstand desiccation and remain viable for long periods. These spores persist during winter on twigs, buds and scales and serve as primary inoculum. Since the covering scales, during summer, are smooth in texture the germ tubes from ascospores or conidia do not get any hold and consequently fail to cause infection. However, the swollen vesicular germ tubes bud off a large number of secondary conidia which, due to their yeast-like consistency adhere to the host buds.

During the rains these conidia may be washed down and carried to other parts of the host. In early spring, when the buds are opening in cool, humid weather these conidia germinate (6). The infection threads fasten themselves to the host surface. The fertilization of the haploid conidia or the binucleate hyphae has taken place by this time. The germ tube presses against the thick cuticle of the host and the fungus slowly grows into it. Soon, the hyphae begin to grow in the leaf tissues, especially in the loose parenchyma. There they ramify as closely septate intercellular hyphae.

Management

Tree and orchard sanitation is important. The tree should be kept free from diseased leaves and twigs. All fallen and diseased leaves, twigs, etc. should be collected and burnt. Good results have been obtained by spraying the trees with lime sulphur and copper fungicides. One spray in autumn and another just before opening of the buds considerably reduce the disease incidence. The spray should be thorough and at the right time. A number of copper fungicides such as Bordeaux mixture (6:6:50) and copper oxychloride (0.25-0.3%) have been recommended. In New Zealand, reduced copper sensitivity in isolates of *T. deformans* was detected and use of chlorothalonil was recommended as an effective alternative to copper fungicides (52). Sharma *et al.* (230) recommended dormant stage (January) spray of 0.05% Bavistin (10 lit per 10 yr old tree) or 0.2% Dithane M-45 (mancozeb). Dormant stage spray was more effective than the flowering stage (February) spray. Mehdi and Shah (173) have reported good control of the disease by fungicide sprays before swelling of buds or at bud burst stage. All fungicides such as hexacap (Captan), Dithane M-45, Bavistin, Topsin, Baycor, Bayleton, and Calixin provided reduction in leaf curl. The best control (90%) was given by Captan (1500 ppm) followed by Dithane M-45 (2500 ppm) giving 87.9% control and Calixin giving 71.6% control. Difolatan may also be used as one applicaion only, before leaf drop is complete. Chlorothalonil (Bravo or Daconil) controls the disease if applied twice, in late fall and in early spring. Tate *et al.* (256) reported that two sprays of ziram, captafol, or thiram given prior to green tip stage provided best control of leaf curl followed by dodine, captan and chlorothalonil. In their study, maneb, mancozeb, cupric hydroxide and copper oxychloride were generally ineffective. The sterol inhibiting fungicides difenoconazole, applied twice during bloom, showed eradicative activity and was as effective as chlorothalonil. Follas and Welsh (76) reported that application of difenoconazole at 8, 10, and 15 g/ 100 lit at 10-14 days interval over the blossom period not only controls brown rot (*Monilinia fructicola*) but leaf curl also. The dosage of 10 g/100 lit gave the optimum control of both diseases. Difenoconazole was superior to flusilazole and as effective as captafol.

In a study of 32 peach cultivars (185) none was found immune to leaf curl. Tolerance was found, irrespective of congenial environmental conditions, in cultivars Bed Will's Early, July Elberta and World's Earliest. The cultivars Sharbati, Early Amber, Sun Gold and Sun Red were highly susceptible. Late blooming varieties generally escape the cool wet weather conducive for infection.

■ COLLAR ROT OF APPLE TREES

Collar rot of apple trees, also known as crown rot, basal rot, or foot rot, is a common disease caused by *Phytophthora cactorum* and probably some other species of *Phytophthora* and *Pythium*. Fruit rot of apple caused by *P. cactorum* had been known since 1875. Association of *P. cactorum* with collar rot was first established by Baines (17). The disease is known to occur in Canada, USA, Netherlands, Germany, Australia, New Zealand and India. It is more serious in orchards on dwarfing rootstock (107). Collar rot also attacks apricot, almonds, cherry, peach, pear, and plum. In India, the occurrence of collar rot had been reported on pear and plum as early as 1939 and on peach in 1942. Its occurrence on apple trees as fruit rot was first reported in 1951 from the Kumaon Hills of Uttar Pradesh and the collar rot phase was later reported from Himachal Pradesh in 1960 (103).

Symptoms

Collar rot appears as cankers on the trunk at or near the soil line. New cankers are not easily detected. At first the infection causes water soaking of the bark that appears as a dark area on the trunk. The dark area enlarges in all directions and, if the girth of the trunk is not large, may encircle it. This results in wilting of the leaves and ultimately the plant. On 5-year or older trees the darkening is generally on one side of the trunk and soon becomes a depressed canker below the level of healthy bark. The cankers are usually oval but sometimes irregular. In early stages the outer bark is firm and intact while the inner bark is slimy and may produce a moist gummy exudate. Old cankers look conspicuously brown. If the tree survives, callus formation may occur on the periphery of the cankers but tissue necrosis in streaks may extend to healthy bark tissue. The cankers may spread to branches and to roots. In roots, the infection always starts from the crown.

On diseased trees, leaves have a tinge of light violet colour which later becomes red. In the first year of infection, on grown-up trees, there is no effect on tree vigour and fruit yield. However, in the next 1-2 years trunk rot may cause death of the tree. The fungus may cause fruit rot also.

The Causal Organism

The hyphae of *Phytophthora cactorum* (Labertand Cohn) Schroet. are normally less than 6 μm thick but they are irregularly swollen though without characteristic hyphal swellings. The simple, sympodial sporangiophores have a short stalk in moist air but an elongated stalk in water. They are slender (0.5-1.0 μm thick) with slight swelling at the base of each branch. The cross wall separating the sporangiophore is 3-4 μm down the pedicel and bears a plug. The sporangia are papillate with short (0.5 μm) pedicel, spherical, broadly ellipsoidal, oval, or obpyriform, occasionally intercalary, and 36-50 × 28-35 μm in size. Each sporangium produces 40-90 zoospores.

P. cactorum is a homothallic species. In the host, after the death of tissues, only oospores are seen. Sex organs are abundant on most media. Antheridia are nearly always paragynous, rarely amphigynous, spherical to irregularly clavate, and 15 × 13 μm in size. Oogonia are spherical or tapering to the base. Their wall is thin, hyaline, or slightly yellow. They measure 19-38 (mostly 25-32) μm in diameter. Oospores are rapidly formed in the host but not in culture. They are always aplerotic and 20-26 μm in diameter. The colourless wall is 2 μm thick. The fungus has several physiological variants. Variability occurs between isolates from different hosts and within the apple isolate. Minimum temperature for growth is 2°C, optimum 25°(20°-28°) C and maximum 30°C.

Collar rot affected apple soil contains many other Pythiaceous fungi such as *Phytophthora parasitica*, *P. syringae*, *P. cinnamomi*, *P. cambivora*, *P. citricola*, *P. megasperma*, *P. drechsleri*, *Pythium ultimum*, *Pythium debaryanum*, *Pythium vexans* and *Pythium polytylum* (167, 206). They may contribute to crown or collar rot of apple. Their pathogenicity on young apple seedlings has been proved.

Disease Cycle and Environmental Relations

Oospores and mycelium in the stem cankers serve as a source of primary inoculum. However, for unaffected orchards the unbudded apple rootstocks from propagation nurseries are the major source of primary inoculum (124). Sporangia formed in the stem cankers cause the secondary infections. Sporangia falling on ground release zoospores. The zoospores or sporangia are disseminated by wind and raindrop splashes to aerial parts of the trees where they can initiate fruit rot.

Oospores survive for several months in moist as well as dry soil. Survival up to 2 years is reported (106). Oospore population in soil is maximum during spring and declines in summer and autumn. Light is essential for oospore germination. Mature oospores require soil moisture near saturation and temperature between 4°and 25°C for germination. Optimum temperature for germination is 20°C. There is no germination at 28°C (17). Maximum production of sporangia is reported to be at 15°C. In natural soil they are produced at 10°-25°C. For Indian isolates, a temperature range of 20°-25°C with high soil moisture (>75% WHC) is most effective for production of sporangia (207). At same temperature range oospore production is favoured by low moisture content (< 50-70% WHC). The zoospores can survive in soil for 20 days under different soil conditions. Their survival is better in relatively drier soils and at 7°-20°C (207). Best recovery of zoospores from soil is at 14°C. Mycelium can survive in natural soil for more than a month at low temperature but only up to 10 days at higher temperatures and under high soil moisture conditions (170, 243). The optimum temperature for growth of mycelium is 28°C. Rana and Gupta (205, 207) also reported that under conditions of low soil temperature and low soil moisture the mycelium could survive for 32 days. Lysis of hyphae increased with increase in temperature and soil moisture. Carbon and nitrogen amendments of soil increase lysis of mycelium which is less in highly acidic soils. The fungus could grow up to 5 mm in soil in 5 days. Growth is inhibited by application of ammonium nitrate to soil. In a study of thermal sensitivity of three species of *Phytophthora* and the effect of soil solarization, Juarez-Palacios *et al.* (127) reported that *P. cactorum* is killed in infested soil within 30 min at 45°C. In field studies, solarized soil reached a maximum temperature of 45°C at 15 cm depth and 33°C at the 45 cm depth compared to 31°and 28°C in non-solarized soil. *P. cactorum* was killed within 2 weeks at the 15 cm depth.

Most of the infections occur in humid and cold weather during spring on the lower part of the trunk, branches and fruits on them. However, cankers are seen only at the end of the season. Periods of soil saturation or flooding are usually needed to induce development of Phytophthora crown and root rot (174). Excessive wet conditions and flooding (heavy soils) increase chances of infection. The pathogen enters the host through intact as well as cracked bark. Wounds caused by frost, cultural operations, removal of grasses and insects act as infection sites. Chances of infection increase with increase in age of the tree. Young apple plants are less susceptible (juvenile resistance) but seedling rot can occur. Susceptibility is maximum during flowering stage of the tree and minimum during leaf fall stage in autumn (38). With the onset of winter rate of infection slows down and callus formation occurs around the cankers. New growth on old trees is usually resistant to infection. Scion bark is resistant during dormancy. Infected fruits usually rot in the stores.

The rate of lesion development after infection depends primarily on the prevailing temperature and moisture, pH, cultivar, scion-rootstock interaction, season, and age of root or shoot. A temperature range of 18°-22°C with maximum lesion development at 18 °C has been found most conducive for disease development under Indian conditions. Temperatures of 20°-25°C are a little inhibitory for disease development. The inter- and intra-cellular mycelium resulting from infection soon destroys the bark tissues. Depending on girth of the trunk, necrotic ring around the trunk can form in 3-4 months.

Management

Sanitary precautions recommended for other apple tree disease are necessary in this disease also. Rootstocks susceptible to *Phytophthora* infection (208) should not be used for raising plants for infested soils. Malling (M) and Malling Merton (MM) series of rootstocks M16, MM 101, MM 104, and MM 114 are highly susceptible to Indian isolates of the pathogen. MM 111 and M 25 have more resistance although MM 111 has been found highly susceptible at some places (103, 174). Crab apple (*Malus baccata*) seedlings which are mostly used as rootstock in India are inferior to most of the M and MM rootstocks in resistance (208). Resistance or susceptibility of rootstock varies from region to region. Rootstock MM 111 is susceptible in British Columbia and Ohio but not in Michigan and Pennsylvania (USA). M 2, M 7, M 26 and MM 111 which are resistant under British conditions are susceptible in parts of North America (cf. 103). Resistance of rootstocks is attributed to high phenolic content.

Budding grafting level influences disease incidence. In a study reported by Agarwala (2) zero mortality was recorded when budding height was 23 cm. The inoculum in soil moves with water and poor drainage conditions predispose tree trunk to infection. Therefore, proper arrangement for drainage and prevention of movement of water from diseased to healthy trees is essential. There have been studies to demonstrate that chemical or physical barriers around diseased trees can reduce movement of inoculum with water in soil.

In the early stages of disease development when cankers have not yet penetrated deep and have not encircled the trunk, removal of affected tissue with a sharp knife, cleaning of the wounds, and application of copper or Bordeaux paste on the wound or disinfection of the wound by flame before applying paste has been recommended.

For chemical control of collar rot through soil treatment, Dexon, Difolatan, zineb, Agallol, Dithane D-14, and mancozeb had been tried and found effective (106, 168, 169, 202). Metalaxyl (Ridomil 25 WP), fosetyl-Al (Aliette 80 WP), mancozeb 80 WP and zineb 75 WP effectively inactivate the fungus in soil up to a depth of 8-10 cm. Ridomil and Aliette check the infection in young seedlings and debarked twigs (203). Pre-plant root treatment to reduce incidence of *Phytophthora* species in dormant apple rootstocks was reported by Jeffers (123). Roots were soaked for one hour in aqueous suspension of 1000 mg/L metalaxyl, copper hydroxide, or fosetyl-Al before planting. Metalaxyl gave the best result followed by copper hydroxide. Other fungicides tried by him were etridiazole and oxadixyl. Tidball and Linderman (269) had earlier reported good control of the disease through application of metalaxyl and fosetyl-Al at the rate of 8.9 kg a.i. and 3.4 kg. a.i. per hectare, respectively. Utkhede (275) and Utkhede and Smith (276) have reported reduced crown and root rot, tree mortality, and increased fruit yield by fosetyl-Al applied as foliar spray (95 g a.i./tree) or as soil drench (91 g a.i./tree in 5 lit water).

Utkhede (273, 274, 275) and Utkhede and Smith (276) reported good biocontrol of the disease with the bacterium *Enterobacter aerogenes* applied at the rate of one million colony forming units (CFU) per tree. *Bacillus subtilis* had earlier been reported as a possible biocontrol agent for *P. cactorum* (273). The disease control was as good as with fungicides. The bacterial antagonists are not significantly affected by metalaxyl, fosetyl-Al or mancozeb and thus fungicides and antagonists can be applied together for synergistic effect (273). *Trichoderma* and *Gliocladium* species also are potential biocontrol agents against Phytophthora root and crown rot of apple.

■ **WHITE ROOT ROT OF APPLE AND PEAR**

White root rot affects apple, pear, and many stone fruit trees. In Himachal Pradesh (India) the disease is present in almost all the apple orchards and total annual loss is estimated at Rs. 6 million or US $ 150,000 (4). At high altitudes the orchards may have up to 25% affected trees. In the Kumaon Hills also the disease is quite prevalent. The disease appears to be a serious problem only in India because there are no reports of its economic importance from other apple growing regions of the world. This may be due to soil conditions specific for the apple belt in India. Susceptibility of different rootstocks in India is reported by Gupta and Sharma (104).

Symptoms

Symptoms of the disease become visible a couple of years after the infection of roots has occurred. The foliage shows symptoms of nutritional deficiency caused by any root disease. There are fewer leaves on the tree and these turn yellow and fall prematurely. Growth of branches is slow. Fruits are small-sized and fewer in number. Examination of roots reveals more specific symptoms. White hairy growth of fungus is the characteristic feature. It may be present on the root surface or beneath its epidermis. Usually this growth is in the form of white strands or flat ribbon-like structures. Later, this fungal growth turns brown and less conspicuous. Numerous, minute, black sclerotia or perithecia develop on the dead bark or on the exposed xylem. Due to rot of the roots the tree becomes weak and its growth stops. Gradually, the tree dies. Perithecia and synnemata of the conidial stage develop only on rare occasions.

The Causal Organism

White root rot of fruit trees is caused by *Dematophora necatrix*. Its sexual (perithecial) stage is *Rosellinia necatrix* in Ascomycotina (Xylariaceae). The fungus causes white root rot of apple, grapevine, pear, almond, apricot, mulberry, etc. The mycelium is hyaline, branched, septate and knotted. The hyphae are slightly swollen at septa thus giving a constricted appearance. The mycelial web can be seen on affected roots as mass of white strands but it disappears when roots are dead. Then, sometimes, perithecia appear in clusters. These are globose, black, short pedicellate at the base, ostiolate, and 1-2 mm in diameter. They are embedded in a ropy subiculum of brown, septate hyphae which forms a crust on the host surface at the base of the perithecium. The hyphae of the subiculum are of two types: some are uniform in thickness, 5-8 μm wide, and others exhibit characteristic pyriform swellings, 2-3 times the diameter of the hyphae, and formed immediately above the septum. Asci are cylindrical, long stalked, with a single membrane, 8-spored, and measure 250-380 × 8-12 μm. They have an apical apparatus for dehiscence of spores. The ascospores are straight or curved, dark brown, 30-50 × 5-8 μm in size. When conidiophores are formed, they are produced independently or in association with perithecia on brown ropey synnemata which project straight outwards as rigid columns. The synnemata are up to 1.5 mm high. The stipe is 40-300 μm thick and is composed of flexuous, intertwined, repeatedly branched threads, 2.0-3.5 μm thick. The apex of these branches contains the conidiogenous cell. Conidia are solitary, simple, ellipsoid or obovoid, hyaline to pale brown, 1-celled, smooth and measure 3.0-4.5 × 2.0-2.5 μm.

Disease Cycle

Infected root pieces left in soil are the main source of survival of the fungus. Where perithecia or sclerotia are formed they may also serve as a source of primary inoculum. In warm and wet weather, exploring hyphae develop from these sources, spread in the soil, and reach healthy roots to cause infection. The pyriform swellings at the tip of the hyphae send infection threads into the host root by direct penetration. Infection can occur on old as well as young trees. The infection first occurs on new rootlets (149). Then the mycelium grows, destroys the cortical tissue, and reaches the main root. Cortex and phloem are destroyed but the xylem remains unaffected. When the root is destroyed the fungus moves through soil to fresh roots. The hyphal pieces can also be spread by irrigation or rain water. On old trees symptoms on aerial parts are seen 2-3 years after infection, depending on the amount of roots destroyed (149, 150).

For growth of the fungus, temperatures of 14°-17°C are ideal. Five years to 20 years old trees have been found to be highly susceptible (4). Poor drainage and slightly acidic soils favour the disease. Generally, the disease is more prevalent in orchards planted after removing forest trees, especially oak.

Management

Certain precautions before starting new plantations can avoid the white root rot disease. As far as possible, apple orchards should not be established by removing forest trees, especially oak. The nurseries and main plantation should have good drainage. The nursery soil should be first thoroughly dried in sun and then zineb, at the rate of 100 g/sq. m of bed, should be mixed with the soil. Roots of seedlings should also be treated with 0.5% zineb. In the main orchard, soil should be fumigated with carbon disulphide or 3% formaldehyde at least 3 weeks before planting.

If the disease is detected on new trees, they can be carefully uprooted, infected roots removed, and a fungicidal paste applied to cut surfaces. These trees can be replanted in properly treated fresh pits. The same treatment can be given to standing trees by exposing the roots. Removal of soil around the base of the tree and application of zineb or pentachloronitrobenzene as drench also helps in reducing the disease. Carbendazim has also been used against *Dematophora necatrix* (102). If infection is found on the trunk, the bark of the affected portion can be removed and a fungicidal paste applied to the exposed surface. Badly affected trees should be uprooted and all root pieces should be removed. Fumigation of the site can be done with carbon disulphide. *In vitro* studies have shown that the fungus is sensitive to Aureofungin, Brassicol (PCNB) and Dikar.

- **BROWN ROT OF POME AND STONE FRUITS**

The term "brown rot" refers primarily to the discolouration and decay of maturing fruits on trees of apple (*Malus* spp.), pear (*Pyrus communis*), peach and nectarine (*Prunus persica*), plum (*Prunus domestica*), apricot (*Prunus armeniaca*) and almond (*Prunus amygdalus*), etc. In technical terms and in its restricted sense the term "brown rot " is used to denote the type of decay of pome and stone fruits brought about by one or the other of three closely related fungi (*Monilinia* spp.) which invade the fruit and produce their spore bearing structures on the surface in the form of grey or yellowish cushion-like tufts (291). The decay is not like the soft decay generally referred to as postharvest soft rot. Each of the three species of the pathogen can infect any of the host trees so that if any one of these crops is infected, it may serve as a source of infection for the others.

The brown rot was first reported by Pearson in 1796 in Germany on pear, plum and peach. In England the disease was reported in 1836. It is now present in almost all the areas of the world where these fruits are grown including India where the disease was reported in 1940. However, there is some geographical difference in the occurrence of a particular species in different parts of the world (291). The losses caused by brown rot fungi are enormous. In addition to loss of fruits before harvest and during storage, there is destruction of blossoms also. In severe infections and in the absence of effective control measures, 50-75 % of fruits may rot in the orchard and the remainder may become infected before they reach the market.

Symptoms

The first symptoms of the disease are seen on the blossoms. Brown spots appear on petals, stamens or pistils. They spread rapidly.involving the entire flower and floral stalk. In humid weather the infected portion is covered with the greyish-brown conidial mass of the fungus. Later, these organs shrivel and dry up, the shrivelled brown mass clinging to the twig for some time. This blossom infection may cause reduced fruit set or lead to twig cankers. Production of conidia from infected flowers and their spread to ripening fruits is generally regarded the most destructive consequence of the blossom blight phase of brown rot. Small, sunken, elliptical, brown cankers develop around the infected floral stalks on the twigs bearing the infected flowers. This causes twig blight when the stem is encircled by the canker. In humid weather, gum and also grey tufts of conidia appear on the bark surface. The twig cankers also result in reduced fruit setting.

Fruit symptoms appear when the fruits are approaching maturity. They may also develop on fallen fruits. Small, circular, brown spots appear and spread in all directions. In humid weather they are soon covered with ash-coloured tufts of conidia which break through the skin. These tufts are either scattered or arranged in concentric rings on the fruit surface. Light influences the development of these conidial tufts. Each day a new ring is formed. One large and several small rotten areas may be present on the fruit. The fungus permeates throughout the flesh of the fruit. The loss of water through ruptured skin and through absorption by the fungal growth causes the fruit to soon dry, shrivel and mummify. Thus, there is no wet rot but dry rot of the fruit and mummification is the final result. The mummies may remain hanging on the tree or may fall on the ground. The rotten fruits that fall on the ground also finally get converted into mummies. Sometimes, small cankers develop on twigs or branches bearing the infected fruits.

The post-harvest brown rot or apple black described in India also has similar symptoms (133). Circular spots develop at the site of lenticels and grow in all directions. They become darker and finally black. The fruits remain firm having rubber-like consistency.

The Casual Organisms

The brown rot of pome and stone fruits is caused by the following three species of *Monilinia* (291)

Monilinia spp.	Synonym	Conidial stage
M. fructigena (Aderh. and Ruhl.) Honey	*Sclerotinia fructigena* Aderh. and Ruhl.	*Monilia fructigena* Pers
M. laxa (Aderh. Ruhl.) Honey	*Sclerotinia laxa* Ader. & Ruhl.	*Monilia cinerea* Bon. *Monilia laxa* (Ehr.) Sacc. and Vogl.
M. fructicola (Wint.) Honey	*Sclerotinia fructicola* (Wint.) Rehm. *S. americana* (Wormald) Norton and Ezekiel	*Monilia cinerea* f. *americana* Wormald

Not all the three species occur in every country or geographic region. *M. fructigena* is common in Europe and England, South Africa, Japan, Manchuria, Uzbekistan, Turkey and is reported from Kashmir and Himachal Pradesh in India. It is not found in North and South Americas. *M. laxa* is found in the western part of United States, Europe, Japan, Manchuria, Turkey, Uzbekistan, and probably in Kashmir (India). In Himachal Pradesh it is reported to cause post-harvest brown rot or apple black (133) *M. fructicola* is found throughout North America, some parts of South America, Australia, and New Zealand. It is not found in Europe, Japan and neighbouring areas.

The distinction between *Sclerotinia* and *Monilinia* is that while in the former true, well defined sclerotia are formed, in the latter stromatized host tissues (pseudosclerotia) are formed in the mummies. The morphological differences between the three species are not great. The conidia of *M. fructigena* are somewhat longer than those of the other two species and in mass they look yellow to buff in colour than ashen-grey. The apothecia of *M. fructicola* are larger than those of the other two species. Ascospores of *M. fructigena* are sharply pointed at each end while in *M. fructicola* the ends of ascospores are rounded. The distinction between *M. fructicola* and .*M. laxa* are still less.

The mycelium is branched, hyaline and multinucleate. Soon after the mycelium has reached a certain stage in its growth, it produces short, hyaline conidiophores which occur in dense, velvety tufts (sporodochia) on the substrate which may be a fruit, flower, or even twig. These conidiophores bear long branched chains of hyaline, ovate or lemon-shaped, and 1-celled conidia. These conidia are held in the chain by disjunctors and are easily separated in wet atmosphere and are blown away by wind.

After invasion of the fruit the mycelium spreads rapidly eventually permeating the entire fruit which shrivels and mummifies. The mummies contain the pseudosclerotia or stromata which give rise to stipes with apothecial fundaments at the apex. These fundaments develop into apothecial cups. The asci are cylindrical and contain 8 ellipsoid, 1-celled, hyaline ascospores. When the asci are ripe, the spores are ejected with force forming a whitish cloud over the apothecia. Microconidia or spermatia are also formed on fruit mummies. They serve as agents of spermatization to initiate formation of apothecia on the pseudosclerotia. *M. fructigena* and *M. laxa* produce apothecia very rarely and appear to be heterothallic while *M. fructicola* is homothallic and readily forms apothecia.

The sporodochia of *M. fructigena* are buff-coloured and about 3-4 mm in diameter. In *M. laxa* and *M. fructicola* they are grey and usually half the diameter of those in *M. fructigena*. Size of conidia of *M. fructigena* is about 12-34 × 9-15 (av. about 22 × 13) μm. In *M. laxa*, conidia are of two types, the winter conidia measuring 5-19 × 4-12 (av. about 11.5 × 8) μm and summer conidia measuring 8-23 × 7-16 (av. about 19 × 13.5) μm. In *M. fructicola* the conidia measure 10-27 × 7-17 μm. In the apothecial stage, asci of *M. fructigena* measure 112-180 × 9-12 μm and those of *M. laxa* 121-188 × 7.5-11.8 μm. The asci of *M. fructicola* are 102-215 × 6-13 μm in size. The ascospores are largest in *M. fructigena* (9-12.5 × 5-6.8 μm). In *M. laxa* they are 7-19 × 4.5-8.5 μm and in *M. fructicola* 6-15 × 4-8.5 μm (291).

Disease Cycle

These fungi survive as mycelium in mummified fruits on the trees and in cankers of affected twigs or as pseudosclerotia in mummies on the ground (7, 11). In the spring, the mycelium in mummified fruits on the trees and in the twig cankers produces new conidia while the pseudosclerotia in mummified fruits buried in soil produce several apothecia which release the ascospores. The conidia and ascospores dispersed by wind cause blossom infection. These spores are also dispersed by rainwater and raindrop splashes and by insects. They germinate within a few hours and cause infection of the host. In flower infection the stigmas appear to be the organs most susceptible to attack although all parts may be infected.

The mycelium resulting from primary infection, especially in humid weather, produces short hyphae which grow together, push upward through the epidermis, and form numerous sporodochia on the rotten shrivelled floral parts from which new masses of conidia are produced for secondary infection. The mycelium travels from floral parts into the fruit spurs and twigs. In the twigs the mycelium causes disintegration and collapse of the cells around the fruit spur, and a depressed, reddish brown, shield-shaped canker develops. Further growth of the canker encircles the twig which dies. The surface of these cankers also is soon covered with conidial masses which serve as secondary inoculum.

Infection does not occur directly. Injury to the host surface by various means permits infection. Insects visiting the blossoms are a source of wounds on the host surface. Susceptibility of the fruit to infection increases with its maturity. In general, immature stone fruits are less susceptible than mature fruits but in peaches the fruit before pit hardening is as susceptible as mature fruit (30). Fruit-to-fruit contact enhances chances of spread of brown rot. In prunes (*Prunus domestica*), 43 to 69% of brown rot caused by *M. laxa* and *M. fructicola* occurred in clustered fruit as opposed to solitary fruit (176). After infection the mycelium at first grows intercellularly, disintegrates tissues by enzymic action, produces conidia on the rotten tissue and finally permeates throughout the fruit. The disjointed cells are held together by hyphal threads. Loss of moisture causes drying and mummification. Apart from insects, birds also cause injury to the maturing fruits and thus help in infection.

Management

Complete control of the blossom blight phase including prevention of sporulation is essential for managing the brown rot of fruits. This is achieved by 2-4 sprays of effective fungicides from the time of pink stage of blossom buds until petal fall. The effective fungicides are captan, chlorothalonil (Bravo), dichlone (Phygon), sulphur, thiram, benomyl (Benlate), thiopahante methyl (Topsin M), triforine and many of the new protectant and systemic fungicides such as iprodione (Rovral), vinclozolin (Ornalin), bitertanol (Baycor) and etaconazole (Vangard). Vinclozolin (449 μg/ml), iprodione (449 μg/ml), propiconazole (44 μg/ml) and tebuconazole (67 μg/ml) are reported to be most effective post-infection and antisporulant fungicides against blossom blight of cherry caused by *M. fructicola*. Efficacy of iprodione against brown rot blossom blight of almond (192) and brown rot of peach fruit (190) has been shown. It has good penetrating activity in to the fruit mesocarp (1). *Monilinia* species develop resistance to systemic fungicides, hence, these are recommended for use in combination with protectant fungicides such as benomyl + dichloran. Ethanol, at concentrations more than 30%, improves the efficacy of benomyl + dichloran treatments for control of post-harvest rot of peach fruits caused by not only *M. fructicola* but also *Rhizopus stolonifer* and *Colletotrichum gloeosporioides* (73).

For reducing sources of primary inoculum infected blossoms or cankers should be removed as early as possible. This reduces inoculum available fo, fruit infection later on in the season. Application of fungicides a few weeks before harvest controls infection of ripening fruits. The applications are continued weekly or biweekly until just before harvest. Control of insects reduces injury to floral parts and fruits. On sweet cherries, fungicides have been applied through overhead sprinkler irrigation system against *Monilinia* (191).

For reducing infection of fruits at harvest and during storage the fruits should be picked and handled with care to avoid surface injury. All fruits showing brown rot should be discarded. Post-harvest rot can be prevented by dipping the fruits before storage in a solution of dichloran + benomyl or dichloran + triforine + benomyl for one min. In peaches immersion of fruits in hot water at 52°C for 2.5 min or exposure of fruits to hot air at 54°C in 80% relative humidity for 15 min has been found effective against post-harvest rot caused by *M. fructicola* (20). Peach fruit dip in 10 % ethanol at 50°C for 2.5 min or 20% ethanol at 46°C for 1.5 min is reported to check postharvest brown rot and Rhizopus rot (164). Gamma radiation of peaches checks development of postharvest brown rot (26). Radiation treatments are very costly and the facility is not available everywhere.

Experimentally, biological control of storage rot has been demonstrated. Fruits coated with a suspension of cells of the bacterium *Bacillus subtilis* are protected against infection by *Monilinia* (197). Storage decay of apple fruits is controlled by similar treatment with *Sporobolomyces roseus* (122).

■ **BOTRYOSPHAERIA WHITE ROT, ROOT ROT AND STEM BROWN**

The species of *Botryosphaeria* cause fruit rot of apples in addition to leaf spots and stem canker. *Botryosphaeria ribis* causes white rot or bot rot of apple fruits in warmer apple growing areas of the world including the northern apple belt of India (133). In addition to fruit rot the fungus causes a serious canker disease (stem brown disease) also on apple, peach, pecan and many other trees. Association of *Botryosphaeria* spp. with gummosis of peach trees is also reported (198). The fungus has a very wide host range and is reported to attack 34 genera in 20 plant families although the pathogenicity varies from host to host. The stem brown disease causes 10-15 % tree mortality in the Kumaon Hills of U.P. and in Himachal Pradesh in India.

Symptoms

In stem canker or stem brown disease, water-soaked spots develop on the branches and twigs. These spots look oily. In the first year of infection there is watery exudation on and around the spots which develop into a sunken, violet canker with a papery bark. In the following winter, the bark sheds exposing a gelatinous layer of tissue. During the next spring, due to development of pycnidia, this layer shows a number of raised areas. Small twigs in the upper portion of the tree are more susceptible than the thicker branches in the lower portion. The cankers increase in size during summer and sometimes encircle the twig or the branch. In susceptible varieties of apple die-back of twigs occurs.

Fruit rot is characterized by a soft watery rot which rapidly affects the entire fruit under warm conditions. Two types of decay are seen. In external rot, light coloured sunken spots are seen on the fruit surface. The flesh of the fruit becomes sticky but no pycnidia of the fungus develop on the fruit surface. In the other type, there is no visible external symptoms but fruits kept at room temperature turn soft and decay rapidly. Red skinned varieties of apple lose colour as a result of infection.

The Causal Organism

The stem brown or bot rot disease is caused by *Botryosphaeria ribis* Gross and Dug. [syn. *Botryosphaeria dothidea* (Mough.:Fr.) Cess. and De Not.] of Ascomycotina. The conidial stage of the fungus is known as *Fusicoccum aesculi* Corda. The fungus has two distinct imperfect stages at different periods of the life cycle. In young lesions or on withered tips of twigs *Macrophoma*-type pycnidia are formed. These are globular and 175-250

μm in diameter. The macroconidia are fusoid, aseptate, multinucleate, hyaline, and measure 45-75×16-25 μm. This is the *Fusicoccum aesculi* conidial stage. Later in the season, stromatic bodies are formed along the stem in which *Dothiorella*-type pycnidia and microconidia are produced. These pycnidia function as spermogonia and the microconidia as spermatia. They induce the perithecial stage (*Botryosphaeria ribis*). The perithecia (pseudothecia) are black and the ostiole is much more pronounced than the ostiole of pycnidia. This gives a rough surface to the cankers. Asci are clavate with filiform paraphyses and measure 100-110 × 16-20 μm. They contain 8 ascospores which are 1-celled, multinucleate, ovoid or fusoid, and measure 16-23 × 7-10 μm.

Disease Cycle

The fungus survives in mummified fruits and in dead bark of the tree and these serve as a source of primary inoculum (252). Ascospores and conidia are abundantly produced throughout the growing season and are discharged and dispersed during rains (251, 252). Conidia are generally more abundant than ascospores. In apples viable spores are exuded by pycnidia in cankers for several years (65). In gummosis of peach trees caused by *B. ribis* (198) conidia play a major role in the natural spread of the disease (285, 286). Possibly, the infection occurs throughout the growing season but in fruits it remains latent until fruits begin to ripen (253). However, Kohn and Hendrix (150) had reported that fruit infection is rare unless the soluble solids in the fruit are greater than 10.5%. The optimum temperature for mycelial growth is approximately 30°C (150). Optimum temperature for germination of conidia varies with isolates and is between 26.7° and 29.5°C. Optimum relative humidity is 95-100 %. No germination of conidia occurs at 8.8°C. Ascospores germinate best at 24.6 °C over a relative humidity range of 92-100% (195, 253).

Management

Sanitary precautions and cleanliness of the tree are important for management of the canker. The infected twigs and branches should be cut off 10-15 cm below the point of infection and the cut ends protected by a fungicidal paste. In India, a fungicidal paste consisting of read lead, copper carbonate and lanolin has been successfully used. In the early stages of infection on thick branches or the main trunk, the affected area may be scrapped off and the wound disinfected and protected by the fungicidal paste. An ancient practice observed in the apple orchards of the Kumaon Hills of U.P. (India) is application of fresh cowdung and mud mixture on the cut ends and scrapped portion of the trunk and branches. Often, it provides as good protection as the fungicidal paste until natural healing of the wounds. Cowdung contains undigested fibre, epithelial cells, bile pigments and salts, nitrogen, phosphorus, potassium, sulphur, micronutrients, and intestinal bacteria and mucus. As a substrate for microbial activity, it might be a means of biological control. Spray of 0.2% captan or other dicarboximide fungicides gives protection to the twigs and branches.

■ BOTRYOSPHAERIA BLACK ROT AND TWIG CANKER

Black rot of apple and pear fruits, frogeye leaf spot and twig canker, all caused by the same fungus (*Botryosphaeria obtusa*) can cause considerable damage to trees as well as to fruits in storage especially in warm, humid areas (13). In frogeye leaf spot, many small, violet spots of indefinite margins appear on the leaf blade. These spots are commonly seen 1-3 weeks after petal fall but can also appear earlier. With increase in size to 3-6 mm diameter, the central portion of the spots turns brown to pale brown, surrounded by violet margins. Finally, the spots have a light brown to grey centre surrounded by dark brown area and violet margin.

In canker of twigs, the infected bark is slightly depressed and reddish brown in colour. This infected area slowly increases in size and turns darker. Cracks may appear at the margin restricting further growth of the canker for some time but growth is soon resumed and another crack may be formed. The process continues. Maximum growth of cankers occurs during early rainy season (July-August in India). After a year of infection the bark may separate from the wood. Callus formation may occur and partially cover the canker. On the surface of the canker small blister-like fruit bodies of the fungus may or may not appear.

In fruit rot, generally only one small, brown spot develops on the fruit. Apart from the taste, this differentiates black rot from bitter rot. The colour of the spot changes to black. These spots increase in dimensions through successive concentric, reddish brown rings. In storage, the rotting portion of the fruit becomes leathery. Decayed fruits neither have a bitter taste nor emit any foul smell. Fungus pycnidia develop as small pustules on the rotting portion. Infected fruits drop from the tree.

The Causal Organism

Botryosphaeria obtusa (Schw.) Shoemaker [syn. *Physalospora obtusa*] is a saprophyte growing on dead bark of apple and other trees. The active parasitic phase is its conidial stage, *Sphaeropsis malorum* Peck. The mycelium is branched, septate, intercellular, thick, at first hyaline, becoming olivaceous to dark brown. Pycnidia are generally globose or subglobose, ostiolate, scattered, and measure 200-300 μm in diameter. The ovoid conidia (pycnidiospores) are hyaline, becoming greenish brown or dark brown at maturity, aseptate, but become 1-septate at the time of germination, and mostly 16-36 × 7-10 μm or more in size. Average size is 25×12 μm. Perithecia are similar to pycnidia in construction, size and shape. Asci are broadly clavate, 90-120 × 17-23 μm in size and thick-wallled. Ascospores are ellipsoidal, aseptate, sometimes becoming 1-septate, hyaline to greenish yellow, and 25-33 × 7-12 μm in size.

Disease Cycle

Mycelium of the fungus survives in dead bark, blighted twigs, and decayed and dried fruits. Perithecia are also formed on dead bark and twigs. Conidia and ascospores from these sources serve as primary inoculum. When pycnidia get wet they release conidia in a several mm long cirrus containing up to 1500 spores. In contact with water the spores are set free from the cirrus. In apple viable spores continue to be exuded from pycnidia in cankers for several years (69). Maximum dispersal of spores occurs during blossoming but some dispersal continues throughout the summer. Wind, water and insects are agents of dispersal. Under favourable conditions of temperature and relative humidity, the spores germinate and cause infection. Free water is essential for good germination of conidia. Only some germination can occur at RH as low as 92 % (12). Ascospore germination is equally good at 98-99 % RH.

Optimum temperature for spore germination is 24°-27°C at which infection is also common. During blossoming and immediately after, infection of leaves is very common in wet weather. Infection of bark and fruits is always through wounds and on leaves through stomata on the upper surface. Incubation period on leaves is 2-4 days, on bark 2-9 days, and on fruits symptoms appear within 2 days of inoculation. The pathogen reaches the xylem and pith after infecting leaves and bark. Fire blight affected trees show more infection of *B. obtusa*.

Management

The control measures are similar to those of white rot. Infected blossoms should be removed from about 15-30 cm below the point of infection and cut ends given fungicidal protection. All decayed and mummified fruits should be collected and burnt. Regular sprays of 0.2 % captan, 0.15 % difolatan or folpet are recommended for control of foliar and twig infection. One dormant spray of Bordeaux mixture (6:6:50) has also been found effective. Rovral (0.2 %), thiabendazole (Tecto) 0.1 % or Bavistin

(0.1 %) used as 5-min fruit dip are effective in controlling postharvest decay (black rot) up to 21 days (53). Arauz and Sutton (13) have reported that sprays of demethylation inhibitor (DMI) fungicides tebuconazole, flusilazole and penconazole are effective in controlling the disease when applied 7 days before inoculation. Only tebuconazole (Elite 45 DF) at 101 μg a.i./ml as post-infection application provided satisfactory protection against and eradication of *B. obtusa*, particularly in foliage. It could be used in combination with benomyl or mancozeb. Benomyl and tebuconazole reduce colonization of fruits. Foliage is protected up to 14 days. Flusilazole and penconazole also provide protection to foliage for 7 days. Models have been developed that use temperature and wetness duration to predict periods favourable for infection of apple fruits and foliage by *B. obtusa* and *B. ribis* (12, 195). These help in determining the appropriate timing for spray of fungicides such as benomyl and tebuconazole. If necessary, insect control measures should also be taken. Tree vigour should be maintained by proper fertilization and watering. The disease is more serious on weak trees.

■ **BITTER ROT OF APPLES**

Bitter rot is mainly a disease of fruits on the trees where it may cause cankers also. The disease was first described in England in 1856 and is found in all apple growing regions of the world. The disease is more prevalent on yellow skinned varieties than on red skinned varieties (237). In warm humid weather enormous losses may be caused by destruction of an entire crop of apples just a few weeks before harvest. The pathogen causing the bitter rot of apples attacks pear fruits also.

Symptoms

Symptoms on fruits generally appear when they are full grown, and sometimes when the fruit is half grown. Initiation of the rot is seen as minute, light brown spots which rapidly increase in size to become circular and slightly sunken in the centre. At first, the spots are smooth and dark brown or black. When they reach a size of 1-2 cm in diameter numerous, slightly raised cushions appear in the spots, mostly near the centre, and some extending outward toward the edge of the spot. The tissue below these cushions turns dark coloured. In humid weather the cushions produce creamy, salmon-pink masses of spores. Often these spore masses are in concentric rings in the spot. As the rotted area increases more rings of spore masses appear. In old spots these spore masses are not seen and the tissue becomes dark brown to black, wrinkled and sunken. The rot also spreads towards the apple core forming a cone of watery rotted tissue that may or may not be bitter. Normally, many spots develop on the fruit, enlarging, coalescing and ultimately covering the entire fruit. Such fruits may drop down and mummify on the ground or they mummify and remain clinging to the twigs (257, 258). Sometimes, infection of fruits may occur just before harvest. In such cases minute red to violet spots develop on the fruit. Usually these spots do not appreciably expand during cold storage. But when the infected fruits are brought out for the market and kept at room temperature rotting starts. Ethylene production occurs in affected fruits.

Cankers on woody parts are rare. They may develop only on 2-3 year old twigs. The bark becomes depressed in an oval shape. The xylem below the lesion is killed. In old cankers parallel cracks may develop on the bark. Callus formation around the cankers stops their further growth.

The Causal Organism

Many fungi are reported to cause rot but the fungus causing bitter rot is known as *Glomerella cingulata*. The conidial stage is *Gloeosporium fructigenum*. A bitter rot of pear caused by *Colletotrichum gloeosporioides* (perfect stage *Glomerella cingulata*) is also reported.

The mycelium consists of slender, sometimes septate, hyphae which are hyaline at first but later become olive coloured. Acervuli develop beneath the epidermis which is ruptured when conidiophores and conidia develop and push the epidermis upward. Conidial mass is creamy and looks pink but individually the conidia are hyaline, unicellular, and oblong. Their size is highly variable, being 10-35 × 3.5-7 μm. Before germination the conidium becomes 1-septate. On germination, 1-2 germ tubes are produced which form appressoria. Perithecial stage is found occasionally on fruits. Only some strains of the fungus produce perithecia. The perithecia develop on stroma of 2-8 mm diameter. They are spherical or subspherical, usually beaked, and measure 125-250 μm in diameter. The clavate asci measure 50-110 × 8 μm. Each ascus contains 8 ascospores which are slightly curved, hyaline, 1-celled, and 35-50 × 12-22 μm in size. The fungus has many strains. One strain can infect apple leaves, survive and sporulate. This strain produces typical symptoms on fruits without sporulation.

Disease Cycle

The fungus survives on mummified fruits on the ground or on the trees. It also survives in cankers on branches and in cracks in the bark. Conidia can survive for long in acervuli. These spores are dispersed by wind, raindrop splashes, and sometimes by insects and birds. Infection of apples on the trees occurs when the fruits are ripening (July-August in India). Presence of moisture on the fruit surface enables spores to germinate quickly and form appressoria. The epidermis is dissolved by enzymic action. Infection can also occur through wounds and lenticels or stomata. In the bitter rot of pears caused by *Colletotrichum gloeosporioides* disease intensity and time required for appearance of symptoms depend on inoculum load (234). The fungus enters the uninjured fruits through lenticels. Hundred per cent infection occurs through the stalk end after removal of stalk and through wounds (235). The infection hyphae enter the fruit within 48 hours of inoculation and colonization of tissues occurs within 96 h. Acervuli develop after 10 days. Green fruits have resistance to infection (231). The fungus grows slowly at 15°-21°C. At temperatures below 15°C conidial formation and germination is considerably reduced. High humidity on the host surface is essential for infection. A temperature of 25°C and 80 % relative humidity is considered ideal (50).

Management

Sanitation and canopy management recommended for brown rot and Botryosphaeria rots are measures applicable to this disease also. Pre-harvest sprays of fungicides at 10-15 d interval reduce infection of ripening fruits. Captan (0.2 %), maneb (0.2 %) or ziram (0.2 %) are recommended fungicides. In China an anti-transpirant compound named *gau-zhi-mo* has been used as spray on trees for good control of bitter rot (105). The compound does not interfere with carbon oxygen diffusion, is not affected by rains, and remains intact on host surface for 15 days.

Immersion of fruits in 2% trisodium phosphate solution before storage has been found effective against post-harvest rot (50). Kaul (131) reported control of post-harvest rot caused by *Glomerella cingulata* by a 3-min dip in 2000 ppm sodium orthophenylphenate solution. Exposure of fruits to hot air at 45°C under 100% relative humidity for 15 min gives control of bitter rot but fruit quality deteriorates (20).

Fruits should be stored at low temperature. Temperature of 0-5°C completely check infection of *G. cingulata* and promote healing of wounds while at 20°C wound healing is suppressed (136).

■ POST HARVEST DECAY OF POME AND STONE FRUITS

The incipient or latent infection of fruits in the orchards and contaminants which lodge on the fruit during picking, harvesting, grading, packing, transportation, storage and various other handling operations before the fruits reach the consumer lead to post-harvest storage and market diseases of fruits. In addition to brown rot, bitter rot and *Botryosphaeria* rot that attack pome and stone fruits, there are many more fungal pathogens that attack these fruits after harvest. Of these some are economically very important. Kaul and Munjal (133) recorded 21 different fungi causing fruit rot of apple in Himachal Pradesh (India). These and others recorded elsewhere, in addition to *Monilinia* spp., *Glomerella cingulata* and *Botryosphaeria* spp., include *Penicillium expansum, Trichothecium roseum, Rhizopus stolonifer, Mucor piriformis, Alternaria alternata, Botryosphaeria stevensii, Aspergillus niger, Botrytis cinerea, Geotrichum candidum, Cladosporium herbarum, Phytophthora cactorum* and *Phoma mali*. Prevalence of rots in Himachal Pradesh, based on average loss, varied from 10.3 to 18% with a mean loss of 14.2%. These decay fungi are of worldwide occurrence and cause similar losses in other countries also.

Blue Mold, Black Mold, Whisker's Rot and Grey Mold Rot

Blue mold rot caused by *Penicillium expansum* Thom is the most common rot of apples having a prevalence of 45.6% in Himachal Pradesh (133). The prevalence is more than the combined prevalence of all other rots. The blue mold rot appears as soft watery spots on fruit surface. The decayed portion is sharply separated from the undecayed tissue. These spots of variable size and shape may occur on any part of the fruit. At first they are shallow but later rapidly extend deeper and by the time the rot reaches the core one third to half of the fruit is affected. The decayed internal tissue is watery and glossy in appearance. The spots do not become sunken. On the surface of the fruit white mycelial growth followed by development of bluish tufts of conidiophores and conidia appears. The rotted fruits emit a musty odour. *Penicillium* and its epidemiology is described under citrus molds. The black mold rot caused by *Aspergillus niger* van Tieghem also looks and develops similarly but the growth soon turns black when conidiophores and conidia develop.

The whisker's rot (described under strawberry diseases) is caused by *Rhizopus stolonifer* (Ehr.:Fr.) Lind. In apples the affected area of the fruit first looks watersoaked and soft and gives a sour smell. Then the skin turns brown and can be easily separated without disturbing the internal tissue. If the skin is not broken the internal tissues lose moisture and the fruit shrinks and mummifies. However, more frequently, the softened skin ruptures during handling of the fruit or under pressure of other fruits in the packing cases and a dirty yellow liquid oozes out. This liquid wets other healthy fruits. The affected fruits rapidly get covered by mycelium extending from within with grey sporangiophores bearing black sporangia at their tips. Occasionally, there is no fungal growth on the surface but inside sparse mycelial growth with black sporangia is seen. The soft watery rot rapidly decays the entire fruit and the spores released from fungal growth infect other healthy fruits in the packing cases. While in early stages of the rot the fruits give off a mildly pleasant smell, soon bacteria and yeasts move in and a sour odour develops. *Rhizopus* can occasionally attack trees also (244).

The grey mold rot caused by *Botrytis cinerea* (described under strawberry fruit rots) and Alternaria black rot caused by *Alternaria alternata* (described under citrus fruit rots) are relatively less prevalent in the apple belt of India (5). The grey mold rot is a serious problem in temperate climate countries. It is assuming greater importance in Himachal Pradesh (India) where 1.2-9.9% incidence in different districts is recorded (19, 232). Its occurrence is common in high hilly areas.

The management of these mold rots basically requires rigid sanitary and canopy management precautions as recommended for brown rot and Botryosphaeria rots. The weather must be dry during harvest, every care should be taken to avoid injury to the fruit skin, the wounded fruits should be culled out, and chemical treatments of fruits should be given within 24-36 hours of harvest. Pre-harvest sprays of fungicides have been recommended for these rots. A pre-harvest spray of captan in peach, cherries and pears is a good safeguard against Penicillium and Aspergillus rots. Spray of Botran (DCNA), one week before harvest checks the Rhizopus rot of peaches. Significant control of grey mold rot of pears by two pre-harvest sprays (30 and 10 days before harvest) of 500 $\mu g/ml$ carbendazim is reported (19, 232). There was no grey mold rot for 19 days at room temperature and for 30 days in cold storage. Thiabendazole and thiophanate methyl were also effective.

Large number of fungicides have been used as post-harvest dip for pome and stone fruits against various rots (129). Fruit dip in 500 ppm benomyl or 1000 ppm thiabendazole for 10-15 sec was recommended by Edney (71) against the blue mold of apples. Benomyl is now not recommended for fruit treatment and has been withdrawn. Dipping of fruits in 1.25% sodium carbonate or 5000 ppm 2-aminobutane for 10-15 sec has also been found effective against blue mold. Fruit dip in captan, though effective, is generally avoided in peaches because it leaves a visible residue on fruits. Iprodione (Rovral) is effective against grey mold and brown rot of sweet cherries and peach fruits. It penetrates the mesocarp of the fruit tissue (1, 192). Chib *et al.* (53) had reported that a 5-min dip of apples in iprodione at 0.2% or carbendazim at 0.1% completely inhibits blue mold (*P. expansum*) for 7 days at 20°-25°C and does not allow complete rotting for 21 days. Fruit dip in 1000-2000 ppm DCNA (Botran) within 12 hours of harvest checks Rhizopus rot. A 5-min dip of apples in 0.2%

difolatan, 0.2% iprodione or 0.1% thiabendazole prevents Rhizopus rot for 7 days and there is significant check up to 21 days (53). However, there are other reports suggesting inefficacy of thiabendazole against *Rhizopus*.

Sulphur dioxide fumigation of apples by release of the gas from sodium metabisulphite has been found useful as a post-harvest treatment (135). Fumigation with low molecular weight aldehydes is effective against post-harvest decay of sweet cherries (166). Sitton and Patterson (242) have reported that controlled atmosphere with carbon dioxide levels above 8% reduces decay by blue mold and grey mold in apple cultivars McIntosh, Delicious and Red Delicious. Use of fungicide-impregnated paper wrappers provide cover to fruit surface and safeguard from bruises and depressions which provide avenues for entry of rot causing organisms. For pome and stone fruits paper wraps impregnated with Botran or sodium orthophenylphenate have been recommended (134). This protects fruits against Penicillium rot, bitter rot, brown rot, and Rhizopus rot (159). In addition to wrapping of fruits, the sheets should be placed at the bottom and top of the packing case. Apples coated with pungent mustard oil withstand onslaught of these molds. Paraffin wax is slightly better for protection against *P. expansum* and *Rhizopus stolonifer* (134).

Temperature of storage (130) and certain temperature treatments of harvested fruits are important aspects of management of fruit decay (132). Storage of Golden Delicious apples at 0°C completely checks blue mold rot but at 5°C some infection does occur. Wound healing at 0°- 5°C aids in prevention of infection and spread of the fungus (136). Chandrani and Kaul (51) have suggested hydrocooling (heat removal or chilling) of peaches in ice water (0°- 2°C) for 30 min for control of Rhizopus rot. For better results, 30 μg/ml DCNA (Botran) was added to the water. The fruits should then be stored at 5°-10°C. At 15°C the hydrocooled peaches decayed completely. According to Sharma (231) storage of pears at 0°C is the most effective way of protection against rot caused by *B. cinerea, P. expansum, Monilinia laxa* and *Glomerella cingulata*. Fruits remained free from decay until 45 days and only 2.2% rot occurred after 60 d. At 5°C decay was checked fairly well with 6.8% rot after 2 months. At 10°C rot started after 15 days and after 60 days 10% rot was recorded. Storage at temperatures above 10°C caused rapid decay, maximum rotting (29.7%) occurring at 30°C of storage. In heat therapy for peaches, 2.5 min dip in hot water at 52°C or in moist (80% relative humidity) hot air at 54°C for 15 min control *R. stolonifer* (20).

Calcium deficiency in fruits results in easy tissue disintegration by decay causing fungi and bacteria. Experimentally, it has been demonstrated that post-harvest pressure infiltration of calcium chloride solution reduces lesion size of decay caused by *Monilinia fructicola* in peach and *Penicillium expansum* in apples (58, 59). Dipping of fruits in calcium chloride solution of up to 12% concentration causes only a little increase in the calcium content of cells. Impregnation under vacuum or pressure does considerably increase the calcium content of cells to a level where decay is significantly reduced (58, 59). Calcium gets integrated into the cell walls and hinders fungal pectolytic enzyme activity associated with intercellular pectic substances. Commercial application of pressure infiltration for apple was put on pilot test in the early 1990s (59). In a study reported by Biggs *et al.* (31) in 1997, when inoculum of *Monilinia fructicola* was sprayed on detached peach fruits the incidence and severity of brown rot were least on fruits that had been dipped in a solution of calcium propionate or calcium silicate. When inoculum was applied as a localized drop to wounded fruits that had been dipped in a solution containing 1200 mg of Ca per litre, brown rot was least for fruits treated with calcium hydroxide or calcium oxide.

In biological control, fungal and bacterial antagonists as alternative to fungicides have been identified and their efficacy demonstrated (117-120, 122, 171, 250, 293). Treatment of pome and stone fruits with cell suspension of *Bacillus subtilis* controls brown rot (*M. fructicola*), *Enterobacter cloacae* controls soft rot caused by *Rhizopus stolonifer*, the fungus *Acremonium brevae*, the bacterium *Pseudomonas cepacia* and several yeasts are reported to control post-harvest decay of apples caused by *P. expansum* and *B. cinerea*. Janisiewicz and Marchi (121) obtained good control of blue mold (*P. expansum*) and grey mold (*Botrytis cinerea*) on pears by dipping fruits in cell suspension of a saprophytic strain of *Pseudomonas syringae*. In some tests they obtained 100% control. McLaughlin *et al.* (171) have reported that the yeast species *Kloeckera apiculata* and *Candida guilliermondii* (*Debaromyces hansenii*) are effective in controlling Rhizopus rot of peach and grey mold of apples. The control of these two molds was enhanced when the yeast suspension was prepared in 2% calcium chloride solution.

■ PINK MOLD ROT

In the apple belt of north India, pink mold rot of apples is prevalent to the extent of 12.6% (133) and is next only to blue mold rot (*Penicillium expansum*) and bitter rot (*Glomerella cingulata*). The disease starts from any injury on the fruit surface. The earliest symptom of this rot appears as light brown spots, about 4 mm in diameter, on the fruit. With increase in the size of spots, the central portion becomes light while the margins remain darker brown. In advanced stage of the rot, under suitable environmental conditions, white mycelial growth of the fungus with light pink masses of conidia on a discoid base appear in the lighter area. The orange-pink colouration and powdery texture develop with age. Later, chocolate brown sunken areas of irregular outline, varying in diameter from 1.25 to 2 cm or more are seen on the fruit. The rotting radiates unevenly towards the core of the fruit. The seed cavities show abundant pink spore masses. Eventually, the whole fruit is involved. The decay is somewhat of brown, firm rot type and at no stage of its development watery rot is seen. However, in the markets secondary invaders (*Aspergillus, Penicillium, Pestalotia*, etc.) quickly establish in the spots and a soft watery dark brown to black rot results. The affected fruits have a bitter taste.

The pink mold rot is caused by the conidial fungus *Trichothecium roseum* Link. Colonies of the fungus on culture media are initially white but later turn pink. Conidiophores are erect, simple, aseptate or sparsely septate, hyaline, and bear conidia in clusters at the apex which is broader than the remainder of the conidiophore. The basal cell of the conidium is truncate or pointed at the point of attachment. The conidia measure 12.4 - 22.3 × 6.2-12.0 μm. The basal cell is 6.2-12.4 μm long while the terminal cell is 6.3-10.6 μm long (252). These conidia are set free from the conidiophores by air currents. The stronger the air current, in the range of 1.7-10 m/sec, the greater the number of spores liberated. Humidity of the air has a very significant effect. Many more spores are set free into a dry rather than in a damp air stream. This is common in all dry spore molds.

Specific control measures are not known but treatments recommended for other molds can reduce the pink mold rot also.

■ MUCOR ROT OF POME AND STONE FRUITS

Mucor rot of apple, pear, and peaches, caused by *Mucor piriformis*, is restricted to cool climate countries of the northern hemisphere although its presence as a rot causing organism of apple was reported in Himachal Pradesh (India) also (133). This rot is significant as it occurs even at 0°C in cold storages and the pathogen is active only at low temperatures at which other post-harvest rot pathogens cease to be active. Mucor rots have not been considered important but the rot caused by *M. piriformis* is reported to be occasionally destructive in certain parts of USA in apples, pears, peaches, nectarines and strawberries. In subtropical regions it is destructive on guava.

In apples and pears infection occurs through stem end, the calyx or puncture wounds. Infected tissues become soft, watery and light brown. Usually, sporangiophores protrude through cracks in the skin or emerge through lenticels. At 0°C infected fruits completely decay in two months of storage, releasing large quantities of juice. Well rotted apple or pear fruits emit a characteristic alcoholic odour.

In peaches and nectarines lesions are circular, light brown, soft, watery and are seen covered with masses of shiny, erect sporangiophores that appear at breaks in the fruit skin or emerge through the lenticels. In cold storage, these fruits sometimes show a narrow band of clean, water-soaked tissue at the lesion margin and usually give off a pleasant aromatic odour. The diseased internal tissue is light brown, very soft, and watery and separates easily from the healthy tissue.

The rots caused by *Mucor* and *Rhizopus* look similar in the early stages. Both produce soft rot by macerating enzymes (polygalacturonase activity). In an advanced stage of the rot, the sporangial growth of *Mucor* is brown while that of *Rhizopus* is black. Under high relative humidity the sporangia of *M. piriformis* absorb moisture, the wall dissolves, and the sporangium becomes a spore drop. Thus, the spores are not disseminated by wind which occurs in *Rhizopus*. The Mucor rot develops very rapidly at 0°C while Rhizopus rot can develop at 20°- 24°C but not at 0°C.

Mucor piriformis A. Fischer was first described in 1892 and was reported as a cause of fruit rot in 1895. The aseptate mycelium consists of hyphae interwoven with stolons but no rhizoids are produced. Grey collumellate sporangia are produced on single or branched sporangiophores which may be short or tall. Young sporangia are milky white or yellowish, later turning grey. They are globose with spiny walls and measure 264-283 μm in diameter. The sporangiospores are ellipsoidal (8.6-9.2 × 7.8-8.1 μm) to subspherical (7.3-7.8 × 4.6-5.3 μm) and smooth. The size of sporangia and spores varies on different hosts. The average diameter of sporangium in one isolate from apricot was found to be 236 μm while in another isolate from peach it was 154 μm (168). Chlamydospores (gemmae) are produced but are not common. The fungus is heterothallic. The mating types (+ and -) occur in separate infections but are brought together by insects (nitidulid beetles and vinegar flies) visiting the decaying fruits. The zygospores are spherical to subglobose and measure 151-180 × 127-156 μm (177).

The fungus grows well and sporulates extensively at 0°-24°C, optimum being 21°C. Mycelium grows slowly at 26°C which is the maximum. Spores germinate at 1°C to 24°C but not at 27°C. Thermal death point of the mycelium is 43°- 46°C and for spores 52°-55°C (177).

The fungus is soil-borne, colonizing the organic matter such as fallen fruits and leaves. More than 75 % spores in soil are present in t top 2 cm. Fallen fruits are infected through contact with infested soil or by fruit-to-fruit spread involving birds and insects. Rains wash down the spores into the soil or splash them on to other fruits. The fungus cannot compete with other soil microorganisms at 20°C or above but is a strong competitor at temperatures below 15°-20°C. Survival of the fungus is best in cool, dry soil. Populations increase in soil only when nutrient substrate (fallen fruits and leaves) is available and cool temperatures with high soil moisture levels prevail. The fungus can survive on infected fruits for 19-20 months and for a year on endocarps of the stone fruits buried in soil (176). In addition to fallen fruits and leaves, the spores can also germinate, grow and sporulate on weeds present in the orchard when temperatures are low and antagonists are limited.

Infested soil and debris serve as source of primary inoculum for contamination and infection of fruits during picking and packing. Since the spores of the fungus are embedded in mucilaginous matrix they rely primarily on raindrop splashes and insects such as *Drosophila melanogaster* and birds for dispersal. *Mucor piriformis*, other species of *Mucor, Rhizopus stolonifer, Monilinia fructicola, Cladosporium* spp., and *Penicillium* spp. have been isolated from *Carpophilus* spp. and *Drosophila melanogaster*. *M. piriformis* can persist on *D. melanogaster* for 15 days and on *C. hemipterus* for 11 days (178). These insects acquire the propagules and transfer them to 75-100 % of injured fruits.

As in most typical post-harvest rots, *M. piriformis* also requires wound on the fruit for infection. On pears and apples, the fungus is associated with the stem punctures and animal wounds. Any type of blemish on fruit skin may permit infection. In cold storages, at 1°C, initial post-harvest infection usually occurs via wounds and secondary spread is by infected fruits contacting healthy fruits or by dissemination of spores in exuded juice dripping from infected fruits. Spores are so virulent that a single spore reaching a wound initiates rot. Pear fruits become particularly vulnerable to infection when they are nearly ripe and over-mature fruits are especially susceptible to decay. Green stone fruits are resistant to infection.

Fungicides generally recommended against post-harvest decay of pome and stone fruits are ineffective against Mucor rot. *Mucor* is highly resistant to benzimidazoles (245) and dicarboximides which have been used for control of post-harvest fruit decay caused by other fungi. Sanitation is the most important step for control of Mucor rot because the rot results from unsanitary conditions. Fallen leaves and fruits should be removed from the floor of the orchard. This reduces the inoculum. The wooden bins or crates that are used in the orchard for picking fruits and transportation easily get contaminated by orchard soil infested with spores of the fungus. These should be thoroughly washed, preferably with chlorinated water, or steamed and internally covered with paper or plastic pads to avoid bruising of fruits. In general, contact between fruits and soil should be avoided at each stage of handling operations. Fallen or wounded fruits should be segregated and not packed with clean fruits.

Pre-storage heating of pear fruits at 27°C for 2 days has been found to reduce Mucor rot in some varieties. This does not affect storage life or the quality of the fruit. The fruits should be dry and wrapped in paper before storage. Pears dipped in hot water at 47°C for 30 min carry very low level of inoculum and there is less or no Mucor rot. This can be done before storage (20).

■ FIRE BLIGHT OF APPLE AND PEAR

The name *fire blight* was first used in 1817 to describe sudden browning of leaves as if they had passed through a hot flame and causing a morbid matter to exude from the pores of the bark. Fire blight is the most destructive disease of pear trees. It also causes damage to apple and quince. The host range of the causal bacterium includes plants in at least 174 species in about 40 genera, all in the family Rosaceae (265). The disease damages susceptible hosts by killing flowers and twigs and by girdling of large branches and trunks resulting in death of the trees. In certain years, blossom blight may result in total loss of the crop. Twig blight eliminates fruiting spurs. The tree structure is disturbed when the main branches are killed. Fruit infection occurs any time until harvest although fruits gain resistance as they mature. Loss in the nursery occurs when the blight kills the young plants. Young trees in the nursery or in the orchard may be killed by a single infection in one season. It occasionally kills mature trees especially when the lower trunk, collar or roots are infected.

Fire blight had been indigenous to North America where it coexisted with wild hosts of the bacterium. It attracted attention when improved methods of apple and pear cultivation were introduced and the disease was found to be a problem when trees were pruned, fertilized and stimulated to grow rapidly. It was the first plant disease reported by T.J. Burrill in 1882 to be caused by bacteria. For decades after its discovery, the disease spread slowly due to geographic and botanical barriers. However, it has continued to spread, especially since 1950s, and is now known to occur in 19 countries in 4 continents. It is present in all of North America and parts of Central America and in Europe, Near East and New Zealand. In New Zealand the disease was found in 1911 but Australia which is in close proximity imposed strict quarantine in 1924 and has remained disease-free (268). Fire blight was reported in England in 1958 and by 1986 it had spread to most of the European countries and to Egypt and Israel. The disease occurs in north western part of India also. It is suspected to be present in the eastern part of China, Japan, South Africa, Russia, Italy and Turkey. However, at many places it has been confused with blossom blight caused by *Pseudomonas syringae* pv. *syringae*.

Symptoms

Infection occurs on flowers, twigs, fruits, branches and the trunk. Usually the first symptoms appear on flowers. The most obvious symptom is the rapid wilt and browning or blackening of infected blossoms or leaves (7, 11). Individual flowers or the entire cluster become water-soaked and turn slightly darker green. White to clear droplets of ooze form on the affected parts. The flowers turn brownish to black, shrivel, and fall or remain hanging on the tree. The bacterial ooze may turn amber or black and dry, leaving a slightly shiny area. The infection progresses on to the leaves on the same spur or on nearby twigs, starting as brown black blotches along the midrib and main veins or along the margins and between the veins. As the blackening progresses, the leaves curl and shrivel, hang down and usually remain clinging to the curled blighted twigs.

The current year's growth and especially the water sprouts (suckers) are usually infected directly and wilt from the tip downward. Symptoms of water-soaking, ooze and wilting are similar to flower infection. The bark turns brownish black and is soft at first but later shrinks and hardens. The wilted tip is hooked or crooked and the leaves turn black and cling to the twig. The infection progresses very rapidly in succulent growth and may advance 10-30 cm in one day.

Under optimal conditions of environments and in susceptible hosts, the blossom or twig infections may progress into branches or trunk, killing large portions of the tree. These infections on branches and trunk form cankers. If the canker girdles the branch, portion above the point of infection dies. If the branch is not completely girdled by the canker, the infection becomes dormant or inactive canker with sunken, sometimes cracked, margins. Cankers without cracks on the periphery may contain viable bacteria surviving on the margins. The interior of the bark adjacent to cankers is often marked with reddish-brown streaks that extend several centimetres into the healthy wood.

Fruit infection occurs through the pedicel or directly through lenticels. Older fruits are generally resistant but even harvested fruits can become infected. Small, immature fruits become water-soaked, then turn brown, shrivel and mummify. Moisture loss from infected fruits is fairly rapid, resulting in these fruit mummies which hang on the trees for months. Finally, they turn dark brown to black in pear and reddish brown in apple.

Under humid conditions, an extremely sticky and stringy ooze is often produced on the surface of any infected part. This ooze may drip and produce long, thin strands (aerial strands). Upon drying these brittle strands break into small fragments that may become wind-borne and serve as inoculum (139).

The Causal Organism

The causal organism of fire blight, *Erwinia amylovora* (Burrill) Winslow *et al.*, the type species of the genus, belongs to the group of erwinias which form neither pectic enzymes nor yellow pigment but do cause necrosis or wilt symptoms through toxins. The bacterial cells, measuring 1.0-3.0 × 0.5-1.2 μm are typically motile by peritrichous flagella. The bacterium grows best at 21°-28°C with a minimum growth temperature of 3°-12°C and maximum 35°-37°C. The generation time under optimal conditions for growth is 70-90 min. Colonies on 5 % sucrose nutrient agar are typically white, domed, shining, mucoid, with radial striations, and a dense flocculent centre or central ring after 2-3 days at 27°C. The G+C content of DNA is 53.6-54.1 mole per cent.

There is a relationship between the extracellular polysaccharide (EPS) capsule of *E. amylovora* and virulence. Isolates without EPS are usually avirulent and often have a rough colony rather than smooth colonies of the virulent strains (32). Bennett (25) suggested that EPS alone is not the virulence determinant. Billing (32) reported that a cell-leakage inducing agent and EPS both were necessary for virulence. Toxin (s) have been suggested as the physiological basis for symptom induction by *E. amylovora*. Goodman *et al.* (85) had reported purification of amylovorin, a host-specific phytotoxic polysaccharide from ooze produced on affected parts.

Disease Cycle

The bacteria mostly survive at the margins of cankers formed in the previous season (23), most often in cankers on large branches and seldom on twigs less than 1 cm in diameter. They also survive in cankers on other hosts and possibly in buds and healthy wood tissue. They have also been isolated from leaves, flowers (35), twigs (82), buds (34) and fruits (70, 279) of pear and apple with no symptoms and on healthy pear scion wood collected from diseased pear trees (180).

The role of epiphytic populations of the bacterium in epidemiology of fire blight has been emphasized in many studies. The presence of bacteria on apparently healthy pear flowers, fruits, leaves, and inactive cankers was demonstrated by Miller and Schroth (180). Occurrence of fire blight outbreaks could be directly related to the presence of epiphytic populations on pear flowers. In many cases, the populations exceeded 0.1 to 1 million colony forming units per flower two weeks prior to appearance of the disease (266). Stigma of the pistil of pear and apple flowers plays a major role in harbouring the epiphytic populations of the bacterium (264). Norelli and Beer (188) found that infections were less frequent when inoculum concentrations on pear and apple flowers were less than 1000 to 10000 CFU per flower. The nectar of the flowers helps in very rapid multiplication of bacteria on the surface.

In spring, the bacteria in cankers become active, multiply and spread into the adjoining healthy bark. During humid and wet weather, these bacterial masses absorb moisture, increase in volume, a part of them exudes or oozes through lenticels or cracks onto the surface (220). The ooze consists of plant sap, millions of bacteria and bacterial metabolites. Usually the ooze is seen when the pear blossoms are just opening. The sweet sticky ooze attracts insects, ants and flies which carry the bacteria on their body to healthy flowers.

The secondary spread occurs through dissemination of bacteria-containing ooze by insects, raindrop splashes and birds. Wind helps in dispersal of aerial strands and aerosols, the minute droplets of water containing bacteria (172). Long distance dissemination is caused by infected budding and grafting material, crates used for packing and transportation of fruits and by infected fruits. Epiphytic populations on these sources also disseminate the bacteria to long distances.

The bacteria enter the host tissue through natural openings such as nectaries, hydathodes and lenticels or through wounds and bruises caused by insects, hail and driving rains. In apple the bacteria penetrate through stigmas, anthers and receptacle walls (213). After entry into the flower the bacteria multiply rapidly. Bacterial metabolites (toxins) cause collapse of cell walls, plasmolysis (84), and sometimes formation of cavities in floral parts which are filled with bacterial cells. The bacteria move intercellularly and through macerated cells of the flower pedicel. They progress to the spur and may invade the leaves. Highly susceptible varieties or succulent growth allow the pathogen to progress into large branches in which cankers are formed. Although infection of leaves may take place through stomata and hydathodes, most leaf infections occur from floral infection. Young tender twigs may be infected by bacteria through their lenticels and through wounds. They may also be infected through flower and leaf infections. In the twig, the bacteria travel intercellularly, or if they enter the injured xylem vessels, they travel over short distances through xylem.

The influence of environment is undoubtedly the most important factor in fire blight epidemics. The blight develops most rapidly at temperatures between 18°and 30°C, optimum for *in vitro* growth of the bacterium is 21°-28°C. The number of days with temperatures above 18°C at the time of bloom coupled with high humidity are implicated in the rate of disease development. Apart from temperature and high humidity, rain also plays an important role in many ways. It transfers the bacteria on the same tree, from the top to lower portions, from tree to tree, produces aerosols and, when with high winds, causes injuries for infection.

Management

The early control measures centered around sanitation and cultural techniques. With the advent of chemical control, especially use of antibiotics, growers have relied heavily on sprays during the bloom period. Despite these efforts, fire blight is still a threatening disease that cannot be controlled by a few simple measures (265). Satisfactory control can be obtained only by utilizing an integrated programme of management that includes many cultural and chemical methods.

1. Quarantine

Once the disease enters an area it becomes permanently established in that area and all efforts to eradicate it prove unsuccessful. Quarantine is essential to prevent entry of the pathogen into an unaffected area through planting material and plant products. These regulatory measures have kept the disease away from many countries. Often the bacterium is present on the fruit surface and on surface of crates used for transportation of fruits. This necessitates precautions in importation of fruits from places where the disease is known to be present. Importation of hosts from such countries should also be avoided.

2. Resistance

No pear or apple varieties are immune to fire blight when conditions are favourable and abundant inoculum is present, but the degree of susceptibility differs. Thus, apple cultivars of Red Delicious group which are inherently resistant do not normally require other control measures. There are many other apple varieties which have considerable resistance to the disease (9, 278). Pear cultivars are much more susceptible to fire blight than apple. Even cultivars which show some resistance are susceptible in favourable conditions.

3. Cultural management practices

Fire blight is less serious and fruits of better quality are produced when orchards are planted on well-drained sites with light soil and soil with pH 6-6.5. Sprinkler irrigation plays the same role as rain in epidemiology of fire blight. This system of irrigation should be avoided where the disease is a problem. Tree nutrition has been studied extensively and a definite relationship has been shown between susceptibility of trees and nitrogen, potassium, phosphorus and some micronutrients. Increased nitrogen causes more succulent growth in the tree and, therefore, more sites for infection. Use of nitrogen is recommended only to meet the shortage of the nutrient as determined by soil and leaf analysis. Ammonium nitrogen should not be applied because its nitrification and availability occurs towards the warmer period of the growing season when the disease also becomes serious.

Pruning should not be very heavy because the vigorous vegetative growth after pruning is very susceptible. Overwintering cankers should be removed during pruning in the dormant stage pruning. It reduces the disease in the following season. Cuts should be made about 30 cm below the point of visible infection. Sterilization of tools between cuts during pruning is also important. The tools, after each cut, should be wiped with a sponge or cloth soaked in 10% sodium hypochlorite solution. The same solution can be used to disinfect the cuts on the branches. Exposure of scions of pome fruit trees infected with *E. amylovora* to 43°C for 23 hours kills the bacteria without causing tissue damage to the plants.

4. Chemical management practices

Satisfactory control of fire blight with chemicals can be obtained only in combination with the above mentioned measures. Sprays of fungicides at silver tip to green tip stage of flower buds considerably reduce disease incidence (8, 14). Bordeaux mixture was one of the first compounds recommended for use as a blossom blight control spray in 1929. However, when conditions were highly conducive for fire blight, it was not very effective. It is still recommended and is being used at a concentration of 2:6:100 in many areas. Where disease pressure is not high it provides adequate protection. Other copper compounds are also effective. Copper is phytotoxic and causes fruit russet. Sprays under quick drying conditions are recommended to reduce chances of russeting.

The discovery of antibiotics for blossom spray was a significant improvement in the control of fire blight and without the problem of phytotoxicity associated with copper compounds. In the early 1950s, streptomycin sprays (100 ppm) were found to give excellent control when applied at the proper time (15, 138). The antibiotic is effective on open blossoms for 2-3 days and, therefore, application must be repeated every 3-5 days to protect newly opened blossoms. For better absorption of the antibiotic by the tissue night sprays have been recommended (197). Oxytetracycline was included in the early formulations of streptomycin (83) but it did not improve control over streptomycin alone.

In 1972 streptomycin-resistant strains of the pathogen were detected (45, 54, 183, 221). These were found where streptomycin was used exclusively and repeatedly for fire blight control and also where it had not been used routinely. In view of this, it is now recommended that streptomycin and tetracycline both should be used to avoid the resistant strains. While mutants resistant to streptomycin are frequent, resistance to oxytetracycline has not been found (158).

For effective control of fire blight the number of antibiotic or fungicide sprays is very high, being repeated every 3-5 days. Sometimes 18 sprays are required. This can be avoided by prediction of possible time of infection and applying fewer but timely sprays. After the primary inoculum from the cankers reaches the flowers it must multiply very rapidly before it can cause infection. Such rapid multiplication occurs only when periods of high humidity or abundant rainfall coincide with warm temperatures (above 18°C). This observation has been used to develop fire blight forecasting systems (182, 267). Computerized models are being used for specific areas to forecast the disease appearance and advise start of spray operations (280).

5. Biological control

The possibility of biological control of fire blight has been explored with bacteriophages and antagonistic bacteria (24). So far not much success has been achieved. However, certain strains of *Erwinia herbicola*, commonly present as epiphyte on plant surfaces, and *Pseudomonas fluorescens* have given encouraging results when used as cell suspension sprays. Honey bees disseminate the antagonistic bacteria to flowers more effectively than manual sprays (268).

■ CROWN GALL DISEASE

The crown gall disease was first discovered in 1870 in the United States but the bacterial association was first established in 1907 in Italy. The disease is now found in whole of North America, Europe, South Africa, Asia, Australia and New Zealand and affects 40 species in 61 families of woody and herbaceous plants (68). In India, crown gall is reported on apple, pear, peach, cherry, plum and almond. In addition, it occurs on apricot and grapevines in many countries.

Symptoms

The galls first appear as small outgrowths on the stem and roots, particularly near the soil line. They are single or in clusters. Their size is variable, being 0.5 to 10 cm or more in diameter. In the early stages of development the tumours are generally soft and round. With increase in size their surface becomes convoluted. Initially they look white or light pink, later the colour changes to dark brown or black due to death and decay of the peripheral cells. Sometimes, there is no distinct line of demarcation between the tumour and the healthy stem. They appear just as swellings encircling the stem. But often they lie outside, close to the host surface and connected only by a narrow neck of tissue. Many tumours remain spongy throughout and may crumble or get detached. Others continue to grow, reach a size of upto 30 cm, become hard and woody, knobby and knotty. Partial rotting of the galls may occur but the area regenerates. Although the tumours are most common on stem and roots near the soil surface, they may also develop on branches, petioles, and on leaf veins.

The damage to plants is in many ways. When nursery stock is affected it is not saleable. The grown up trees and vines (in grapevine) remain stunted with small and lustreless yellowish leaves. Yield of fruits is reduced. Crown gall affected trees are susceptible to adverse conditions, particularly cold (winter injury).

The Causal Organism

Crown gall is caused by *Agrobacterium tumefaciens* (Smith and Townsend) Conn. The history of the genus *Agrobacterium* can be traced to the discovery of a bacterium-induced plant tumour in 1907 by E.F. Smith and C.O. Townsend who named it *Bacterium tumefaciens*. H.J. Conn created the genus *Agrobacterium* in 1942 and gave the bacterium its present name.

Agrobacterium belongs to family Rhizobiacea. Species of *Agrobacterium* which produce galls are known as *A. tumefaciens* while species similar in almost every respect to *A. tumefaciens* but not pathogenic is called *A. radiobacter*. The species responsible for hairy root of nursery stock is called *A. rhizogenes*. The species *A. rubi* closely resembles *A. rhizogenes* but produces cane galls on raspberry. Some believe it to be same as *A. tumefaciens*. There was also a proposal that since there is no morphological and biochemical difference between *A. tumefaciens* and *A. radiobacter* except that the former is pathogenic while the latter is saprophytic, the two should be merged and the pathogenic form should be called *A. radiobacter* pv. *tumefaciens*. Similarly, since there is not much difference between biotype 2 of *A. tumefaciens* or *A. radiobacter* and *A. rhizogenes* the latter should be called *A. radiobacter* pv. *rhizogenes*.

In *Agrobacterium* the cells are rod-shaped with few peritrichous flagella. They are Gram-negative, aerobic and produce a considerable amount of extracellular slime on the carbohydrate containing media. There are 3 biotypes of *A. tumefaciens* and *A. radiobacter* which differ in colony size and colour on specific media, maximum growth temperature and other biochemical characters (137, 156). Biotype 1 is most common and most pathogenic. None of these biotypes in *A. radiobacter* is pathogenic. The G+C content of the DNA in *A. tumefaciens* is 60.2-60.8 mole % and that of *A. rhizogenes* 61.4 mole %.

A. tumefaciens became important when the role of plasmids (a self-replicating, extrachromosomal, hereditary, circular DNA found in many bacteria and some fungi, which is not required for survival of the organism) in its pathogenicity was discovered (156). The isolates of the species contain one or more large plasmids some of which have the tumour-inducing genes (Ti genes). When the bacterium containing such plasmids enters the host it may transfer part of the plasmid with Ti genes onto the host DNA where it integrates with host genes which start the uncontrolled cell growth resulting in gall or tumour formation. The presence of the bacterium during tumour growth is not required. Some isolates of *A. tumefaciens* and also the species *A. radiobacter* possess plasmids but lack the Ti genes. They are not pathogenic. The virulence and host range of *A. tumefaciens* is, thus, determined by its plasmids possessing Ti genes not the nuclear chromosome (157, 186, 263).

Disease Cycle

Agrobacterium species constitute a common component of the soil microflora. They are most commonly found in close association with plant roots, in plant galls induced by pathogenic strains, and in the soil surrounding galled plants (141, 143). Plant galls are an excellent source of *A. radiobacter* (non-pathogenic) and pathogenic strains of *A. tumefaciens* (156). These bacteria are thousand times more in rhizosphere than in non-rhizosphere soil. The number of pathogenic forms is often less than 1 % of the total agrobacteria isolated (141, 218, 219), although occasionally it may be as high as 50% (143).

The bacteria survive in soil as saprophyte for several years and enter the host through some wound, during pruning, planting of cuttings, etc. Once inside the host tissue the bacteria occur primarily intercellularly. They insert the Ti part of the plasmid into cells of the fresh wounds. The cells have been conditioned by the recent wound to be receptive for the foreign DNA. They take in several copies of the T-DNA which become incorporated in several places along the various chromosomes of the host plant. These cells are now transformed and express the genes of T-DNA along with their own DNA. Transformed cells produce opines which can be utilized only by the T-DNA carrying bacteria, and also increased amounts of indole acetic acid (NAA), variable amounts of cytokinins and various enzymes.

As a result of the above mentioned transformation of the host genetic material one or more groups of hyperplastic cells appear in the cortex or in the cambial layer depending on the depth of the wound. These cells may contain one to several nuclei. They divide very rapidly, produce cells that show no differentiation or orientation. In 10-14 d after infection a small swelling can be seen. As the irregular division and enlargement of cells continues unchecked the swellings enlarge and develop in to a young tumour. Bacteria are absent from the centre of the tumour but may be present in intercellular spaces on the periphery. By this time certain cells have differentiated into vessels and tracheids but these have no connection with the vascular system of the host. As the tumour cells increase in size and number, they exert pressure on the surrounding tissues which are distorted and crushed. Crushing of xylem vessels reduces the flow of water and nutrients to the upper parts of the tree by as much as 20 % of the normal (110).

The young tumours which are smooth and soft have no epidermis and are easily invaded by saprophytic microbes and insects. This results in the decay of peripheral cells of the tumours which become black. The decay of peripheral cells releases the bacteria present in them into the soil. As the tumours further enlarge they become woody and hard. When they are unable to obtain as much water as required for their further growth they stop growing in size, decay sets in, and necrotic tissues separate from the body of the tumour. Mostly some portion of the tumour remains alive and forms additional tumours in the same or the following season.

In addition to the primary tumours formed at the site of infection, secondary tumours are formed in infections of very young and expanding host tissue. Sometimes, these secondary tumours develop on scars of fallen leaves or on unwounded parts of the stem, petiole and even on the leaf midrib and larger veins at some distance from the site of the primary tumour. These secondary tumours have no bacteria in them. Their starting point seems to be the xylem of the vascular bundles. When pieces of such secondary tumours are grafted on healthy plants, they develop into large tumours. This shows that the presence of bacteria is required only in the beginning to trigger the malignancy, later, the tumour cells continue to grow on their own.

Management

The regulatory measures are a must even if other control measures are available. Nursery stock showing infection should be rejected. Susceptible nursery stock should not be planted in soil known to contain the bacterium. Infested fields should be planted with maize or other grain crops for several years before they are used for planting fruit trees. Since the bacterium enters only through wounds, injury to roots and the lower stem should be avoided during cultivation. Root chewing insects should be controlled by application of soil insecticides and by keeping the nursery bed raised. Nursery stock should be budded rather than grafted because the disease is more common in grafts. Antiseptic tape should be used for the grafts. Some control of existing galls was reported by painting them with commercially available hydrocarbons (216) which selectively kill the gall tissue but the method is not widely practised. Penicillin and vancomycin have been used for partial control (37, 38). Algatoll paint (20 parts in 80 parts methyl alcohol) had been recommended (14) against crown gall of almond.

Crown gall has provided an example of commercial biological control of a plant disease (142, 217). A particular strain of *A. radiobacter* (K84) carries a bacteriocin producing gene in its plasmids. It produces the bacteriocin Agrocin 84 which is highly effective against many pathogenic strains of *A. tumefaciens* (142, 144, 184, 187). The Agrocin 84 producing strains of *A. radiobacter* are highly antagonistic to *A. tumefaciens* and have been used in suspension for dip treatment of nursery stock, seedlings and seeds. Cells of the antagonist can be stored dry for long periods on suitable carriers (vermiculite, peat, carboxymethyl cellulose) at 4°C if the inoculum concentration on the carrier is sufficiently high. The antagonist establishes on surface of the roots where it produces the bacteriocin and prevents gall formation by *A. tumefaciens*. However, applicability of this biological control is limited to only those strains of *A. tumefaciens* which are susceptible to the bacteriocin. Development of strains resistant to Agrocin 84 has been reported (60). Rapid and prior colonization of the host surface by the bacteriocin producing strain is a prerequisite for successful control.

A more recent innovative development in biological control of crown gall is the commercial release of a derivative biocontrol strain (K1026) which has been improved by recombinant DNA technology (145). This happens to be the first genetically engineered biocontrol agent released for commercial use.

■ BACTERIAL CANKER AND GUMMOSIS OF STONE FRUIT TREES

The most important bacterial diseases of deciduous fruit trees caused by *Pseudomonas* spp. are bacterial blossom blight or blast of pear and bacterial canker of stone fruit trees. Bacterial blossom blight was first investigated in England but has been reported from many other countries and continents. It had been confused with fire blight of pear in many countries. Canker was first investigated in Germany in 1907 although it had been reported earlier in 1902. It occurs in all major stone fruit growing regions of the world. It primarily affects stone fruit trees. In India, the disease was observed on apricot in Himachal Pradesh and the Kumaon Hills of Uttar Pradesh in 1971. In most countries blossom blight and canker are considered to be caused by the same bacterium (*Pseudomonas syringae*).

The disease develops when the host is stressed or dormant. Often the pathogen may be intimately associated with the trees but they remain without symptoms. It becomes conspicuous when it appears as cankers on the main trunk and branches and as spots on leaves. It kills the young trees and reduces the yield of old trees. Tree losses from 10 to 75 % have been reported in young orchards (7). The disease also kills leaf and flower buds reducing the yield by 10 to 20 % (7). Fruits may also be attacked reducing their quality.

Symptoms

The type of symptom depends on the host cultivar, rootstock, age of the tree, origin of inoculum and plant part invaded, strain of the pathogen, horticultural practices and nature of pre-disposing factors (62, 107). The characteristic symptom of the disease is formation of cankers accompanied by gum exudation. These cankers usually develop at the base of an infected spur, at the bud union, and in pruning wounds (107, 108). New cankers are typically formed in late winter and early spring. The bark of spurs first shows narrow brown streaks which develop into canker. They then spread mostly upward and sometimes downward and on the sides. The infected areas are slightly sunken and darker brown in colour than the surrounding healthy bark. The underlying cortical tissue is bright orange to brown. Narrow, brown streaks extend into the healthy tissue above and below the canker. In spring when dormancy of the tree is broken (early growing season), gum is produced by the tissue surrounding most cankers. The gum breaks through the bark and runs down on the surface of the limbs. Cankers in which gum is not produced are similar but usually are softer, more moist, sunken, and may have a sour smell. When the trunk or branch is girdled by a canker the leaves above the infected area curl, turn lighter green and then yellow. Within a few weeks the branch or the entire tree above the canker dies. Thus, die-back of terminal shoots or twigs is a common symptom.

On cherry, apricot, and pear dormant bud blast is especially serious. The pathogen may be present in dormant leaf and flower buds which are often killed. Those that survive, open normally in spring but collapse in early summer (107). In cross section, the infected buds show brown areas on the bud scales extending across the base of the bud. The damage to buds becomes most obvious in the orchard during full bloom when infections cause light bloom. Blossom infection is followed by development of cankers on twigs and spurs. Infection of individual flowers is not common and occurs only under highly favourable conditions. The dead flowers remain attached to the tree.

Leaf infections, especially on cherry, appear as water-soaked spots, about 1-2 mm in diameter, which later become brown and dry. Shot holes may be seen later. However, leaf symptoms are limited to some stone fruit cultivars grown in geographic regions with wet summers and high humidity (108). When the fruit is infected, flat, superficial, dark brown spots develop. The spots are 2-3 mm deep, depressed, and may have underlying gum pockets on cherry, while on peach they are 2-10 mm in both diameter and depth. The underlying flesh is dark brown to black and sometimes spongy. In artificial inoculations highly aggressive strains of the pathogen engulf nearly the whole fruit after three days of inoculation at 25°C.

The Causal Organisms

The three pathovars of *Pseudomonas syringae* responsible for above described cankers are: *Pseudomonas syringae* pv. *syringae* van Hall, *Pseudomonas syringae* pv. *morsprunorum* (Wormald) Young *et al.* and *Pseudomonas syringae* pv. *persicae*. *Ps. syringae* pv. *syringae* is globally regarded as the most damaging. The pathovar *morsprunorum* is highly specialized and is pathogenic only to cherry and plum (107). These bacteria are rod-shaped and 0.7-1.2 x1.5-3.0 μm in size. They are Gram-negative and motile by more than one polar flagella. The bacteria are strict aerobes. Optimum temperature for growth is 25°-30°C and no growth occurs at 41°C. The G+C content of the DNA varies from 59.9 to 60.0 mole %.

Most strains of the pathovar *syringae* produce the phytotoxin syringomycin, a polypeptin class of antibiotics, which plays some role in necrosis of tissue and formation of cankers and determines the virulence. Many of the toxin producing strains contain plasmids but their role is not yet clearly known. Many strains are ice-nucleation-active, that is, as epiphytes on leaf surface they serve as nuclei for ice formation at freezing temperatures (155) and, therefore, cause frost injury facilitating infection. The same bacteria produce bacteriocins toxic to non-ice-nucleation-active strains, thus assuring a competitive advantage for themselves (cf. 7).

Disease Cycle

P. syringae survives in cankers, buds, and systemically inside other host tissues with no symptoms (73, 107). In addition, the pathovar *syringae* is reported to survive epiphytically on weeds growing in the orchard (154, 211) and also on host leaves that show no symptoms. Among all the phytopathogenic bacteria, *P. syringae* group appears to be best adapted for epiphytic growth (defined as an increase in pathogen cells on apparently healthy external parts of the shoot). The pathogen is also known to spread from colonized peach blossoms, that are not killed, to developing seeds, which in turn produce infected seedlings and carry the pathogen to the orchard (107). When surviving as dormant cell the bacterium is in a hypobiotic stage when the metabolism is at the minimum. This enables long term survival (73). Although survival in plant debris is possible the soil survival is highly restricted due to poor competitive ability of the bacterium.

Bacteria enter the tree branches and shoots through the bases of infected buds or spurs, through growth cracks at the junction of branches, and also through wounds caused by leaf fall (leaf scars), pruning and other agents such as insects. Infection of buds occurs at the base of the outer bud scales and then spreads throughout the base of the bud. Killing of tissue across the base separates the growing point from rest of the tree. Bacteria can grow downward into the stem and kill stem tissues around the base of the bud. Flower infection is rare but when it occurs the bacteria enter through natural openings and through wounds caused by insects and wind-blown rain. The infections of pear blossoms by *Ps. syringae* have been variously reported through trichomes, calyx cup, nectaries, sepals and pedicels. Under very humid conditions bacteria multiply rapidly on the floral parts and may advance into the spur and twigs causing formation of cankers. Leaf infections usually occur on young succulent leaves in areas with cool and wet springs and during periods of high winds and prolonged leaf wetness. Entry into the leaf is through stomata. Infection of buds, blossoms and young leaves is favoured by frost injury.

In leaf scar infection of cherry trees it has been shown that bacteria present on leaf surface are mobilized in rain and sucked into the broken ends of leaf trace vessels by negative tension in the vascular system of the tree (62, 281). The rate of leaf scar infection depends on the concentration of bacterial cells in the inoculation drop. The leaf scar infections are important only in cherry-*Ps. syringae* pv. *morsprunorum* system. They are not a major avenue of infection in areas where bacterial canker is caused by *Ps. syringae* pv. *syringae* (62).

After entry, the bacteria move intercellularly through the bark into the medullary rays of vascular bundles. In an advanced stage of infection, the parenchyma may disintegrate. Cavities are formed in which bacterial masses are present. Although xylem vessels may be invaded the bacteria do not move far in these vessels. Cankers develop most rapidly soon after the peak cold period is over and before rapid tree growth starts. The side of the tree facing the sun shows more infections because of the warming effect of sunrays. When higher temperatures set in, growth of the cankers is slowed down and the host forms callus tissue around the

cankers. These cankers become inactive. In those cankers in which callus formation is not complete bacteria survive and become a source of primary inoculum in the next season. The ability to form callus depends on varietal reaction of the host, its age and succulence, temperature and humidity during the season, and the type of rootstock.

In leaf infections, intercellular growth of bacteria causes necrosis of tissue and formation of spots. During wet weather bacterial ooze comes out and spreads the bacteria to adjoining and other leaves through contact, insects, and wind or rains. Leaves become resistant as they become old. With the onset of the summer season most of the bacteria inside the leaves and on their surface die but those that survive can initiate new infections on buds and stems.

Management

Bacterial canker is unlikely to be a serious problem provided crucial pre-disposing factors are avoided or counteracted. Trees are particularly susceptible in some soils, in water-logged soils that drain poorly, and during prolonged periods of drought. Thus, the selection of sites for new orchards should take into consideration these facts. Use of marginal land for establishing a new orchard should be avoided.

Only healthy budwood should be used for propagation. Susceptible varieties should be propagated on resistant rootstock and should be grafted as high as possible. Only healthy nursery trees should be planted in the orchard. The disease situation is aggravated if the pathogen has been introduced uniformly into young nursery trees through infected buds or rootstock. These trees might not become diseased in the nurseries where optimal growth conditions can be maintained easily. But when the trees are severely stressed during transplanting they are more prone to disease development until they have become established in their new environment (108).

Chemical control of the canker phase of the disease both in the nursery and in the orchard is based on spray with fixed copper or Bordeaux mixture before the winter season starts (10:15:100) and in spring (6: 9:100) before blossoming. However, wide distribution of copper-resistant strains of *Ps. syringae* pv. *syringae* on cherry blossoms indicates that copper-based fungicides cannot be used indefinitely (107). Streptomycin applied in spring is more effective in reducing leaf spot than Bordeaux mixture but it does not control canker initiation and development. In general, chemical sprays have not proved very effective against the disease.

Cankers on trunks and thick branches can be destroyed by flame (propane torch). The flame is aimed at the canker, especially at its margins for 5-20 sec, until the underlying tissue begins to crackle and char. The

treatment is carried out in early to mid-spring and, if necessary, repeated 2-3 weeks later (109).

■ REFERENCES

1. Adaskaveg, J.E. and J.M. Ogawa. 1994. Penetration of iprodione into mesocarp fruit tissue and suppression of gray mold and brown rot of sweet cherries. *Plant Dis.* **78:** 293.

2. Agarwala, R.K. 1970. Relative importance of the control methods of Phytophthora collar rot disease of apple trees, pp. 632-638. In: *Plant Disease Problems.* S.P. Raychaudhuri *et al.* (eds.). Indian Phytopathol. Soc. New Delhi.

3. Agarwala, R.K. 1985. *Apple Scab.* Directorate of Publications, Haryana Agric. Univ., Hissar (India). 18 p.

4. Agarwala, R.K. and V.C. Sharma. 1966. White root rot of apple in Himachal Pradesh. *Indian Phytopath.* **19:** 82.

5. Agarwala, R.K. and V.C. Sharma. 1968. Storage rot diseases of apple. *Indian Phytopath.* **21:** 294.

6. Agarwala, R.K., K.N. Arora and A. Singh. 1966. Effect of temperature and humidity variation on the development of peach leaf curl in mid-hills and the control. *Indian Phytopath.* **19:** 308.

7. Agrios, G.N. 1988. *Plant Pathology*, 3rd Ed. Academic Press.

8. Aldwinckle, H.S. and S.V. Beer. 1978. Fire blight and its control. *Hortic. Reviews* **1:** 243.

9. Aldwinckle, H.S. and S.V. Beer. 1979. Recent progress in breeding for fire blight resistance in apples and pears in North America. *EPPO Bull.* **9:** 7.

10. Anagnostakis, S.L. and D.E. Aylor. 1991. Efficacy of ascospores of *Venturia inaequalis* in producing scab lesions on apple leaves. *Plant Dis.* **75:** 918.

11. Anderson, H.W. 1956. *Diseases of Fruit Crops.* McGraw Hill.

12. Arauz, L.F. and T.B. Sutton. 1989. Influence of temperature and moisture on germination of ascospores and conidia of *Botryosphaeria obtusa. Phytopathology* **79:** 667.

13. Arauz, L.F. and T.B. Sutton. 1990. Protectant and after infection activity of fungicides against *Botryosphaeria obtusa* on apple. *Plant Dis.* **74:** 1029.

14. Ark, P.A. 1941. Chemical eradication of crown gall on almond trees. *Phytopathology* **31:** 956.

15. Ark, P.A. 1953. Use of streptomycin dust to control fire blight. *Plant Dis. Rep.* **37:** 405.

16. Aylor, D.E. 1998. The aerobiology of apple scab, *Plant Dis.* **82:** 838.

17. Baines, R.C. 1939. Phytophthora trunk canker or collar rot of apple trees. *J. Agric. Res.* **59:** 159.

18. Banihashemi, Z. and J.E. Mitchell. 1976. Factors affecting oospore germination in *Phytophthora cactorum,* the incitant of apple collar rot. *Phytopathology* **66:** 443.

19. Banyal, D.K. and R.L. Sharma. 1996. Efficacy of pre-harvest fungicidal sprays on gray mold rot and fruit quality of pear in storage. *Indian J. Mycol. Pl. Pathol.* **26:** 286.

20. Barkai-Golan, R. and D.J. Phillips. 1991. Post-harvest heat treatment of fresh fruits and vegetables for decay control. *Plant Dis.* **75:** 1085.

21. Barnes, E.H. and E.B. Williams. 1960. A biochemical response of apple tissues to fungus infection. *Phytopathology* **50:** 844.

22. Becker, C.M., T.J. Burr. and C.A. Smith. 1992. Overwintering of conidia of *Venturia inaequalis* in apple buds in New York orchards. *Plant Dis.* **76:** 121.

23. Beer, S.V. and J.L. Norelli. 1977. Fire blight epidemiology : Factors affecting release of *Erwinia amylovora* by cankers. *Phytopathology* **67:** 1119.

24. Beer, S.V., J.R. Rundle and J.L. Norelli. 1984. Recent progress in the development of biological control of fire blight: A review. *Acta Hort.* **151**: 195.

25. Bennett, R.A. 1980. Evidence for two virulence determinants in the fire blight pathogen *Erwinia amylovora. J. Gen. Microbiol.* **116**: 351.

26. Beraha, L., G.B. Ramsey, M.A. Smith and W.R. Wright. 1959. Effect of gamma radiation on brown rot and Rhizopus rot of peaches and the causal organism. *Phytopathology* **49**: 354.

27. Beresford, R.M. and D.W.L. Manktelow. 1994. Economics of reducing fungicide use by weather-based disease forecasts for control of *Venturia inaequalis* in apples. *N.Z. J. Crop Hortic. Sci.* **22**: 113.

28. Berkett, L.P. and K.D. Hickey. 1982. Location of primary infection sites of apple powdery mildew and its significance in management strategies. *Phytopathology* **72**: 705.

29. Berkett, L.P. *et al.* 1988. Relation of application timing to efficacy of triadimefon in controlling apple powdery mildew. *Plant Dis.* **72**: 310.

30. Biggs, A.R. and J. Northover. 1988. Early and late season susceptibility of peach fruits to *Monilinia fructicola. Plant Dis.* **72**: 1070.

31. Biggs, A.R., M.M. El-Kholi, S. El-Neshawy and R. Nickerson. 1997. Effect of calcium salts on growth, polygalacturonase activity, and infection of peach fruit by *Monilinia fructicola. Plant Dis.* **81**: 399.

32. Billing, E. 1984. Studies on avirulent strains of *Erwinia amylovora. Acta Hort.* **151**: 249.

33. Boneti, J.I.S. and Y. Katsurayama. 1999. Chemical control of apple scab under conditions of prolonged leaf wetness. *Phitopatologia Brasileira* **24**: 31.

34. Bonn, W.G. 1979. Fire blight bacteria in symptomless dormant apple and pear buds. *Can. J. Plant Pathol.* **1**: 61.

35. Bonn, W.G. 1981. Monitoring epiphytic *Erwinia amylovora* and the incidence of the blight of apple and pear in southeastern Ontario. *Acta Hort.* **117**: 31.

36. Boyle, A.M. and R.M. Price. 1963. Vancomycin prevents crown gall. *Phytopathology* **53**: 1272.

37. Braun, A.C. and A.M. Boyle. 1945. Application of penicillin to crown gall. *Phytopathology* **35**: 521.

38. Browne, G.T. and S.M. Mircetich. 1996. Effect of months of inoculation on severity of disease caused by *Phytophthora* spp. in apple root crown and excised shoots. *Phytopathology* **86**: 290.

39. Burchill, R.T. 1960. The role of secondary infections in the spread of apple mildew (*Podosphaera leucotricha*). *J. Hortic. Sci.* **35**: 66.

40. Burchill, R.T. 1968. Field and laboratory studies of the effect of urea on ascospore production of *Venturia inaequalis. Ann. Appl. Biol.* **62**: 297.

41. Burchill, R.T. 1978. Powdery mildews of tree crops. In: *The Powdery Mildews*, p. 565. D.M. Spencer (ed.). Academic Press.

42. Burchill, R.T. and C.J. Williamson. 1971. Comparison of some new fungicides for the control of scab and powdery mildew of apple. *Plant Pathology* **20**: 173.

43. Burchill, R.T., K.E. Hutton, J.E. Crosse and M.E. Garrett, 1965. Inhibition of the perfect stage of *Venturia inaequalis* by urea. *Nature (London)* **205**: 520.

44. Burchill, R.T., E.L. Frick, M.E. Cook and A.A. Swait. 1979. Fungitoxic and phytotoxic effect of some surface-active agents applied for the control of apple powdery mildew. *Ann. Appl. Biol.* **91**: 41.

45. Burr, T.J., J.L. Norelli, C.L. Reid, L.K. Capron, L.S. Nelson, H.S. Aldwinckle and W.F. Wilcox. 1993. Streptomycin resistant bacteria associated with fire blight infections. *Plant Dis.* **77**: 63.

46. Butt, D.J. 1972. The timing of spray for the protection of terminal buds on apple shoots from powdery mildew. *Ann. Appl. Biol.* **72**: 239.

47. Butt, D.J. and M.J Jeger. 1986. Components of spore production in apple powdery mildew (*Podosphaera leucotricha*). *Plant Pathology* **35**: 491.

48. Butt, D.J. and X. M. Xu. 1994. Vintem TM- a computerized apple scab warning system for use on farms. *Norwegian J. Agric. Sci. Suppl.* **17**: 247

49. Butt, D.J., A.H.M. Kirby and C.J. Williamson. 1973. Fungitoxic and phytotoxic effects of fungicides controlling powdery mildew of apple. *Ann. Appl. Biol.* **75**: 21.

50. Chand, J.N., M.R. Kondal and R.K. Aggarwal. 1968. Epidemiology and control of bitter rot of apples caused by *Gloeosporium fructigenum*. *Indian Phytopath.* **21**: 257.

51. Chandrani, A.N. and J.L. Kaul. 1997. Effect of hydrocooling with or without fungicides on whisker's rot of July Alberta peaches. *J. Mycol. Pl. Pathol.* **27**: 76.

52. Cheah, L.H., K.G. Tate, A.W. Hunt and M. de Silva. 1993. Decreased sensitivity of *Taphrina deformans* (peach leaf curl) to copper fungicides in Hawkes Bay. *Proc. 46th New Zealnd Plant Prot. Conf., Christchurch.* August 10-12, 1993.

53. Chib, H.S., B.R. Gupta, P.S. Andotra and C.N. Dar. 1983. Evaluation of some fungicides for the control of post-harvest rots of apple through fungicides in Kashmir. *Indian J. Mycol. Pl. Pathol.* **13**: 353.

54. Chiou, C.-S. and A.L. Jones. 1991. The analysis of plasmid mediated streptomycin resistance in *Erwinia amylovora*. *Phytopathology* **81**: 710.

55. Cimanowski, J. and M. Szkolnik. 1985. Post-symptom activity of sterol-inhibitors against apple powdery mildew. *Plant Dis.* **69**: 562.

56. Conway, 1982. Effect of postharvest calcium treatment on decay of Delicious apples. *Plant Dis.* **66**: 402.

57. Conway, W.C., G.M. Greene, K.D. Hickey. 1987. Effect of preharvest and postharvest calcium treatment of peaches on decay caused by *Monilinia fructicola*. *Plant Dis.* **71**: 1084.

58. Conway, W.C., C.E. Sams, R.G. McGuird and A. Kelman. 1992. Calcium treatment of apples and potatoes to reduce post-harvest decay. *Plant Dis.* **76**: 329.

59. Conway, W.S., C.E. Sams, G.A. Brown, W.B. Beavers, R.B. Tobias and L.S. Kennedy. 1994. Pilot test for the commercial use of postharvest pressure infiltration of calcium in to apples to maintain fruit quality in storage. *Hort. Technol.* **4**: 239.

60. Cooksey, D.A. and L.W. Moore. 1982. High frequency spontaneous mutations to agrocin 84 resistance in *Agrobacterium tumefaciens* and *A. rhizogenes*. *Physiol. Plant Pathol.* **20**: 129.

61. Coyier, D.L. 1974. Heterothallism in the apple powdery mildews fungus *Podosphaera leucotricha*. *Phytopathology* **64**: 246.

62. Crosse, J.E. 1966. Epidemiological relations of the pseudomonad pathogens of deciduous fruit trees. *Annu. Rev. Phytopathol.* **4**: 291.

63. Crosse, J.E., C.M.E. Garrett and R.T. Burchill. 1968. Change in the microbial population of apple leaves associated with the inhibition of the perfect stage of *Venturia inaequalis* after urea treatment. *Ann. Appl. Biol.* **61**: 203.

64. Creemers, P., A. Vanmechelen, H. Lyr *et al* (eds). 1999. Fungal diseases of pome fruits in relation with anti-resistance strategies for modern fungicides. *Int. Reingardsbrunn Symp. Germany*, May 1998. **2**: 257.

65. Dahmen, H. and T. Staub. 1992. Protective, curative, and eradicative activity of difenoconazole against *Venturia inaequalis*, *Cercospora arachidicola* and *Alternaria solani*. *Plant Dis.* **76**: 774.

66. Daines, R., D.J. Weber, E.D. Bunderson and T. Roper. 1984. Effect of early sprays on control of powdery mildew fruit russett on apples. *Plant Dis.* **68**: 326.

67. Dar, G.N. and R.N. Kaul. 1981. Powdery mildew disease of apple and its control in Kashmir. *Pestology* **5**: 9.

68. DeCleane, M. and J. Deley. 1976. The host range of crown gall. *Bot. Rev.* **42**: 389.

69. Drake, C.R. 1971. Source and longevity of apple fruit rot inoculum, *Botryosphaeria ribis* and *Physalospora obtusa*, under orchard conditions. *Plant Dis. Rep.* **55**: 122.

70. Dueck, J. 1974. Survival of *Erwinia amylovora* in association with mature apple fruit. *Can. J. Plant Sci.* **54**: 349.

71. Edney, K.L. 1970. Some experiments with thiabendazole and benomyl as post-harvest treatment for the control of storage rots of apple. *Plant Pathology* **19**: 189.

72. Ellis, M.A., D.C. Ferree and D.E. Spring. 1981. Photosynthesis, transpiration, and carbohydrate content of apple leaves infected by *Podosphaera leucotricha*. *Phytopathology* **71**: 392.

73. Endert, E. and D.F. Ritchie. 1984. Overwintering and survival of *Pseudomonas syringae* **pv.** *syringae* and symptom development in peach trees. *Plant Dis.* **68**: 468.

74. Fabreges, C. and P. Lagouarde. 1998. Vision: Fungicide active against scab and powdery mildew of apple. *Phytoma* **505**: 45.

75. Felciano, A., A.J. Felciano, J. Vendrusculo, J.E. Adaskaveg and J.M. Ogawa. 1992. Efficacy of ethanol in postharvest benomyl-DCNA treatment for control of brown rot of peach. *Plant Dis.* **75**: 226.

76. Follas, G. and R.D. Welsh. 1993. Control of leaf curl in stone fruit with difenoconazole. *Proc. 46th New Zealand Plant Prot. Conf., Christ Church.* August 10-12. 1993.

77. Funt, R.C., M.A. Ellis and L.V. Madden. 1990. Economic analysis of protectant and disease-forecast-based fungicide spray programs for control of apple scab and grape black rot in Ohio. *Plant Dis.* **74**: 638.

78. Gadoury, D.M. and W.E. MacHardy. 1982. Effect of temperature on development of pseudothecia of *Venturia inaequalis*. *Plant Dis.* **66**: 464.

79. Gadoury, D.M. and W.E. MacHardy. 1982. Preparation and interpretation of squash mounts of pseudothecia of *Venturia inaequalis*. *Phytopathology* **72**: 92.

80. Gadoury, D.M. and W.E. MacHardy. 1982. A model to estimate maturity of ascospores of *Venturia inaequalis*. *Phytopathology* **72**: 901.

81. Gadoury, D.M., R.C. Seem, D.A. Rosenberger, W.F. Wilcox, W.E. MacHardy and L.P. Berkett. 1992. Disparity between morphological maturity of ascospores and physiological maturity of asci in *Venturia inaequalis*. *Plant Dis.* **76**: 277.

82. Gawda, S.S. and R.N. Goodman. 1970. Movement and persistence of *Erwinia amylovora* in shoot, stem, and root of apple. *Plant Dis. Rep.* **54**: 576.

83. Goodman, R.N. 1954. Fire blight control with agrimycin, a streptomycin-terramycin combination. *Plant Dis. Rep.* **38**: 874.

84. Goodman, R.N. and J.A. White. 1981. Xylem parenchyma plasmolysis and vessel wall disorientation caused by *Erwinia amylovora*. *Phytopathology* **71**: 844.

85. Goodman, R.N., J.S. Huang and P.Y. Huang. 1974. Host-specific phytotoxic polysaccharides from apple tissues infected by *Erwinia amylovora*. *Science* **183**: 1081.

86. Gupta, G.K. 1975. Epidemiology, forecasting and possible control of apple scab (*Venturia inaequalis*). *Pesticides* **9**: 31.

87. Gupta, G.K. 1979. Role of on-season, post-harvest and pre-leaf fall sprays in the control of apple scab (*Venturia inaequalis*). *Indian J. Mycol. Pl. Pathol.* **9**: 139.

88. Gupta, G.K. 1979. Scab of *Cotoneaster bacillaris* in India caused by *Venturia inaequalis* (*Spilocaea* stage). *Plant Dis. Rep.* **63**: 156.

89. Gupta, G.K. 1981. Maturation and discharge of ascospores of *Venturia inaequalis*. *Indian Phytopath* **34**: 502.

90. Gupta, G.K. 1983. Behaviour of fungicides and various spray schedules in the control of apple scab (*Venturia inaequalis*). *Int. J. Trop. Plant Dis.* **1**: 181.

91. Gupta, G.K. 1985. *Apple Scab.* Published by E. Merck (India), Bombay.

92. Gupta, G.K. 1987. Investigations on the effect of urea and fungicides in suppressing the ascigerous stage of apple scab pathogen. *Int. J. Trop. Plant Dis.* **5**: 93.

93. Gupta, G.K. 1992. Apple scab, pp. 1-31. In: *Plant Diseases of International Importance.* Vol. III. *Diseases of fruit crops.* J. Kumar *et al.* (eds.). Prentice Hall.

94. Gupta, G.K. and V.C. Lele. 1976. The scab fungus (*Venturia inaequalis*) on apple twigs in Kashmir Valley. *Curr. Sci.* **45**: 565.

95. Gupta, G.K. and V.C. Lele. 1980. Role of urea in suppression of ascigerous stage and comparative *in vitro* efficacy of fungicides against apple scab. *Indian J. Agric. Sci.* **50**: 167.

96. Gupta, G.K. and V.C. Lele. 1980. Prevalence, distribution and intensity of apple scab in Kashmir Valley. *Indian J. Agric. Sci.* **50**: 45.

97. Gupta, G.K. and V.C. Lele. 1980. Morphology, physiology and epidemiology of the scab fungus, *Venturia inaequalis* (Cooke) Wint in Kashmir Valley. *Indian J. Agric. Sci.* **50**: 51.

98. Gupta, G.K. and J. Kumar. 1985. Studies on the curative (after infection) and eradicant (post-symptom) activity of fungicides against apple leaf infection by *Venturia inaequalis*. *Pesticides* **19**:18.

99. Gupta, G.K. and K.D.Verma. 1985. Post-symptom antisporulant activity of fungicides on apple scab foliage lesions. *Indian J. Agric. Sci.* **55**: 381.

100. Gupta, G.K. and K.D. Verma. 1986. Studies on the keeping quality of apple fruits infected with scab (*Venturia inaequalis*) in late summer season, pp. 307-309. In: *Adv. Res. Trop. Fruits* (ed.) T. R. Chadha. UHF, Solan, India.

101. Gupta, S.K. and G.K. Gupta. 1991. Eradicative activity of fungicides against apple powdery mildew. *Indian J. Mycol. Plant Pathol.* **21**: 70.

102. Gupta, V.K. 1977. Possible use of carbendazim in the control of Dematophora root rot of apple. *Indian Phytopath.* **30**: 527.

103. Gupta, V.K. 1999. *Phytophthora cactorum*—A threatening pathogen of apple. *Indian Phytopath.* **52**: 105.

104. Gupta, V.K. and K.D. Sharma. 1978. Comparative susceptibility of apple rootstocks to *Dematophora necatrix*. *Indian Phytopath.* **31**: 377.

105. Han, J.-S. 1990. Use of antisporulant epidermal coatings for plant protection in China. *Plant Dis.* **74**: 263.

106. Harris, D.C. 1991. The Phytophthora disease of apple. *J. Hort. Sci.* **66**: 513.

107. Hattingh, M.J. and I.M.M. Roos. 1992. Bacterial canker of stone fruits, pp. 394-404. In: *Plant Diseases of International Importance.* Vol. III. *Diseases of fruit crops.* J. Kumar *et al.*(eds.). Prentice Hall, New Jersey.

108. Hattingh, M.J., I.M.M. Roos and E.L. Mansvelt. 1989. Infection and systemic invasion of deciduous fruit trees by *Pseudomonas syringae* in South Africa. *Plant Dis.* **73**: 784.

109. Hawkins, J.E. 1976. A cauterization method for the control of cankers caused by *Pseudomonas syringae* in stone fruit trees. *Plant Dis. Rep.* **60**: 60.

110. Hebburn, A.G. 1982. The biology of the crown gall—A plant tumor induced by *Agrobacterium tumefaciens*, pp. 101-113. In: *Bacteria and Plants.* M. Rhodes-Roberts and F.A. Skinner (eds.). Academic Press, New York.

111. Heyes, C.C. and J. H. Andrews. 1983. Antagonism of *Athelia bombacina* and *Chaetomium globosum* to the apple scab pathogen, *Venturia inaequalis*. *Phytopathology* **73**: 650.

112. Hildebrand, P.D., C.L. Lockhart, R.L. Newery and R.G. Ross. 1989. Resistance of *Ventuiria inaequalis* to bitertanol and other demethylation-inhibiting fungicides. *Can. J. Plant Pathol.* **10**: 311.

113. Hill, S.A. 1975. The importance of wood scab caused by *Ventura inaequalis* as a source of infection of apple leaves in the spring. *Phytopath. Z.* **82**: 216.

114. Hirst, J.M. and O.J. Stedman. 1961. The epidemiology of apple scab (*Venturia inaequalis*). Frequency of air-borne spores in orchards. *Ann. Appl. Biol.* **49**: 290.

115. Hirst, J.M. and O.J. Stedman. 1962. Epidemiology of apple scab (*Venturia inaequalis*). II. Observations on the liberation of ascospores. *Ann. Appl. Biol.* **50**: 525.

116. James, J.R. and T.B. Sutton. 1962. Environmental factors influencing pseudothecial development and ascospore maturation of *Venturia inaequalis*. *Phytopathology* **52**: 1073.
117. Janisiewicz, W.J. 1987. Post-harvest biological control of blue mold of apples. *Phytopathology* **77**: 481.
118. Janisiewicz, W.J. 1988. Biological control of diseases of fruits, pp. 153-165. In: *Biocontrol of Plant Diseases*. K.G. Mukerji and K.L. Garg (eds.). CRC Press, Boca Raton, Fl., U.S.A.
119. Janisiewicz, W.J. 1988. Biological control of post-harvest diseases of apples with antagonist mixtures. *Phytopathology* **78**: 194.
120. Janisiewicz, W.J. and J. Roitman. 1988. Biological control of blue mold and gray mold on apple and pear with *Pseudomonas cepacia*. *Phytopathology* **78**: 1097.
121. Janisiewicz, W.J. and A. Marchi. 1992. Control of storage rot of various pear cultivars with a saprophytic strain of *Pseudomonas syringae*. *Plant Dis*. **76**: 555.
122. Janisiewickz, W.J., D.L. Peterson and R. Bors. 1994. Control of storage decay of apples with *Sporobolomyces roseus*. *Plant Dis*. **78**: 466.
123. Jeffers, S.N. 1992. Pre-plant root treatments to reduce incidence of *Phytophthora* species on dormant apple rootstocks. *Plant Dis*. **76**: 12.
124. Jeffers, S.N. and H.S. Aldwinckle. 1988. Phytophthora crown rot of apple trees: Sources of *Phytophthora cactorum* and *P. cambivora* as primary inoculum. *Phytopathology* **78**: 328.
125. Jesper, M.J. and D.J. Butt. 1983. The effects of weather during perennation on epidemics of apple mildew and scab. *EPPO Bull*. **13**: 79 .
126. Jones, A.L., G.R. Ehret, E.L. Hadidi, M.J. Zabik, J.N. Cash, and J.W. Johnson. 1993. Potential for zero residue disease control programs for fresh and processed apples using sulphur, fenarimol and myclobutanil. *Plant Dis*. **77**: 1114.
127. Juarez-Palacios, C., R. Feli-Gastelum, R.J. Wakeman, E.J. Poplomats and J.E. DeVey. 1991. Thermal sensitivity of three species of *Phytophthora* and the effect of soil solarization on their survival. *Plant Dis*. **75**: 1160.
128. Kapoor, J.N. 1967. *Podosphaera leucotricha*. CMI Description of Pathogenic Fungi and Bacteria.
129. Kaul, J.L. 1982. Comparative effectiveness of systemic fungicides for the control of post-harvest fungal rots of apple. *Indian Phytopath*. **35**: 315.
130. Kaul, J.L. 1984. Effect of temperature on the development of fungal rots of apple. *Indian Phytopath*. **37**: 573.
131. Kaul, J.L. 1986. Efficacy of biphenyl and sodium orthphenylphenate in controlling fungal rots of fruits. *Indian Phytopath*. **39**: 285.
132. Kaul, J.L. and R.L. Munjal. 1980. Effectiveness of hot water treatments in controlling fungal rots of apple. *Indian Phytopath*. **33**: 484.
133. Kaul, J.L. and R.L. Munjal. 1981. Post-harvest fungal diseases of apple in Himachal Pradesh. *Indian Phytopath*. **34**: 80.
134. Kaul, J.L. and R.L. Munjal. 1982. Fruit wrappers and skin coating for control of post-harvest decay of apple. *Indian J. Mycol. Plant Pathol*. **12**: 179.
135. Kaul, J.L. and R.L. Munjal. 1982. Effectiveness of ammonia and sulphur dioxide fumigation to control certain post-harvest diseases of apple. *Indian J. Mycol. Pl. Pathol*. **12**: 211.
136. Kaul, J.L. and R.L. Sharma. 1996. Lesion development and wound healing in Golden Delicious apples at different temperatures in relation to blue mold and bitter rot. *Indian J. Mycol. Pl. Pathol*. **26**: 97.
137. Keane, P.J., A. Kerr and P.B. New. 1970. Crown gall of stone fruits. II. Identification and nomenclature of *Agrobacterium* isolates. *Australian J. Biol. Sci*. **23**: 585.
138. Keil, H.L. and R.W. Wilson. 1968. Relation of streptomycin sprays to fire blight control and apple fruit residue. *Plant Dis. Rep*. **52**: 259.

139. Keil, H.L. and T. van der Zwet. 1972. Aerial strands of *Erwinia amylovora* structure and enhanced production by pesticide oil. *Phytopathology* **62**: 355.

140. Kellor, W. and Scheinpflug. 1987. Fungal resistance to sterol biosynthesis inhibitors: A new challenge. *Plant Dis.* **71**: 1066.

141. Kerr, A. 1969. Crown gall of stone fruits. I. Isolation of *Agrobacterium tumefaciens* and related species. *Australian J. Biol. Sci.* **22**: 111.

142. Kerr, A. 1972. Biological control of crown gall: seed inoculation. *J. Appl. Bacteriol.* **35**: 493.

143. Kerr, A. 1974. Soil microbiological studies on *Agrobacterium radiobacter* and biological control of crown gall. *Soil Sci.* **118**: 168.

144. Kerr, A. 1980. Biological control of crown gall through production of agrocin 84. *Plant Dis.* **64**: 25.

145. Kerr, A. 1989. Commercial release of a genetically engineered bacterium for the control of crown gall. *Agric. Sci.* **2**: 41.

146. Ketskhoveli, E.B. 1976. Effectiveness of curative pruning against powdery mildew of apple, *Podosphaera leucotricha* (Ell. & Ev.) Salmon. *Rev. Plant Pathol.* **55**: 525.

147. Khosla, K.K. and G.K. Gupta. 1992. Curative activity of sterol-inhibiting fungicides against leaf infection of apple by *Venturia inaequalis*. *Indian J. Plant Prot.* **20**: 50.

148. Knauf-Beiter, G., H. Dahmen, U. Heyes and T. Staub. 1995. Activity of cyprodonil: optimal treatment timing and site of action. *Plant Dis.* **79**: 1198.

149. Kohn, F.C. Jr. and F.F. Hendrix. 1982. Temperature, free moisture, and inoculum concentration effects on the incidence of white rot of apple. *Phytopathology* **72**: 313.

150. Kohn, F.C. Jr. and F.F. Hendrix. 1983. Influence of sugar content and pH on development of white rot of apples. *Plant Dis.* **67**: 410.

151. Kumar, J. and G.K. Gupta. 1986. Influence of host response and climatic factors on the development of conidial stage of apple scab fungus, *Venturia inaequalis*. *Indian J. Mycol. Pl. Pathol.* **16**: 123.

152. Kumar, J. and G.K. Gupta. 1986. Influence of NPK fertilizers on the susceptibility of apple plants to scab pathogen (*Venturia inaequalis*). *Indian Phytopath.* **39**: 480.

153. Kunz, S., B. Lutz, H. Deising and K. Mendgen. 1998. Assessment of sensitivities to anilinopyrimidine and strobilurin fungicides in populations of the apple scab fungus *Venturia inaequalis*. *J. Phytopathol.* **146**: 231.

154. Latorre, B.A. and A.L. Jones. 1979. Evaluation of weeds and plant refuse as potential source of inoculum of *Pseudomonas syringae* in bacterial canker of cherry. *Phytopathology* **69**: 1122.

155. Lindow, S.E. 1983. The role of bacterial ice nucleation in frost injury to plants. *Annu. Rev. Phytopathol.* **21**: 263.

156. Lippincott, J.A., B.B. Lippincott, and M.P. Starr. 1981. The genus *Agrobacterium*, pp. 842-855. In: *The Prokaryotes* Vol. I. M.P. Starr *et al.* (eds.) Springer-Verlag, Berlin.

157. Loper, J.E. and C.L. Kado. 1979. Host range conferred by the virulence specifying plasmids of *Agrobacterium tumefaciens*. *J. Bacteriol.* **139**: 591.

158. Loper, J.E., M.D. Henkels, R.G. Roberts, G.G. Grove, M.J. Willett and T.J. Smith. 1991. Evaluation of streptomycin, oxytetracycline and copper resistance in *Erwinia amylovora* isolated from pear orchards in Washington State. *Plant Dis.* **75**: 287.

159. Luepschem, N.S. 1964. Effectiveness of 2,6-dichloro-4 nitroaniline impregnated peach wraps in reducing Rhizopus decay losses. *Phytopathology* **54**: 1219.

160. MacHardy, W.E. 1996. *Apple Scab: Biology, Epidemiology and Management*. 545 pp.

161. MacHardy, W.E. and D.M. Gibson. 1985. Forecasting the seasonal maturation of ascospores of *Venturia inaequalis*. *Phytopathology* **75**: 185.

162. MacHardy, W.E. and D.M. Gadoury. 1986. Patterns of ascospore discharge by *Venturia inaequalis*, *Phytopathology* **76**: 985.

163. MacHardy, W.E. and D.M. Gadoury. 1989. A revision of Mills's criteria for predicting apple scab infection periods. *Phytopathology* **79**: 304.

164. Margosan, D.A., J.L. Smilanick, G. F. Simmons and D.J. Henson. 1997. Combination of hot water and ethanol to control postharvest decay of peaches and nectarines. *Plant Dis.* **81**: 1405.

165. Margot, P., F. Haggenberger, J. Amrein and B. Weiss. 1998. CGA 279202: a new broad spectrum strobilurin fungicide. *Brighton Crop Protection Conf.* 1998. Vol 2: 375.

166. Matheis, J.P. and R.G. Roberts. 1993. Fumigation of sweet cherry (*Prunus avium*) fruit with low molecular weight aldehydes for post-harvest decay control. *Plant Dis.* **77**: 810.

167. Matheron, M.E., D.J. Young and J.C. Matejka. 1988. Phytophthora root and crown rot of apple trees in Arizona. *Plant Dis.* **72**: 481.

168. McIntosh, D.L. 1969. Suitability of several fungicides for treating soil infested with *Phytophthora cactorum* to prevent crown rot disease of apple trees. *Plant Dis. Rep.* **53**: 182.

169. McIntosh, D.L. 1971. Dilution plates used to evaluate initial and residual toxicity of fungicides in soil to zoospores of *Phytophthora cactorum*, the cause of crown rot of apple trees. *Plant Dis. Rep.* **55**: 213.

170. McIntosh, D.L. 1972. Effect of soil water suction, soil temperature, carbon and nitrogen amendments and host rootlets on survival in soil of zoospores of *Phytophthora cactorum*. *Can. J. Bot.* **50**: 269.

171. McLaughlin, R.J., C.L. Wilson, S. Droby, R. Ben-Arie and E. Chalutz. 1992. Biological control of post-harvest diseases of grapes, peach, and apple with the yeasts *Kloeckera apiculata* and *Candida guilliermondii*. *Plant Dis.* **76**: 470.

172. McManus, P.S. and A.L. Jones. 1994. Role of wind driven rains, aerosols, and contaminated budwood in incidence and spatial pattern of fire blight in apple nursery. *Plant Dis.* **78**: 1059.

173. Mehdi, A. and A.M. Shah. 1994. Efficacy of fungicides in controlling peach leaf curl disease. *Indian Phytopath.* **47**: 427.

174. Merwin, I.A., W.F. Wilcox and W.C. Stiles. 1992. Influence of orchard ground cover management on the development of Phytophthora crown and root rot of apple. *Plant Dis.* **76**: 199.

175. Michailides, T.J. 1991. Characterization and comparative studies of *Mucor* isolates from stone fruits from California and Chile. *Plant Dis.* **75**: 373.

176. Michailides, T.J. and D.P. Morgan. 1997. Influence of fruit-to-fruit contact on the susceptibility of French prune to infection by *Monilinia fructicola*. *Plant Dis.* **81**: 1416.

177. Michailides, T.J. and R.A. Spotts. 1990. Postharvest diseases of pome and stone fruits caused by *Mucor piriformis* in the Pacific Northwest and California. *Plant Dis.* **74**: 537.

178. Michailides, T.J. and R.A. Spotts. 1990. Transmission of *Mucor piriformis* to fruits of *Prunus persica* by *Carpophilus* spp. and *Drosophila melanogaster*. *Plant Dis.* **74**: 287.

179. Miller, P.M. and S. Rich. 1968. Reducing discharge of *Venturia inaequalis* ascospores by composting overwintered leaves. *Plant Dis. Rep.* **52**: 728.

180. Miller, T.D. and M.N. Schroth. 1972. Monitoring epiphytic population of *Erwinia amylovora* on pear with a selective medium. *Phytopathology* **62**: 1175.

181. Mills, W.D. 1944. Efficient use of sulphur dust and sprays during rains to control apple scab. *Ext. Bull. Cornell Exp. Sta.* 630.

182. Mills, W.D. 1955. Fire blight development on apple in western New York. *Plant Dis. Rep.* **39**: 206.

183. Moller, W.J., M.N. Schroth and S.V. Thomson. 1981. The scenario of the fire blight and streptomycin resistance. *Plant Dis.* **65**: 563.

184. Moore, L.W. and G. Warren. 1979. *Agrabacterium radiobacter* strain 84 and biological control of crown gall. *Annu. Rev. Phytopathol.* **17**: 163.

185. Nautiyal, M.A., S. Prakash and A. Kumar. 1988. Screening of peach cultivars for their resistance against peach leaf curl. *Indian J. Agric. Sci.* **58**: 575.

186. Nester, E.W. and T. Koshuge. 1981. Plasmids specifying plant hyperplasia. *Annu. Rev. Phytopathol.* **17**: 163.

187. New, P.B. and A. Kerr. 1972. Biological control of crown gall: field measurements and glasshouse experiments. *J. Gen. Microbiol.* **35**: 279.

188. Norelli, J.L. and S.V. Beer. 1984. Factors affecting the development of fire blight blossom infections. *Acta Hortic.* **151**: 37.

189. Northover, J. 1975. Captafol (single application technique, SAT) simplifies early season control of apple scab. *Plant Dis. Rep.* **59**: 357.

190. Northover, J. and K.E. Schneider. 1993. Activity of plant oils on diseases caused by *Podosphaera leucotricha, Venturia inaequalis* and *Albugo occidentalis*. *Plant Dis.* **77**: 152.

191. Ogawa, J.M. *et al.* 1975. *Monilinia* life cycle on sweet cherries and its control by overhead sprinkler fungicide applications. *Plant Dis. Rep.* **59**: 876.

192. Osorio, J.M., J.E. Adaskaveg and J.M. Ogawa. 1993. Comparative efficacy and systemic activity of iprodione and the new experimental anilide E—0858 for control of brown rot of peach fruit. *Plant Dis.* **77**: 1140.

193. Osorio, J.M., J.E. Adaskaveg and J.M. Ogawa. 1994. Inhibition of mycelial growth of *Monilinia* species and suppression and control of brown rot blossom blight of almond with iprodione and E—0858. *Plant Dis.* **78**: 712.

194. Palmiter, D.H. 1934. Variability in monoconidial culture of *Venturia inaequalis*. *Phytopathology* **24**: 22.

195. Parker, K.C. and T.B. Sutton. 1993. Effect of temperature and wetness duration on apple fruit infection and eradicant activity of fungicides against *Botryosphaeria dothidea*. *Plant Dis.* **77**: 181.

196. Parker, K.C. and T.B. Sutton. 1993. Susceptibility of apple fruit to *Botryosphaeria dothidea* and isolate variation. *Plant Dis.* **77**: 385.

197. Powell, G. 1967. Night application for improved effectiveness of streptomycin in preventing fire blight disease. *Plant Dis. Rep.* **51**: 605.

198. Pusey, P.L. 1993. Role of *Botryosphaeria* species in peach tree gummosis on the basis of differential isolation from outer and inner bark. *Plant Dis.* **77**: 170.

199. Pusey, L.P. and C.L. Wilson. 1984. Post-harvest biological control of stone fruit brown rot by *Bacillus subtilis*. *Plant Dis.* **68**: 753.

200. Puttoo, B.L., S. Gulamuddin, M.L. Koul, M.A. Fazil and S.N.Teng. 1976. Comparative efficacy of systemic and non-systemic fungicides in controlling apple scab (*Venturia inaequalis*) under field conditions. *Pesticides* 10 **(2)**: 51.

201. Qasba, G.N., G.N. Dar and A.M. Shah. 1982. Scab on *Cotoneaster aitchinsoni* in India caused by *Venturia inaequalis* (*Spilocaea pomi*). *Indian Phytopath.* **35**: 698.

202. Rana, K.S. and V.K. Gupta. 1981. Cause and control of collar rot of apple in India. *Indian Phytopath.* **34**: 17.

203. Rana, K.S. and V.K. Gupta. 1983. *In vitro* and *in vivo* efficacy of systemic and protectant fungicides against *Phytophthora cactorum*. *Indian J. Mycol. Pl. Pathol.* **13**: 272.

204. Rana, K.S. and V.K. Gupta. 1983. Effect of different soil factors and amendments on the production of sporangia and oospores in *Phytophthora cactorum*. *Proc. Indian Nat. Sci. Acad.* **49B**: 706.

205. Rana, K.S. and V.K. Gupta. 1983. Effect of soil factors and amendments on the lysis and growth of *Phytophthora cactorum*. *Indian J. Mycol. Pl. Pathol.* **13**: 307.

206. Rana, K.S. and V.K. Gupta. 1984. Occurrence of Pythiaceous fungi in collar rot affected apple soils of Himachal Pradesh. *Indian Phytopath.* **37**: 39.

207. Rana, K.S. and V.K. Gupta. 1984. Survival of *Phytophthora cactorum* zoospores in relation to soil factors and amendments. *Indian Phytopath.* **37**: 477.

208. Rana, K.S. and V.K. Gupta. 1984. Relative susceptibility of apple rootstock to *Phytophthora cactorum. Indian Phytopath.* **37**: 530.

209. Rana, K.S. and V.K. Gupta. 1984. Effect of fungicides on the viability of *Phytophthora cactorum* propagules in soil. *Phytopath. Z.* **110**: 245.

210. Revathi, R., P. Annamalai and D. Lalithakumari. 1992. Laboratory mutant of *Venturia inaequalis* resistant to sterol inhibitors. *Indian Phytopath.* **45**: 331.

211. Roos, I.M.M. and M.J. Hattingh. 1986. Weeds in orchards as potential source of inoculum for bacterial canker of stone fruits. *Phytophylactica* **18**: 5.

212. Roosje, G.S. 1963. Research on apple and pear scab in the Netherlands from 1938 until 1961. *Neth. J. Pl. Pathol.* **69**: 132.

213. Rosen, H.R. 1936. Mode of penetration and progressive invasion of fire blight bacteria into apple and pear blossoms. *Ark. Agr. Exp. Sta. Bull.* **331**.

214. Sanogo, S. and D.E. Aylor. 1997. Infection efficiency of *Venturia inaequalis* ascospores as affected by apple flower bud developmental stage. *Plant Dis.* **81**: 661.

215. Schnabel, G., L. Parisi and M. Blanke. 1998. Testing new fungicides to reduce apple scab. *Acta Hortic.* **466**: 83.

216. Schroth, M.N. and D.C. Hildebrand. 1968. A chemotherapeutic treatment for selectively eradicating crown gall and olive knot neoplasma. *Phytopathology* **58**: 848.

217. Schroth, M.N. and W.J. Moller. 1976. Crown gall controlled in the field with a non-pathogenic bacterium. *Plant Dis. Rep.* **60**: 275.

218. Schroth, M.N., S.V. Thomson and D.C. Hildebrand. 1971. Epidemiology and control of fire blight. *Annu. Rev. Phytopathol.* **12**: 389.

219. Schroth, M.N., A.R. Weinhold, A.H. McCain, D.C. Hildebrand and N. Ross. 1971. Biology and control of *Agrobacterium tumefaciens. Hilgardia* **40**: 537.

220. Schroth, M.N., S.V. Thomson and D.C. Hildebrand. 1974. Epidemiology and control of fire blight. *Annu. Rev. Phytopathol.* **12**: 389.

221. Schroth, M.N., S.V. Thomson and W.J. Moller. 1978. Streptomycin resistance in *Erwinia amylovora. Phytopathology* **68**: 565.

222. Schwabe, W.F.S. 1979. Changes in the susceptibility of apple leaves as influenced by age. *Phytopathology* **69**: 53.

223. Schwabe, W.F.S. 1984. Changes in the susceptibility of developing apple fruit to *Venturia inaequalis. Phytopathology* **74**: 118.

224. Schwabe, W.F.S. 1997. The effect of different wetting regimes on the protective action of apple scab fungicides. *Deciduous Fruit Grower* **47**: 262.

225. Sharma, I.D., A. Nath and S.K. Patyal. 1996. Estimation of N-dodecylguanidine acetate (dodine) residues in apple (*Malus domestica*). *Pesticides Res. J.* **8**: 191.

226. Sharma, J.N. 1995. Efficacy of fungicidal spray schedules for the control of apple scab. *Indian J. Mycol. Pl. Pathol.* **25**: 350.

227. Sharma, J.N. and V.K. Gupta. 1995. Studies on apple scab forecasting in Himachal Pradesh. *Indian Phytopath.* **48**: 225.

228. Sharma, J.N. and J.L. Kaul. 1997. Studies on the development of storage scab in apple and its management. *Indian Phytopath.* **50**: 396.

229. Sharma, J.N. and K.D. Verma. 1996. Curative and eradicant activity of some sterol-inhibiting fungicides against *Venturia inaequalis* causing apple scab. *Indian J. Plant Prot.* **24**: 76.

230. Sharma, R.C., V.K. Gupta and R.C. Garg. 1990. Dormant stage application of fungicides for the management of peach leaf curl. *Indian Phytopath.* **43**: 259.

231. Sharma, R.L. 1996. Effect of temperature in the management of post-harvest diseases of pears. *Indian J. Mycol. Pl. Pathol.* **26**: 233.

232. Sharma, R.L. and D.K. Banyal. 1996. Incidence of gray mold rot of pears in Himachal Pradesh. *Indian J. Mycol. Pl. Pathol.* **26**: 232.

233. Sharma, R.L., J. Kumar and V. Rao. 1988. Performance of some scab resistant cultivars of apple in Kullu Valley, Himachal Pradesh. *J. Tree Sci.* **7**: 45.

234. Sharma, S., R. K. Agarwala and N. P. Dahroo. 1986. Population levels of *Colletotrichum gloeosporioides* in relation to symptom development in pear fruits. *Indian Phytopath.* **39**: 120.

235. Sharma, S., R.K. Agarwala and J.C. Kaushik. 1986. Mode of infection and histopathological changes induced by *Glomerella cingulata* in pear fruits. *Indian Phytopath.* **39**: 469.

236. Sharma, S.K. and V.K. Gupta. 1980. Efficacy of fungicides in relation to time of application against powdery mildew of apple. *Indian J. Mycol. Pl. Pathol.* **10**: 174.

237. Sharma, S.K. and V.K. Gupta. 1994. Influence of fungicides on spore germination, sporulation and control of apple powdery mildew. *Indian J. Mycol. Pl. Pathol.* **24**: 24.

238. Sharma, S.K., S.K. Gupta and D.S. Kaith. 1992. Efficacy of fungicides in relation to number of sprays against powdery mildew of apple. *Indian J. Mycol. Pl. Pathol.* **22**: 267.

239. Sharma, V.P. and S.K. Sharma. 1991. Effect of apple scab spray schedule on powdery mildew and Alternaria leaf spot. *Indian J. Mycol. Pl. Pathol.* **21**: 190.

240. Singh, B.M. and J.N. Chand. 1969. Studies on the resistance of apple fruits to bitter rot caused by *Gloeosporium fructigenum*. *Indian Phytopath.* **22**: 179.

241. Singh, J.R. and E.W. Williams. 1956. Identification of three physiologic races of *Venturia inaequalis*. *Phytopathology* **46**: 190.

242. Sitton, J.W. and M.E. Patterson. 1992. Effect of high carbon dioxide and low oxygen controlled atmosphere on post-harvest decays of apple. *Plant Dis.* **76**: 992.

243. Sneh, B. and D.L. McIntosh. 1974. Studies on the behaviour and survival of *Phytophthora cactorum* in soil. *Can. J. Bot.* **52**: 795.

244. Snowden, A.L. 1990. *Post-harvest Diseases and Disorders of Fruits and Vegetables.* Vol. I.

245. Spotts, R.A. and L.A. Cervantes. 1986. Populations, pathogenicity, and benomyl resistance of *Botrytis* spp., *Penicillium* spp. and *Mucor piriformis* in packing houses. *Plant Dis.* **70**: 106.

246. Spotts, R.A. and P.M. Chen. 1984. Cold hardiness and temperature response of healthy and mildew-infected terminal buds of apple during dormancy. *Phytopathology* **74**: 542.

247. Spotts, R.A., R.P. Covey and P.M. Chen. 1981. Effect of low temperature on survival of apple buds infected with the powdery mildew fungus. *HortScience* **16**: 781.

248. Spotts, R.A., L.A. Cervantes and F.J.A. Niederholzer. 1997. Effect of dolomitic lime on production of asci and pseudothecia of *Venturia inaequalis* and *V. pirina*. *Plant Dis.* **81**: 96.

249. Stanis, V.F. and A.L. Jones. 1985. Reduced sensitivity to sterol inhibiting fungicides in field isolates of *Venturia inaequalis*. *Phytopathology* **75**: 1098.

250. Sugar, D., R.G. Roberts, R. J. Hilton, T. L. Righotti and E.E. Sanches. 1994. Integration of cultural methods with yeast treatment for control of post-harvest decay in pear. *Plant Dis.* **78**: 791.

251. Sutton, T.B. 1981. Production and dispersal of conidia by *Physalospora obtusa* and *Botryosphaeria dothidea* in apple orchards. *Phytopathology* **71**: 584.

252. Sutton, T.B. and J.V. Boyne. 1983. Inoculum availability and pathogenic variation in *Botryosphaeria dothidea* in apple production areas of North Carolina. *Plant Dis.* **67**: 503.

253. Sutton, T. B. and L. F. Arauz. 1991. Influence of temperature and moisture on germination of ascospores and conidia of *Botryosphaeria dothidea*. *Plant Dis.* **75**: 1146.

254. Sutton, T.B. and A.L. Jones. 1979. Analysis of factors affecting dispersal of *Podosphaera leucotricha* conidia. *Phytopathology* **69**: 380.

255. Tandon, R.N. and R.K. Kakkar. 1970. Some new and interesting post-harvest diseases, pp. 301-307. In: *Plant Disease Problems*. S.P. Raychaudhuri *et al.*(eds). IPS New Delhi.

256. Tate, K.G., P.M. Wood and A.J. Papay. 1996. Field evaluation of fungicides for control of peach leaf curl (*Taphrina deformans*). *Proc. 47th New Zealand Plant Protection Conf.* 9-11 Aug. 1994.

257. Taylor, J. 1971. A necrotic leaf blotch and fruit rot of apple caused by a strain of *Glomerella cingulata*. *Phytopathology* **61**: 221.

258. Taylor, J. 1971. Epidemiology and symptomatology of apple bitter rot. *Phytopathology* **61**: 1028.

259. Thakur, V.S. and G.K. Gupta. 1990. Evaluation of pre-symptom activity of fungicides on symptom expression, conidia production and viability of *Venturia inaequalis*. *Indian Phytopath.* **43**: 520.

260. Thakur, V.S. and G.K. Gupta. 1991. Studies on the effect of sterol inhibitors on apples scab, *Venturia inaequalis*, pathogenesis. *Indian J. Plant Prot.* **19**: 185.

261. Thakur, V.S. and K. Khosla. 1999. Relevance of Mills' infection periods to apple scab (*Venturia inaequalis*) prediction and rescheduling fungicide application in Himachal Pradesh. *Indian J. Agric. Sci.* **69**: 152.

262. Thakur, V.S., G.K. Gupta and R. Malhotra. 1995. Antisporulant activity of ergosterol biosynthesis-inhibiting fungicides in the management of apple scab, *Venturia inaequalis*. *Indian Phytopath.* **48**: 35.

263. Thomashow, M.F. *et al*. 1980. Host range of *Agrobacterium tumefaciens* is determined by the Ti-plasmid. *Nature* (London) **283**: 794.

264. Thomson, S.V. 1986. The role of the stigma in fire blight infections. *Phytopathology* **76**: 476.

265. Thomson, S.V. 1992. Fire blight of apple and pear, pp. 32-65. In: *Plant diseases of international importance*. Vol. III. *Diseases of fruit crops*. J. Kumar *et al*. (eds). Prentice-Hall.

266. Thomson, S.V., M.N. Schroth, W.J. Moller and E.O. Reil. 1975. Occurrence of fire blight of pears in relation to weather and epiphytic populations of *Erwinia amylovora*. *Phytopathology* **65**: 353.

267. Thomson, S.V., M.N. Schroth, W.J. Moller and W.O. Reil. 1982. A forecasting model for fire blight of pear. *Plant Dis.* **66**: 576.

268. Thomson, S.V., D.R. Hansen, K.M. Flint and J.D. Vanderberg. 1992. Dissemination of bacteria antagonistic to *Erwinia amylovora* by honey bees. *Plant Dis.* **76**: 1052.

269. Tidball, C.J. and R.G. Linderman. 1990. Phytophthora root and stem rot of apple rootstocks from stool beds. *Plant Dis.* **74**: 141.

270. Tomerlin, J.R. and A.I. Jones. 1983. Development of apples scab in the orchard and during cold storage. *Plant Dis.* **67**: 147.

271. Tweedy, B.G. and D. Powell. 1960. Alternaria rot of apple. *Phytopathology* **50**: 657.

272. Ulaya, G. and W. Koller. 1999. Baseline sensitivities of *Venturia inaequalis* populations to the strobilurin fungicide kresoxim methyl. *Plant Dis.* **83**: 274.

273. Utkhede, R.S. 1984. Antagonism of isolates of *Bacillus subtilis* to *Phytophthora cactorum*. *Can. J. Bot.* **62**: 1032.

274. Utkhede. R.S. 1986. Biology and control of apple crown rot caused by *Phytophthora cactorum*. A Review. *Phytoprotection* **67**: 1.

275. Utkhede, R.S. 1987. Chemical and biological control of crown and root rot of apple caused by *Phytophthora cactorum*. *Can. J. Plant Pathol.* **9**: 295.

276. Utkhede, R.S. and E.M. Smith. 1991. Effects of fosetyl-Al, metalaxyl and *Enterobacter aerogenes* on crown and root rot of apple caused by *Phytophthora cactorum* in British Columbia. *Plant Dis* **75**: 406.

277. Utkhede, R.S. and E.M. Smith. 1995. Effect of nitrogen form and application method on incidence and severity of Phytophthora crown and root rot of apple trees. *European J. Plant Pathol.* **101**: 283.

278. van der Zwet, T. and H.L. Keil. 1979. Fire blight: A bacterial disease of Rosaceous plants. *U.S. Dep. Agric. Handbook* **510**.

279. van der Zwet, T., S.V. Thomson, R.P. Covey and W.G. Bonn. 1990. Population of *Erwinia amylovora* on external and internal apple fruit tissues. *Plant Dis.* **74**: 711.

280. van der Zwet, T., A.R. Biggs, R. Heflebower and G. W. Lightner. 1994. Evaluation of MARYBLYT computer model for predicting blossom blight on apple in West Virginia and Maryland. *Plant Dis.* **78**: 225.

281. Vigouroux, A. 1989. Ingress and spread of *Pseudomonas* in stems of peach and apricot promoted by frost-related water-soaking of tissues. *Plant Dis.* **74**: 854.

282. Verma, K.D. and S.K. Gupta. 1992. Effect of pre-leaf-fall spray of urea and fungicides in suppressing the ascigerous stage of apple scab pathogen in Himachal Pradesh. *Plant Dis. Res.* **7**: 68.

283. Verma, K.D. and J.N. Sharma. 1997. Cultural and morphological variability in *Venturia inaequalis*, the apple scab pathogen. *J. Mycol. Pl. Pathol.* **27**: 252- 254.

284. Warner, J. 1990. *Phytoprotection* 71: 1.

285. Weaver, D.J. 1974. A gummosis disease of peach trees caused by *Botryosphaeria dothidea*. *Phytopathology* **64**: 1429.

286. Weaver, D.J. 1979. Role of conidia of *Botryosphaeria dothidea* in the natural spread of peach tree gummosis. *Phytopathology* **69**: 330.

287. Wicks, T. 1974. Tolerance of apple scab fungus to bezimidazole fungicides. *Plant Dis. Rep.* **58**: 886.

288. Wieneke, J., R.P. Covey Jr. and N. Benson. 1971. Influence of powdery mildew infection on S and Ca accumulation in leaves of apple seedlings. *Phytopathology* **61**: 1100.

289. Wilcox, W.F., D.I. Wasson and J. Kovach. 1992. Development and evaluation of integrated reduced spray program using sterol methylation inhibition fungicides for control of primary scab. *Plant Dis.* **76**: 669.

290. Wilson, C.E. and M. Wisniewski. 1989. Biological control of post-harvest diseases. *Annu. Rev. Phytopathol.* **27**: 425.

291. Wormald, H. 1954. *The Brown rot Diseases of Fruit Trees*. H.M.S. Office, London. Tech. Bull. No. 3, 113 pp.

292. Yoder, K.S. 1992. Powdery mildew of apple, pp. 66-89. In: *Plant Diseases of International Importance*. Vol. III. *Diseases of Fruit Crops*. J. Kumar *et al.* (eds.) Prentice Hall.

293. Yohalem, D.S., E.V. Nordheim and J.H. Andrews. 1996. The effect of water extract of spent mushroom compost on apple scab in the field. *Phytopathology* **86**: 914.

Mango Diseases

■ **BACTERIAL LEAF SPOT, BLIGHT AND CANKER OF MANGO**

Bacterial blight and canker disease of mango has been known to occur in India for a long time, perhaps as early as 1881, but was first reported in 1948 from Pune in Maharashtra. Doidge (20) had reported a similar disease from South Africa. In India, the disease is present throughout the country but is particularly serious in the northern states. It affects leaves, petioles, trunk and thick branches, and fruits and causes leaf spots and cankers. In warm, humid weather of the mango season it causes significant loss to foliage. Fruit drop to the extent of 10-15% is also reported. The disease incidence and disease severity have varied from 0.52 to 42.0% and 15 to 90%, respectively, in different states of India (35). Upto 80% loss may occur in areas with high winds and heavy rains (16). Different strains of the bacterium cause different predominant symptoms. The leaf spot isolate and fruit or twig canker isolates are different.

Symptoms

Groups of minute, water-soaked lesions appear towards the tip of the leaf. They increase in size to about 1-4 mm and turn brown to black in colour. These spots are surrounded by a chlorotic halo. They are delimited by the leaf veins Large necrotic patches may be formed by coalescing of several spots. These patches sometimes dry up with decrease of atmospheric humidity. The leaf symptoms are very conspicuous on young newly formed leaves. Brown spots are seen on these leaves which curl when there is heavy infection. In warm and very humid weather, but when there are no heavy rains, drops of amber coloured exudation can be seen on these leaves. The spots are often rough and raised due to heavy bacterial exudation. The young leaves invariably shed and can be seen on the orchard floor. When a greater portion of the lamina of older leaves is affected these leaves also fall down. Petioles, fruits and tender stems are also infected. On young fruits, water-soaked lesions appear

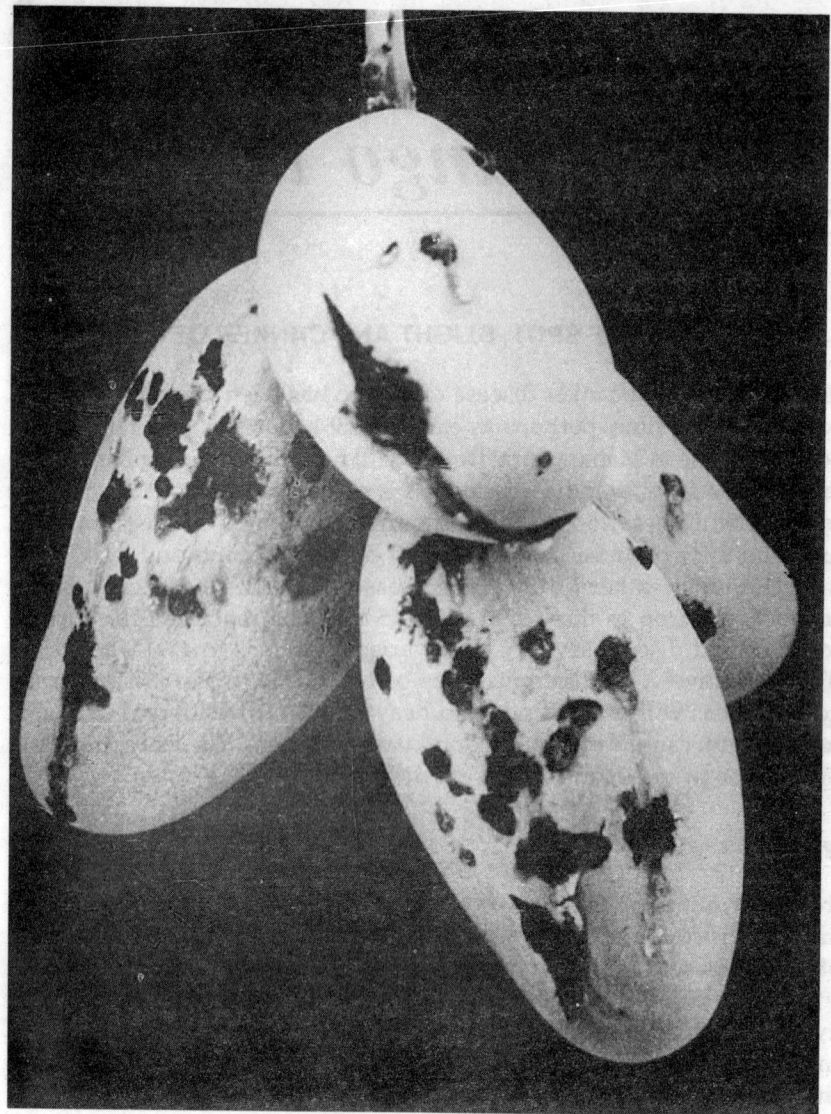

Fig. 14. Bacterial spots on mango fruits.
Courtesy: Dr. Ram Kishun

and turn dark brown to black. Cracks may appear in the skin of the fruit. Badly affected fruits drop prematurely. The fruits with cracked skin invite other secondary pathogens and post-harvest rot of fruits may occur. Even

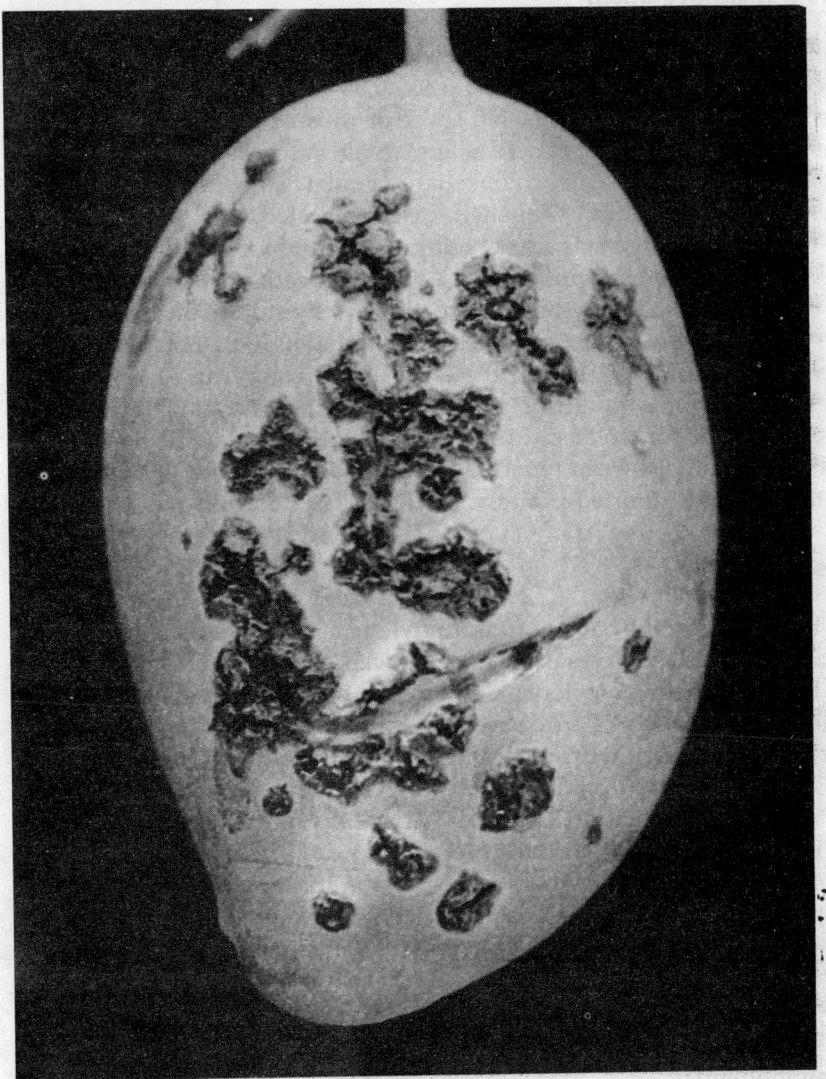

Fig. 15. Bacterial spots on mango fruits. Cracks on the skin.
Courtesy: Dr. Ram Kishun

on the tree, such fruits with cracks invite ants and insects and half mature fruits undergo rotting and shed. Xylem plugging and phloem distortion also occurs.

The Causal Organism

Xanthomonas campestris pv. *mangiferae-indicae* (Patel *et al.*) Robbs *et al.* The bacterial cells are rod-shaped and 0.36 - 0.54 × 1.44 μm in size. The cells are single or in short chains. No spore or capsule are formed. The cells are motile by a single polar flagellum. Stain reaction is Gram-negative. On potato dextrose agar, colonies are circular, smooth, glistening and creamy white. Gelatin is liquefied, casein digested, and litmus reduced by the bacterium. Nitrates are not reduced to nitrite. Optimum temperature for growth of the bacterium is 27°C. Thermal death point is about 55°C.

On the basis of biochemical and pathogenicity studies on 20 isolates of the bacterium from different states of south India, 10 pathotypes or races have been identified (16). In these studies, the cultivar Alphonso was highly susceptible to all the isolates while the cultivar Bangalore was resistant to most isolates. Cultural and biochemical variations in the bacterium had also been reported by Venugopal *et al.* (95). Three or four pathotypes are reported from the northern states of India.

Disease Cycle

In the orchards, the bacterium is a phylloplane resident throughout the year (47). It also survives in cankers on twigs and smaller branches, stone of the fruit, and as phylloplane microflora on weeds. It spreads rapidly during rains. Long-distance spread is caused by infected planting stock. The bacteria enter the fruits through bruises and other types of injuries. Disease development is favoured by high humidity (RH above 90 %) and a temperature range of 25°-30°C. Maximum infection of fruits occurs when minimum and maximum (night and day) temperatures are 22°C and 25°C. Rainfall is a major weather factor affecting fruit infection (47). High wind velocity is also a favourable factor for the disease. Many insects such as ashy weevil (*Myllocerus discolor* var. *variegata*), leaf webber (*Orthega vadrusalia*) and bugs (*Canteconidia furcellata*) mechanically transmit the bacteria on their legs and mouth parts (34).

Management

Although no specific control measures are recommended, spray of the antibiotic streptocycline is reported to give some control (36, 50). In north India, Mishra (49) has recommended 2 sprays of 300 ppm streptocycline with 0.3 % copper oxychloride in May at 10 days interval. *Bacillus* spp. such as *B. coagulans*, are reported as potential biocontrol agents (35).

Sinha and Hoda (78) screened a large number of mango varieties for resistance to the disease. None was resistant. Cultivars Langra, Gulabkhas, Kesar and Mankurud were moderately resistant. Bombai Green, Hemasagar, Fazli, Swarnarekha and Anupam were susceptible. In south India, severe natural infection in cultivars Mulgao, Alphanso, Neelam, Rumani, Bangalore and Baneshan was reported but Bombai Green was found resistant. Mishra (49) screened 212 mango varieties and found 95 of them free from the disease. Other scientists have reported Bombai Green, Fazri and Swarnarekha as resistant cultivars. Absence of standardized method for screening and presence of variability in the pathogen seem to give variable results of screening trials.

■ POWDERY MILDEW OF MANGO

Powdery mildew is a destructive disease of mango in India, Egypt and probably other parts of the world. In India, it is common in U.P. in the north and in Maharashtra, Karnataka and Tamil Nadu in the south. The disease generally occurs during December to March or longer depending on climatic conditions in the area. Usually its appearance coincides with blossom stage of the tree and blossom blight is the most destructive phase of the disease. In addition to loss of blossoms, infection of leaves in some areas causes defoliation. Even fruits may be affected and shed. Losses ranging from 20 to 90 % have been reported (54). Although a serious disease, little work has been done on it.

Symptoms

In the plains, powdery mildew on leaves is not common. Symptoms appear on the inflorescence. The sepals are more prone to infection than other floral parts. White or grey powdery growth appears on the panicles and soon the flowers are blighted, turn brown to black and either shed or remain as dry masses in the inflorescence. The number of flowers is considerably reduced and consequently there is less fruiting. Fruits formed in diseased panicles fail to grow and shed before reaching pea-size. Young leaves and shoots on trees growing in the cool mid-hills of north India show a similar powdery growth. These leaves also dry up and shed. If infection occurs after fruit set, the fruits soon fall down. The fruits are often malformed and off-coloured. Symptoms of die-back may also occur.

The Causal Organism

Powdery mildew of mango is caused by *Oidium mangiferae* Berth. The fungus is morphologically similar to some species of *Erysiphe* except for the absence of a known perithecial stage which has not been found anywhere in the world. The mycelium consists of septate, hyaline, ectophytic hyphae measuring 4.1 to 8.2 μm in diameter. Haustoria are lobate. Conidia (oidia) are oblong, single-celled, hyaline, thin-walled and formed in a chain. They measure $25 - 48.9 \times 16 - 23.9$ μm. The germ tube from the conidium forms appressorium on the host surface from which infection peg penetrates the epidermis to form haustorium.

Disease Cycle

The fungus survives as mycelium and conidia on the tree branches and also on other annual and perennial hosts. In favourable weather conidia formed on the mycelium are dispersed by wind and spread the disease. The optimum temperature for germination of conidia is 22°C, maximum 32°C and minimum 9°C. The conidia are quickly destroyed in dry weather and under strong sunlight. Generally, cold nights, light rains or foggy weather at the time of flowering favour initiation of the disease. In the mango belt of north India, the disease destroys mango panicles during the second half of March when average minimum temperatures are around 15°-17°C and the relative humidity 65-85%. High wind velocity favours spread of powdery mildew.

Management

Spray of fungicides is the best management strategy for mango powdery mildew (22, 23, 82). One spray should be given before flowering followed by 2-3 sprays at fortnightly intervals. Pre-infection or prophyllactic sprays are better. They can reduce the disease by half. Sprays after the disease has appeared give about 25-30 % control. Sulphur based fungicides have been most commonly recommended. These include Karathane (0.25 - 0.5 kg/ha), Sulfex and lime sulphur. One spray of 0.2 % Sulfex as soon as the disease is detected is reported to be effective, cheap and non-hazardous (84). Datar (15) compared many fungicides and found Topsin M (thiophanate methyl) at 0.5 % as the best, followed by Karathane at 0.2 %. The first spray was given in the first week of January after emergence of inflorescence and four subsequent sprays were given at 10-days interval. A single foliar spray of 0.1 M solution of phosphates is reported to induce systemic protection and suppresses mildew lesions on the leaves.

In addition to fungicide sprays, removal and burning of affected leaves, blossoms, twigs, etc. are recommended as sanitary precautions. Datar (14) screened 26 Indian varieties of mango against powdery mildew and found only the variety Totapari to possess moderate resistance. Other varieties were either moderately or highly susceptible. The varieties Dashehari and Langra were in the highly susceptible category.

■ MANGO ANTHRACNOSE AND DIE-BACK

The anthracnose of mango fruits in association with leaf spots, blossom blight, wither-tip or die-back, all caused by *Colletotrichum gloeosporioides* Penz., is a widespread disease and has major economic importance in areas where mango is grown commercially. Diseases with the above names have been reported from many countries including India, Philippines, Indonesia, Trinidad, Peru, Hawaii and Portugal (19, 72). In India, the disease was first described in the state of Punjab. In several states, the disease is often a limiting factor in profitable cultivation of mango (80).

The anthracnose occurs on all parts of the tree but is most common on flowers and flower stalks. Young and mature fruits are also infected. If early infection of fruits takes place they fall off. Infected ripe fruits are blemished hence fetch a low price in the market. In storage the anthracnose leads to postharvest storage rot.

Symptoms

The leaf spot and twig anthracnose is characterized by leaf spots, wither-tip and blossom blight. Young leaves are most susceptible. Numerous oval or irregular, brown to black spots appear on the leaf blade restricting further growth of the leaf. Under humid conditions these spots increase rapidly and form irregular, necrotic areas. Sometimes, the dead parts of the blade separate and fall down. Badly affected leaves may also shed. Symptoms of wither-tip or die-back appear at the tip of very young shoots. Black necrotic areas develop on the affected twig which dries from the tip downward, accompanied by defoliation of the branch. Generally, thick branches are not infected but when infected the affected portion is destroyed. In wet and warm weather, acervuli of the fungus develop as pink bodies on the dead and infected parts. Later, these fungus structures turn black. In blossom blight, minute, brown to black lesions appear on floral parts and floral stalks. Affected flowers in the panicle dry and shed. This causes significant loss to fruit setting. In severe infections, the entire inflorescence may be destroyed with no fruiting at all.

On green immature fruits the infection appears as black spots. The tissue beneath the spot becomes hard and ruptures. This may cause fall of the fruit. Anthracnose of ripening fruit is characterized by the development of black spots of varying forms which may be slightly sunken or may show cracks. These spots increase in size as the fruit ripens and often coalesce. Finally, the entire fruit is covered by these spots and shows soft rot symptoms. The spots are often concentrated at the stem-end and sometimes in streaks toward one side of the fruit. This suggests that the disease has spread through spores washed down by rain water from the stem-end. Such fruits either rot on the tree or, more often, carry the disease and rot during transport and storage.

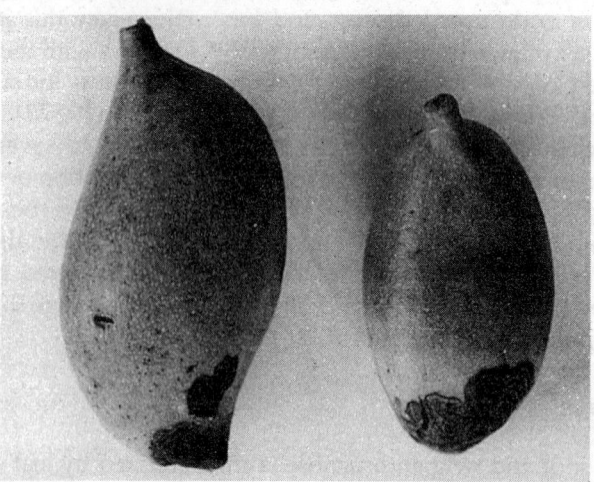

Fig. 16. Mango anthracnose.

The Causal Organism

Colletotrichum gloeosporioides Penz., anamorph of the Ascomycetous fungus *Glomerella cingulata* (Ston.) Spould and Shrenk, is a widely distributed fungus causing leaf spot and anthracnose on citrus, avocado, sugarcane, etc. In addition to *C. gloeosporioides*, many other fungi are reported to be associated with die-back, leaf blight, and fruit rot symptoms on mango. These include *Diplodia natalensis, Botryodiplodia theobromae, Botryosphaeria ribis, Physalospora rhodina, Ceratocystis fimbriata, Macrophoma mangiferae* and *Hendersonula toruloidea* (11, 24, 26, 70, 93, 94). Most of these are only weak parasites.

The mycelium of *C. gloeosporioides* consists of rather narrow, sparsely septate hyphae which are at first hyaline but later take on a slightly dark colour. Acervuli are abundantly formed on the affected host surfaces.

These develop at first as tangled subepidermal masses of hyphae. These structures are elongated or irregular in shape and up to 500 μm in diameter. From the tangled mass of hyphae numerous closely packed conidiophores arise and partially raise the epidermis. The conidiophores are hyaline and unbranched. One or more conidia are formed from the apex of each conidiophore. Setae in acervuli are common on twigs but not on fruits. These structures are 4-8 μm thick, brown, and slightly swollen at the base. They almost completely cover the acervulus. The conidia remain embedded in a viscid fluid. In moist conditions this viscid fluid swells and ruptures the epidermis exposing the conidia. Being embedded in the fluid the conidia are not dispersed by wind currents. They are splashed by raindrops and insects. The conidia are sub-hyaline but look pinkish in mass. They are broadly oval to oblong with rounded ends, aseptate, and sometimes with 1-2 globules. Their size is highly variable, the average being around 12-16 × 4-6 μm. On germination the conidial germ tubes form clavate or irregular appressoria which measure 6-20 × 4-12 μm. The perithecial stage (*Glomerella cingulata*) develops on stromatoid cushions. The perithecia are immersed in the host tissue. They are more or less compounded, subspherical, and with prominent ostiolar hair. Asci are subclavate, often slightly pedicellate, and measure 55-70 μm in length. The ascospores are allantoid, hyaline, aseptate, 3.5-5 μm wide and 12-22 μm long. They are difficult to distinguish from conidia. The fungus is heterothallic and the sexual stage is not common.

Disease Cycle

Diseased twigs, leaves and dried fruits on the tree and on the orchard floor are a prolific source of perennation of the fungus and fresh infections. The pathogen can survive as leaf spots on the tree throughout the year under north Indian conditions. The fungus has a long saprophytic survival ability on dead twigs. Fitzell and Peak (21) had reported that conidia of *C. gloeosporioides* var. *minor* (*G. cingulata*), attacking mango are produced in lesions on leaves, defoliated branch terminals, mummified inflorescence and flower bracts. There is abundant production of conidia during flush growth and flowering of the trees. Prolonged periods of rains at this stage help severe outbreaks of the disease. Ascospores have no role in epidemiology (19). There is also a report that the shot-hole symptoms on leaves are caused by insects and the fungus enters the leaf tissue as a secondary invader. But it produces inoculum for infection of blossoms and fruits.

Conidial production in acervuli is favoured by temperatures of 10° - 30° C and a relative humidity of 95 - 97 %. The fungus does not grow at a relative humidity less than 95 %. Thus, humid and misty conditions, temperatures of 24°-32°C, especially at the time of development of shoots, flowers and fruits, are considered most favourable for infection. In north India, the months of July-August provide such conditions and anthracnose spreads very rapidly. The winter rains preceding or during blossoming also provide such conditions.

On the fruits, most of the infections take place from the start of the blossoming until fruits are more than half grown. Infection pegs from the appressoria enter the fruit through pores in the skin but the infection remains latent and the fungus grows only to a limited extent in the epidermal layers of the fruit. Further growth of the fungus usually occurs during ripening of the fruit. Latent infection can occur through lenticels also. Appressoria have an important role in epidemiology of anthracnose. Most of the appressoria from germ tubes of conidia lodged on the fruit do not germinate immediately but remain firmly attached to the fruit skin as the latent stage of the pathogen. These latent appressoria serve as inoculum that gives rise to anthracnose spots on ripe fruits (8).

Management

Tree and general orchard sanitation are important steps in the management of mango anthracnose. Vigorous growth of trees should be maintained by proper fertilization and watering during the summer. Diseased twigs should be pruned and destroyed along with fallen leaves. Pruning should be followed by sprays of suitable fungicides such as Bordeaux mixture (6:6:50) or copper oxychloride (0.15%). Four to five sprays between January and July (in northern India) have given satisfactory control of the disease. A combination of captan and zineb sprays has been found to be very effective. One pre-blossom spray of 0.3% captan followed by two sprays of 0.2% zineb at 14-21 days intervals are recommended. According to Tandon and Singh (88) mango anthracnose can be effectively checked by giving the trees two sprays of zineb or a copper fungicide at the time of flowering. Subsequent sprays may be started just before the onset of monsoons and continued till harvest at 14 days intervals. These field sprays also reduce the chances of post-harvest anthracnose rot of fruits.

Postharvest chemical treatments of fruits are generally not very effective against decay. Instantaneous dip of fruits in 1000 ppm benomyl or 2000 ppm thiabendazole before storage is reported to give a good control of fruit rot. Hot water treatment has been recommended (48, 63) but the combination of temperature and the time of exposure is to be decided according to the variety of mango. Tandon and Singh (89) have recommended a 15-min immersion of fruits of cultivars Langra and Dassehari in hot water at 50°C. Barkai-Golan and Phillips (3) have listed

5-min immersion in water at 52°C. The treatment does not control stem-end rot. Immersion for 5 min in water at 55°C was recommended by Smoot and Segall (79). In Florida (USA) commercially used treatment involves a drench in water at 53°C for 3-5 min (48). Immersion of fruits in water at a constant temperature of 46°C for 90-115 min also checks anthracnose rot as well as stem-end rot (*Diplodia natalensis*). Forced air treatment at 48°C for 115 min is also effective against anthracnose (48). The hot water treated fruits can be stored safely for 5 wk. At the time of harvest and during transportation and storage special care should be taken to avoid injury to fruits. Storage of fruits should be under dry and cool conditions.

■ TWIG BLIGHT, DIE-BACK AND LEAF BLIGHT

The symptoms of die-back and fruit rot described for anthracnose are often confused with similar symptoms caused by many weak fungal pathogens. These include *Phoma* sp. causing twig blight (32), *Botryodiplodia theobromae* causing die-back of mango trees (11, 94), and *Macrophoma mangiferae* causing leaf blight and seedling and plant mortality in nurseries (26, 93). In Puerto Rico, die-back of mango was associated with the conidial stage of *Physalospora rhodina* (2). The conidial stage of this fungus is *Diplodia natalensis*. In Brazil, a twig blight of mango was attributed to *Ceratocystis fimbriata*. In Florida, the death of mango trees in 1949 was attributable to *Diplodia* sp. In the Republic of Niger (Africa) a destructive die-back of mango has been attributed to *Hendersonula toruloidea*, a weak pathogen that causes damage when the tree is under stress (70). Ramos, *et al.* (67) have described a twig blight of mango caused by the conidial stage of *Botryosphaeria ribis* which they consider to be confined to the Americas. Before there was evidence of fungal etiology of typical die-back symptoms, there had been assumptions that because there was constant association of the nematode *Hemicriconemoides mangiferae* the problem of die-back and decline of trees could be due to nematode infestation of roots. However, in young trees, which also suffered from the disease, there was no evidence of nematode association with the tree roots. Micronutrient deficiency problems were also suspected. Generally manganese and iron were deficient in the diseased plants but it could not be proved that the deficiency of these elements was responsible for die-back and decline.

Symptoms

Kanitkar and Uppal (32) had reported that in the twig blight of mango caused by *Phoma* sp., water-soaked spots appear on the twigs which grow upward but do not encircle the twig. The bark turns dark brown and finally the twig dries. The fungus also causes postharvest stem-end rot of ripe fruits. In the die-back caused by *Botryodiplodia theobromae* (11, 94) the main symptoms are blighting of twigs and branches, especially of grown up trees, and defoliation giving a blighted look to the tree. The bark of the twig, behind the tip, becomes darker in colour which later turns black. As this black area advances, the new green shoots show blighting. On leaves, brown area develops from the midrib and grows along the lateral veins. Eventually, the entire leaf blade turns brown. Within a month all infected leaves fall down. Gummosis on infected twigs and shoots also occurs.

In the blight of mango caused by *Macrophoma mangiferae* (26) symptoms include appearance of round, light brown spots on leaves which turn into dark brown areas, oval or irregular in shape, and surrounded by broad, dark violet, slightly raised margins. Old spots turn ash grey in colour. Black pycnidia of the fungus develop on the spots on both surfaces of the leaf. Spores from pycnidia are deposited at the leaf tip by rain drops and the tip soon dries. Eventually, the necrosis spreads all over the leaf blade. Similar brown necrotic lesions develop on branches and twigs which leads to postharvest stem-end rot of fruits. The same pathogen has been found to cause large scale mortality of nursery seedlings and plants in Punjab (92, 93). Seedlings and young grafted plants suffered from rapid necrosis and drying of leaf blade from tip and margins. Elongated, elliptical, grey lesions were seen around the graft union or crotch angles or at the base of petioles.

In the tip die-back caused by the conidial stage of *Botryosphaeria ribis*, reported from Florida by Ramos *et al.* (67), sometimes referred to as mango decline, the symptoms include terminal and marginal necrosis of the leaves, which ultimately lead to the death of the leaf blade. The die-back starting from twigs gradually progresses to larger branches with eventual reduction in the number of feeder roots. The consequences of poor roots system follow and the tree shows decline. In artificial inoculations of leaves or cut ends of stems, Ramos *et al.* (67) observed progressive die-back of twigs and branches of adult trees or in the main stem apices of seedlings with considerable reduction in growth. In the early stage of infection some branches showed browning of the leaf petiole and midrib, which extended downward as the disease progressed. The terminal leaves of the branch died but remained attached to the twig for some time. When the fungus reached the vascular system of the main stem of seedlings, a complete defoliation of apical branches occurred.

The Pathogens

Phoma, *Macrophoma*, *Diplodia*, *Botryodiplodia* and *Hendersonula* all belong to the family Sphaeropsidaceae of the order Sphaeropsidales of Deuteromycotina. *Macrophoma* was at one stage considered a subgenus of *Phoma* with conidia longer than 15 μm. Sutton (87) had considered it a synonym of *Sphaeropsis*. Most of the species of *Macrophoma* have been transferred to *Macrophomina* or *Sphaeropsis*. *Botryodiplodia* is similar to the genus *Macrophoma* in the immature conidial stage.

Mycelium of *Phoma* sp. is branched, septate, hyaline or pale brown. Pycnidia are immersed or semi-immersed, lenticular or globose, ostiolate, pale or medium brown to black. Conidiophores are small, barrel-shaped or are absent. Conidia are hyaline, aseptate, sometimes 1-septate, thin-walled, ovoid or of various shape. The pycnidia measure 88-248 μm in diameter. The conidia measure 16.5-26.1 $\times$ 4.8-6.9 μm (32). In *Macrophoma mangiferae* the pycnidia are globose or subglobose, ostiolate, light brown, and measure 77-231 μm in diameter. Conidiophores are hyaline, thin-walled and 8-11 $\times$ 1.5-2.0 μm in size. The aseptate, hyaline, and elliptical conidia measure 10.5-24.5 $\times$ 5.3-7.0 μm, the average size being 19.8 $\times$ 6.5 μm (27). Hingorani *et al.* (28, 29, 30) had indicated the possibility of the perfect stage of this species being a *Physalospora*. *Macrophoma*, *Diplodia* and *Botryodiplodia* are considered the conidial stage of *Botryosphaeria* (25).

Diplodia natalensis and *Botryodiplodia theobromae* are described under postharvest decay of mango fruits. *Botryosphaeria ribis* (syn. *Botryosphaeria dothidea*) and its conidial stage *Fusicoccum aesculi* are described under Botryosphaeria white rot of apple. The isolate from mango in Florida (67) produces pigmented hyphae and swollen hyphal cells or chlamydospores. In cultures, the dark mycelium produces stromatic pycnidia with hyaline aseptate conidia. On mango seedlings the aseptate conidia measure 16.8 $\times$ 7.2 μm. The pathogen invades the vascular system of mango.

Disease Cycle

These fungi are weak pathogens parasitizing the plant when it is under stress (water, nutrition and temperature) or has been injured. The pycnidia present on dead twigs, dead bark, and sometimes dried fruits on the tree or on the orchard floor serve as a source of primary inoculum. Infections usually occur through wounds. In *Macrophoma mangiferae*, the fungus can survive on infected leaves for more than a year (27). It also survives as dormant mycelium in the wood and as pycnidia on the infected bark (92). When pycnidia are immersed in water they release conidia within 5 min. The conidia can germinate at any temperature from 10° to 35°C. At

26°C, the conidia of *M. mangiferae* germinate in water within 2-3 hours. Under saturated conditions for 72 hours, 100% infection occurs at 27°-41°C (28). The gummosis caused by *Botryodiplodia* (*Lasiodiplodia*) *theobromae* is more common during the summer than during the rains and in the winter. In the tip die-back caused by *B. ribis* deficiency of iron, zinc and manganese in the tree is reported to be associated with the disease (67). Water stress affects xylem near the cambium and this also helps invasion by the fungus. *B. ribis* appears to be an aggressive pathogen of mango. It invades the vascular system, induces tyloses and dark inclusion bodies in the xylem vessels and causes necrosis of the xylem. The hyphae are present in the xylem vessels (67, 68).

Management

Sanitation and other measures recommended for control of anthracnose generally are sufficient against this disease also. Pruning of affected twigs followed by application of a fungicidal paste on the cut ends reduces inoculum. Under the weather conditions in Punjab (India), the Macrophoma blight of young mango plants could be effectively controlled by 4 sprays of 0.2% captafol, 0.1% carbendazim or 0.1% thiophanate methyl. The sprays were started in the first week of July in a 14-days schedule.

■ GREY BLIGHT OF MANGO

The disease is similar to anthracnose leaf spot and Macrophoma leaf spot in symptoms. Brown spots develop on the leaf blade. These spots are isolated, scattered or in groups and spread all over the leaf surface. In the early stages the spots look light brown with darker margins but in older spots the central portion turns olive-grey. Numerous, minute, dot-like fungal structures (pseudopycnidia or acervuli) appear on this portion of the spot. Shot-hole may be caused when the central dried portion separates and falls down. Sometimes, the pathogen causes storage rot of fruits. The fruits carry the infection while still on the tree. Brown to olive black spots develop when the fruits are kept in stores. Decay leads to complete rotting and shrinking of the fruit (90).

Grey blight of mango is caused by *Pestalotiopsis mangiferae* (P. Henn) Stey (syn. *Pestalotia mangiferae* P. Henn). Like the anthracnose fungus, this species also belongs to Melanconiaceae of Coelomycetes in Deuteromycotina. The mycelium consists of septate, branched, hyaline to light brown, inter- and intracellular hyphae. Acervuli form from a stroma of brown, thin-walled pseudoparenchyma. Conidiophores develop from upper cells of the pseudoparenchyma. They are hyaline, cylindrical, branched and septate in the upper and lower portion. Conidia are 5-celled, oblong-clavate or clavate-fusiform, 18.5 - 27.7 × 8 - 11 μm in size. The middle 3 cells are pigmented while the apical and basal cells are hyaline. There is a difference in the shade of browning in the 3 middle cells which are 15-16 μm long. The septa and walls of the conidium are sometimes black. The apical cell bears 3 coarse, widely divergent, 19-26 μm long setulae (flagella). The pedicel of the conidium is short.

The pathogen survives on infected leaves and attacks the new leaves through a wound or when the tree is weakened by nutritional or water stress. Secondary infections are caused by conidia dispersed by rain and wind. The optimum temperature for growth of the fungus is 20°- 23°C (71). Management strategies adopted for anthracnose are sufficient to control this disease.

■ RED RUST OF MANGO LEAVES

This is an algal disease very common on mango leaves during the rainy season and for a short period thereafter. This disease is found in all mango growing regions of the world especially in areas where the atmospheric humidity is high and the average rainfall is around 250 cm. In most varieties, the attack on leaves takes place during the rainy season, after fruit harvest. Fruits are generally not attacked. The loss due to these algal spots is not much.

Orange coloured spots are formed by a hairy growth on the leaf surface. On the margins of the spots orange coloured sporangia of the alga are seen. Similar algal growth occurs on twigs also. Rains disturb these sporangia and spores are released. The causal organism is *Cephaleuros mycoides*.

Normally no control measures are required. The alga disappears from trees during the cold season and also during the fruiting season. In case of severe attacks copper based fungicides can be sprayed. Fungicidal control measure for anthracnose should be enough for this disease also.

■ POSTHARVEST FUNGAL ROT OF MANGOES

In addition to fruit rot caused by infection of the mango blight bacterium, *Xanthomonas campestris* pv. *mangiferae-indicae*, and the anthracnose fungus, *Colletotrichum gloeosporioides*, there are a large number of other fungi and some bacteria that are typical postharvest decay agents of mangoes. They cause the damage mostly during transport from the production areas to the cities. The cumulative loss due to decay in mangoes may be

Fig. 17. Red rust (alga) on mango leaf.

8-18 %. Some of the more important of these rots are stem-end rot caused by *Diplodia natalensis* (5, 59) and *Botryodiplodia (Lasiodiplodia) theobromae,* soft rot caused by *Rhizopus arrhizus* and *R. stolonifer,* black mold rot *(Aspergillus niger),* Aspergillus rots caused by *A. fumigatus, A. terreus* and *A. nidulans,* and a soft rot caused by the bacterium *Erwinia carotovora.*

Stem-end Rot

This rot always starts from the stem-end of the fruit at the point of attachment of the pedicel (fruit stalk) with the fruit (83). Dharam Vir (17), from a study of fruits collected from the market, had stated that the infection is spread all over the fruit surface including the stem-end. The spread of spores from the stem-end infections are known to pass the infections to other parts of the fruit. The skin of the fruit around the pedicel becomes dark green and water-soaked and then turns into a brown black circular lesion, surrounded by lighter margin. This lesion may bear pycnidia of the fungus. In wet weather the lesion spreads rapidly covering uniformly about two thirds of the fruit surface. A soft rot of the fruit follows. The internal flesh becomes brown. The fungus can attack mango leaves, twigs (die-back) and flowers (blossom blight).

The causal organisms have been referred as *Diplodia natalensis* Pole-Evans and *Botryodiplodia theobromae* Pat. These two species are very similar and have the same perfect stage, *Physalospora rhodina* Cooke or *Botryosphaeria rhodina* (Cke.) Von Arx in Ascomycotina. According to Sutton (87) the correct generic name to adopt for the tropical and subtropical fruit pathogens popularly known as *Botryodiplodia theobromae* is *Lasiodiplodia theobromae* (Pat.) Gritt. and Maubl. The mycelium is immersed, branches, septate, and dark brown. The pycnidia are scattered or aggregated, black, globose, at first immersed becoming erumpent, unilocular, thick-walled and ostiolate. They measure 150-180 μm in diameter. Conidiophores in the pycnidia are slender, hyaline and branched. The conidia are at first hyaline, then dark brown, thick-walled, aseptate when young, becoming 1-septate, ellipsoid or ovoid. They measure   16.2-30 (26) × 9.8 14.1 (11.4) μm (94). Upto 50 μm long paraphyses are present in the pycnidium.

Lasiodiplodia theobromae is often considered a disease agent of stressed or weakened plants. It has a wide host range that includes fruit trees as well as certain field crops and causes leaf spots, cankers, root rot, fruit rot and seed decay on different hosts. The fungus survives on the trees in cankers or in the bark or dead twigs and also through pycnidia on fallen diseased plant parts. The *Physalospora* stage usually grows as a saprophyte on dead plant debris. Dried banana leaves generally used for packing mangoes are reported to be a source of inoculum of *Diplodia* (1). Latent infection of fruits by stem-end rot fungi can also occur in the orchard (31). *Lasiodiplodia theobromae* invades mango pedicel through wounds and in maturegreen and ripe fruits, the invasion can occur without quiescence. Infection also occurs directly through the skin of ripe fruits (58). There is evidence that infection usually occurs when the fruits are harvested in cloudy, wet weather and the pedicels are removed (desapping) on the floor of the orchard (60, 61). These fungi are present in soil on the orchard floor. Fruits exposed in the orchard for 8 h and then bagged and brought to the store show more stem-end rot than fruits placed in cellophane bags immediately after harvest (57).

The stem-end rot of mangoes is a high temperature disease. The optimum temperature for growth of *L. theobromae* is reported as 24°-30°C while *D. natalensis* grows well at temperatures of 20°-40°C with optimum at 30°C (60, 61). Maximum temperature for infection by *L. theobromae* is

31.5°C and minimum 25.9°C. Conidial production is enhanced by wet weather. Fruit rot is severe in excess rainfall, more than 80 % relative humidity and high temperature conditions.

Black mold and Rhizopus Rots

Aspergillus and Rhizopus rots are important market diseases of mangoes. About 3 to 3.6 % monetary loss to the fruit dealers has been reported due to these two rots (57). Generally, the Aspergillus or black mold rot also starts from the stem-end but when the fruits are bruised and inoculum is present the rot can start from any point on the fruit surface. In the beginning a light brown, circular spot develops. The colour of the lesion may vary with the variety of mango and stage of development of the rot. The spots expand to cover the area around the stem-end. In 3-4 days the diseased portion becomes depressed and later the fruit shrinks. The affected portion is soft. Fungal growth can be seen on the fruit. This growth, after sporulation, appears black. Sometimes fruits are infected on the tree due to wounds. In such cases, first yellow spots appear and enlarge to form grey spots of 1-1.5 mm diameter. Later these spots become dark brown or black.

Mostly the Aspergillus rot is caused by *Aspergillus niger* but other species such as *A. fumigatus* and *A. nidulans* have also been isolated from decaying fruits. *Aspergillus* is the conidial stage of *Eurotium*, an Ascomycetous fungus. The mycelium of *Aspergillus* is septate, hyaline, and much branched. The conidia are produced in chains on a vesicle at the tip of the conidiophore. Enormous numbers of conidia are produced by the fungus which are dry and dispersed in the atmosphere.

Rhizopus soft rot is similar to that described for strawberries but is not as common as Aspergillus and stem-end rots in mangoes although it can be destructive if fruits are badly bruised and are transported in closely packed baskets at high temperature over long distances. *Rhizopus stolonifer*, *R. arrhizus*, and *R. oryzae* are common species. *R. stolonifer* is generally reported from temperate regions but is often destructive on tropical fruits such as mangoes and papaya. The rot can appear from any point of the fruit surface where the fungus spores have lodged on wounds. It starts as water-soaked spots which soon get covered with, first a whitish growth, then black sporangial heads of the fungus. Because of its very rapid growth through stolons it spreads very fast causing the soft rot of fruits. Enormous numbers of spores are produced and these can resist dry conditions and heat. They are present in the atmosphere of the containers and vans. The bruising caused to improperly packed fruits during movement enables these spores to infect all the fruits in the containers and whole lot may rot.

Management of Postharvest Decay of Mangoes

Avoidance and sometimes post-harvest chemical or physical treatments are the major strategies for management of decay of perishable fruits like mangoes which are eaten uncooked (18). Since the fungi causing stem-end and other postharvest rots are basically saprophytes or weak parasites they need an injured host surface to get a foothold on the fruit. The spores of *Aspergillus* and *Rhizopus* become part of the atmosphere and are always present in the orchard and godowns and the probability of their landing on bruised or wounded tissues is high. Thus, the following precautions are considered most important.

1. Extreme care should be taken to prevent any injury to fruits at various operations involving picking, grading, packing and transportation.
2. Fruits should not be allowed to over-ripe on the tree. Over-ripe fruits should not be packed with ripening fruits. Unnecessary exposure of fruit to orchard soil should be avoided and the pedicels should be removed only in the storage room.
3. The basket or crates should have sanitized packing paper at the bottom and sufficient padding with sanitized paper should be provided between fruits. The baskets or containers should be disinfected and dried thoroughly.
4. During loading and unloading baskets should not be roughly handled.

Rhizopus species generally do not grow at temperatures below 5° C. If the fruits are to be kept in storage, the temperature should be around 10° C. This considerably slows down most rots and prevents secondary infections.

Normally, if the mango trees have received fungicidal sprays for the control of anthracnose, no pre-harvest chemical treatments are required for prevention of storage decay. Postharvest fruit treatments of mango have been only vaguely described. These include chemical treatments, heat treatments and radiation (62, 81). Fruit dip in 6 % suspension of borax at 43° C against stem-end rot was recommended. This treatment had been in use for a long time for temperate fruits but has limitations. Borax is not soluble in water. The fruits must be rinsed with water to remove borax residue. Borax has been now replaced with sodium ortho-phenylphenate but no reports of its use in mangoes are available. Pathak, *et al.* (64) found thiabendazole effective against stem-end rots. Raoof and Om Prakash (69) had reported that applying a coat of pure mustard oil gives 90 per cent control of Aspergillus rot. Botran is highly effective against *Rhizopus stolonifer* but not against *R. arrhizus*. The hot water treatment (46°C for 90-115 min) recommended for anthracnose rot control is effective against stem-end rot also. In a brief report, Pandey and Om Prakash (55) have suggested 30-min hot water (52°C) treatment of fruits of cultivar Dashehari after harvest to prevent stem-end rot caused by *Lasiodiplodia theobromae*. The fruits could be stored for 26 days (21 d at 12°C and 5 days at ambient temperature for ripening) with no stem-end infection. A 15-min treatment also prevents stem-end rot if fruits are stored at 12° C but not at room temperature. If 0.1 % carbendazim or thiophanate methyl is added to the water, the time of treatment could be reduced to 5 min.

■ BLACK TIP OR MANGO NECROSIS

Black tip, tip rot or mango necrosis is a fruit disease peculiar to India. It does not occur in any other country. It is the most common disease of mangoes in the northern states of the country. South India appears to be free from the disease. Some of the best mango varieties are highly susceptible. As many as 90% of the fruits on trees in close proximity to brick kilns may bear the necrotic lesion and become useless for sale and consumption.

Mango black tip is characterized by necrosis of tissue at the distal end of the fruit. The first symptom is the development of a small etiolated area at the distal end which gradually spreads, turns nearly black, and covers the tip completely. The tip is flattened with the outer skin turning hard and sunken (9). The inner portion is soft and yields a dark brown liquid due to rotting induced by saprophytic bacteria. The disease commonly occurs when fruits are about 6-8 weeks old or when they are reaching ripening stage.

Das Gupta and Verma (10, 12, 13) observed that the disease was most common in orchards in the vicinity of brick kilns. They suggested that the smoke from kilns polluted the air with toxic gases like sulphur dioxide and these gases caused necrosis of tissues of the fruit. In laboratory experiments the disease could be produced by exposing mango fruits to smoke and sulphur dioxide (10). Sulphur dioxide injury to vegetation is well documented. However, the major effects reported are on leaves. These effects are bleaching of the interveinal tissue in acute cases and chlorosis under conditions of low absorption of sulphur dioxide. Mango trees having fruits with necrotic symptoms do not show sulphur dioxide toxicity on leaves. Later, Das Gupta and Sen (12) stated that boron was deficient in the mango fruits exposed to brick kiln fumes. Sprays of borax at 3-4 kg/ 500 lit water at an early stage of fruiting substantially reduced the disease incidence. One or two sprays were recommended by them.

■ MANGO MALFORMATION

Malformation disease of mango has become most threatening to this popular fruit tree all over India. It is one disease in which the etiology is yet not clearly understood. In northern India more than 50 % trees are affected. Although the disease was noticed in the state of Bihar (north India) as early as 1891, attention to it was given since mid-1950s when it started assuming serious proportions in other states of the country. Considering its prevalence and lack of control measures, malformation can be considered as the most destructive disease of mango. The disease is also reported from Pakistan, Middle East, South Africa, Brazil, Central America, Mexico, Israel, and USA (40, 46, 73, 74).

Symptoms

The disease appears in two forms: vegetative malformation and floral malformation (39). One or both may be present on the same tree.

Vegetative malformation

This symptom is more pronounced on seedling plants or new grafts than on old plants. However, it is present on old trees also particularly on trees which have suffered from floral malformation for some time. Vegetative buds in the axils of leaves or at the apex of seedlings swell and produce small shootlets which bear small, scaly leaves. The apical growth of plant stops and, as a result, numerous vegetative buds sprout, producing hypertrophied growth. Often the shootlets and their branches at the top of the seedlings are not distinguishable due to overcrowding and the whole mass of rudimentary leaves gives a bunch-like appearance which has been called bunchy top (53, 74). Thick shootlets arising from the swollen axillary buds may elongate and produce secondary branches which, in turn, elongate further and bear small rudimentary leaves at the internodes. Collectively the whole structure looks like witches' broom. Seedlings infected at an early stage finally die. Those infected at later stages may continue growth. Grown-up trees in the orchard also show similar vegetative malformation in the leaf axils and at the apex of branches. The bunches of small shootlets bearing numerous scaly leaves are seen at the apex of growing shoots crowded with normal leaves, on shoot tips bearing malformed panicles along with malformed buds, or all along the length of the internode. These compact structures soon dry up and remain clinging to the shoots as dry masses. Development of vegetative malformation is often seen to continue throughout the year.

Fig. 18. Mango malformation.

Floral malformation

This symptom appears on bearing trees when they start flowering. The inflorescence is malformed due to greatly enlarged flowers in the panicle. Normal flowers are open, white and soon set fruit. In a severe attack, malformed panicles produce much larger number of flowers than the healthy panicles but most of the flowers remain unopened. These unopened flowers are green and larger in size than the healthy flowers. The ovaries are non-functional (44). Soon the whole bunch crowded with green unopened flowers hangs down due to its own heavier weight. Sometimes, there may be fruit set in the malformed panicle but the fruits are extremely small and shed early. Partial malformation also occurs in the same panicle. Some healthy flowers set normal fruits which reach maturity. When two panicles are on one shoot, the tip of both may not be malformed. Recovery of malformed panicles is rare.

The cause

Symptoms and the effects of mango malformation have been studied exhaustively but the actual cause of the malady has been very controversial. On the basis of the symptoms and physiological changes in the tissues, the disease was considered a physiological disorder (33, 45).

It was suspected a virus disease since no organism had been isolated from the diseased tissue and symptoms resembled leafhopper transmitted virus diseases (75, 76). The malformation was also considered an effect of attack of eriophyid mites (51, 52, 85). None of these reports could provide conclusive proof of association of the cause they suggested with the disease. Hormonal imbalance had prompted the use of indole acetic acid as a control measure and some relief was claimed. No manipulation of macro and micronutrients has been able to reduce disease incidence or lead the plants to recovery. Electron microscope studies have not revealed the presence of virus particles or mycoplasmal bodies. Use of acaricides, similarly, could provide only temporary relief. In more recent reports, indirect effect of one or more fungi internally present in the plant has been claimed as the cause of the disease. Constant association of *Fusarium moniliforme* and *Fusarium oxysporum* with the diseased parts of the tree has been demonstrated (4, 6, 86, 91). However, in the strict sense of Koch's postulates, their pathogenicity has not been yet demonstrated.

The hypothesis in support of the role of *Fusarium* (40) is that conidia of *Fusarium*, particularly *F. moniliforme* var. *subglutinans*, are taken up by wounded mango roots from the soil and are apoplastically transported to the growing points (shoot tips, axillary buds, etc.) where they multiply saprophytically or parasitically and gradually release malformation-inducing-principles (MIP) for an extended period of time. This principle works through growth substance imbalance and conditions the host cells to produce malformed growth. Once the growth phase of the fungus is over due to nutrient depletion it produces secondary toxic metabolites (TP) which are translocated in the plant and result in toxicity symptoms such as reduced growth and necrosis of malformed tissue, seedling necrosis, and loss of flowering. The presence of malformin-like substances in malformed mango tissues was proposed by many scientists (65, 66, 77). The presence of such substances in culture filtrates of *Fusarium* is also documented (56, 41). The malformation in mango was, therefore, attributed to the malformin produced by the fungus present in high density in interaction with the mango tissues.

Plants raised from or on seedlings having vegetative malformation, invariably develop the disease in the orchard. A tree once infected never recovers in spite of drastic pruning of malformation-bearing twigs and branches and in spite of temporary remission of symptoms through sprays of growth substances. Seasonal variations in the disease incidence were reported by Majumdar and Sinha (44) who suggested a correlation between disease incidence and ambient temperature at the time of flowering. There was less incidence of the disease when temperature was artificially raised around the trees during the flowering period and flower buds appearing early show more disease than those appearing late. The incidence of mango malformation is most common where mean temperature during winter is around 16° C. Incidence of the disease is very low in south India where temperatures remain relatively high during most part of the year.

Management

Fungicides, nutritional amendments, application of growth regulators (IAA), and many cultural practices such as root pruning, all have failed to give a lasting control of the mango malformation disease. An integrated schedule consisting of pruning of affected parts, sprays of mangiferin-zinc or copper chelates, and acaricides thrice (pre-bud burst, after emergence, and pre-blossom stages) is reported to give significant reduction in malformation and increase in the number of healthy inflorescences and fruit set (7). The long lasting effect of these treatments has not been reported. These can be probably only additive steps to a more permanent solution. On the basis of information available it can be visualized that the first step in the malformation management strategy should be to raise seedlings for plants and for rootstock for grafts in a nursery where soil is free from fungi suspected to be involved in the disease. Fumigation of nursery soil with fungicides such as vapam may be useful in preparing soil for raising seedlings. In addition to seedlings being free from vegetative malformation, it should also be ensured that scion is selected from trees which are completely free from vegetative and floral malformation. Injury to roots during planting in the orchard should be avoided. Periodical sprays of systemic fungicides and acaricides may be given as additional precautions. Chemicals used in the control of powdery mildew and anthracnose may be sufficient in this respect.

■ GIANT OR LEAFY MISTLETOES

Dendrophthoae is a common phanerogamic parasite of fruit and wasteland and roadside trees. In ancient Sanskrit literature it is mentioned as *Vrikshabhaksh* meaning 'eater of trees'. This describes the damage done by it. In India mango trees are the worst sufferers from this parasite. In northern India 60-90% of the old, tall 'desi' (indigenous) type of mango trees and a large number of other trees are heavily or moderately infected. Where there is no timely eradication of the parasite the entire tree becomes uneconomical in the course of time. The parasite spreads from tree to tree and a major portion of orchard may be affected.

Dendrophthoae spp. (Loranthaceae-Viscoideae) is a warm climate shrub found in India and Malaysia. Its allied genus, *Phoradendron (serotinum)* is found in North America and *Viscum album* in California, Europe and many other countries. These are semi-parasites of tree trunks and thick branches. The leathery and evergreen leaves possess chlorophyll and synthesize carbohydrate constituent of their requirement. Since the parasites attack the aerial parts of the host trees, situated far above the soil level, and since they are devoid of a true root system they are dependent on the host for water and mineral nutrients. Other manufactured food from the host also passes into the parasite. They obtain these by developing suckers (haustoria) which grow into the host tissue and become intimately associated with the vascular elements of the host. The continuous drain on nutrients by the parasite deprives the host of its growth requirements. In due course of time the attacked branch withers (37, 38).

Dendrophthoae falcata, the common species reported in India, is a strongly branched and glabrous shrub. The stem is thick, erect, or flattened at the nodes and appears to arise in clusters at the point of attack. This cluster forms a dense and bushy growth which can be easily spotted on the trees. The point at which the host is attacked and where the haustorium penetrates often swells to form a tumour. These tumours vary in size according to the age of the parasite (42, 43). Sometimes, the parasite, instead of confining its attack to one point, produces a creeping branch which grows closely along the host stem and forms haustoria at intervals. The flowers are borne in clusters. They are long and tubular, usually greenish white or red in color according to species. The fruit is fleshy and contains a solitary seed. It is sweet and eaten by birds, cattle and other animals.

The parasite is spread by dispersal of its seed mostly by birds and to some extent by other animals. The birds are attracted by brilliant colour of the fruit. The pulp is sticky and thus the seeds are easily carried by birds. When seeds are deposited on other trees, at the junction of branches with the trunk, they germinate and give rise to haustoria, establishing the parasite. Droppings of birds containing seeds also help in dissemination of the parasite.

In the early stage of the parasitic development the damage to the tree may not be appreciable but later the parasite increases in vigour and the effects become apparent. Beyond the point of attack fresh growth of the host shoot is stunted. The damage is most marked in the production of new growth by the host. The quality and yield of fruits is considerably lowered. Leaves may be reduced in size and may show an unhealthy green colour, an effect well marked in mango. The effects of the attack also depend on vigour of the host tree. A large tree, if mildly attacked, will not show any effect.

The commonly known method of control of the parasite is to top off the affected branches. It is important that the branches should be cut sufficiently low so that all vestiges of the haustorial system of the parasite are eradicated. In the early stages of the growth of the parasite, it can be easily detached from the host without damaging the latter. If the tumour is on one side of the branch then the wood just below the tumour may be sawed off. Injection of copper sulphate and 2,4-D into affected branches has been found effective in many hosts. A spray of diesel oil emulsion in soap water is also effective in eradicating the parasite from mango trees (42).

■ REFERENCES

1. Alicbusan, R.V. and L.A. Schafer. 1958. Diplodia stem-end rot of mango. *Philipp. Agric.* **42**: 319.
2. Alvarez-Garcia, L. and J. Lopez-Garcia. 1971. Gummosis, die-back and fruit rot disease of mango (*Mangifera indica*) caused by *Physalospora rhodina* (B. & C.) Cke. in Puerto-Rico. *J. Agric. Univ., Puerto Rico* **55**: 435.
3. Barkai-Golan, R. and D.J. Phillips. 1991. Postharvest heat treatment of fresh fruits and vegetables for decay control. *Plant Dis.* **75**: 185.
4. Bhatnagar, S.S. and S.P.S. Beniwal. 1977. Involvement of *Fusarium oxysporum* in causation of mango malformation. *Plant Dis. Rep.* **61**: 894.
5. Chakrabarty, D.K. and D.N. Srivastava. 1964. Stem-end rot of mango and orange fruits incited by *Diplodia natalensis. Curr. Sci.* **33**: 285.
6. Chakrabarti, D.K. and S. Ghosal. 1985. Effect of *Fusarium moniliforme* var. *subglutinans* infection on mangiferin production in the twigs of *Mangifera indica. J. Phytopath.* **113**: 47.
7. Chakrabarti, D.K. and S. Ghosal. 1989. The disease cycle of mango malformation induced by *Fusarium moniliforme* var. *subglutinans* and the curative effects of metal chelates. *J. Phytopath.* **125**: 238.
8. Daquiose, V.R. and T.H. Quimio. 1979. Latent infection in mango caused by *Colletotrichum gloeosporioides. Philipp. Phytopathol.* **15**: 35.
9. Das Gupta, S.N. and G.S. Verma. 1939. Studies in the diseases of *Mangifera indica*. I. Preliminary observations on the necrosis of the mango fruits with special reference to the external symptoms of the disease. *Proc. Indian Acad. Sci.* **9B**: 13.
10. Das Gupta, S.N. and G.S. Verma. 1941. Studies in the diseases of *Mangifera indica*. III. Investigations into the effects of sulphur dioxide on the mango fruits. *Proc. Indian Acad. Sci.* **13B**: 71.
11. Das Gupta, S.N. and A.T. Zacharia. 1945. Studies in the diseases of *Mangifera indica*. V. On the die-back disease of the mango tree. *J. Indian Bot. Soc.* **24**: 101.
12. Das Gupta, S.N. and C. Sen. 1960. Studies in the diseases of *Mangifera indica*. XI. The effect of boron on mango necrosis. *Phytopathology* **50**: 431.
13. Das Gupta, S.N. *et al.* 1956. Studies in the diseases of *Mangifera indica*. IX. Isolation of brick kiln fume constituents causing mango necrosis. *Indian J. Agric. Sci.* **26**: 259.
14. Datar, V.V. 1983. Reaction of mango varieties to powdery mildew incited by *Oidium mangiferae. Indian J. Mycol. Pl. Pathol.* **13**: 111.
15. Datar, V.V. 1986. Management of powdery mildew of mango with fungicides. *Indian Phytopath.* **39**: 271.
16. Dayakar, B.V. and S.S. Gnanamanickam. 1996. Biochemical and pathological variations in strains of *Xanthomonas campestris* pv. *mangiferae-indicae* from south India. *Indian Phytopath.* **49**: 227.

17. Dharam Vir. 1970. Control of Alternaria rot of tomato and Diplodia rot of mango, pp. 316-320. In: *Plant Disease Problems*. S.P. Raychaudhuri *et al.* (eds.). Indian Phytopath. Soc. New Delhi.

18. Dodd, J.C., P. Jeffries and M.J. Jeger. 1989. Management strategies to control latent infection in tropical fruits. *Ann. Appl. Biol.* **20**: 49.

19. Dodd, J.C., A. Estrada and M.J. Jeger. 1992. Epidemiology of *Colletotrichum gloeosporioides* in the tropics, pp. 308-325. In: *Colletotrichum: Biology, Pathology and Control*. I.A. Bailey and M.J. Jeger (eds.). CAB, Wallingford.

20. Doidge, E.M. 1915. A bacterial disease of mango. *Ann. Appl. Biol.* **1**: 1.

21. Fitzell, R.D. and C.M. Peak. 1984. The epidemiology of anthracnose disease of mango: inoculum source, spore production and dispersal. *Ann. Appli. Biol.* **104**: 53.

22. Gafar, K. and M.K. Abdel-mejid. 1970. Chemical control measure for powdery mildew of mango. *Agric. Res. Rev. Cairo* **48**: 10.

23. Gupta, P.C. and J.K. Dang. 1980. Occurrence and control of powdery mildew of mango in Haryana. *Indian Phytopath.* **33**: 631.

24. Halos, P.M. and G.G. Divingracia. 1970. Histopathology of mango fruits infected by *Diplodia natalensis*. *Philipp. Phytopathol.* **6**: 16.

25. Hanlin, R.T. 1989. *Illustrated Genera of Ascomycetes*. Am. Phytopath. Soc. Press, St. Paul, Minn.

26. Hingorani, M.K. and O.P. Sharma. 1856. Blight disease of mango. *Indian Phytopath.* **9**: 195.

27. Hingorani, M.K., O.P. Sharma and H.S. Sohi. 1960. Studies on the blight disease of mango caused by *Macrophoma mangiferae*. *Indian Phytopath.* **13**: 137.

28. Hingorani, M.K., O.P. Sharma and N.J. Singh. 1961. Pycnidia formation in *Macrophoma mangiferae*, the causal organism of blight disease of mango. *Indian Phytopath.* **14**: 48.

29. Hingorani, M.K., T.S. Reddy, N.J. Singh and H.S. Sohi. 1961. The perithecia producing mutant of *Macrophoma mangiferae*. *Indian Phytopath.* **14**: 98.

30. Hingorani, M.K., T.S. Reddy and N.J. Singh. 1961. Comparative studies of *Macrophoma mangiferae* and its ultra-violet light induced mutant. *Indian Phytopath.* **14**: 139.

31. Johnson, C.I., A.W. Cooke and A.J. Mead. 1993. Infection and quiescence of mango stem-end rot pathogens. *Acta Horticulturae* **341**: 329.

32. Kanitkar, U.K. and B.N. Uppal. 1939. Twig blight and fruit rot of mango. *Curr. Sci.* **8**: 470.

33. Khan, M.D. and A.H. Khan. 1963. Some chemical studies on malformation of mango inflorescence in West Pakistan. *Punjab Hort. J.* **3**: 229.

34. Kishun, R. 1986. Role of insects in transmission and survival of *Xanthomonas campestrus* pv. *mangiferae-indicae*. *Indian Phytopath.* **39**: 509.

35. Kishun, R. 1997. Mango bacterial canker disease in India and its management. *Abstr. Nat. Symp. Recent Advances in Diagnosis and management of Important Plant Diseases*. IPS Zonal Meeting, Kanpur (India), 19-20 Dec. 1997.

36. Kishun, R. and H.S. Sohi. 1984. Control of bacterial canker of mango by chemicals. *Pesticides* **18**: 32.

37. Kuijt, J. 1969. *The Biology of Parasitic Flowering Plants*. Univ. California Press, Berkeley.

38. Kuijt, J. 1977. Haustoria of phanerogamic parasites. *Annu. Rev. Phytopathol.* **17**: 91.

39. Kumar, J. and S.P.S. Beniwal. 1987. Vegetative and floral malformation: Two symptoms of the same disease on mango. *FAO Plant Prot. Bull.* **35**: 21.

40. Kumar, J. and S.P.S. Beniwal. 1992. Mango malformation, pp. 357-393. In: *Plant Diseases of International Importance*. Vol. III. *Diseases of Fruit Crops*. J Kumar *et al.* (eds). Prentice-Hall, New Jersey.

41. Kumar, N. and S. Ram. 1999. Presence of malformin-like substances in the culture filtrates of *Fusarium* species. *Indian Phytopath.* **52**: 134.

42. Kumar, N.C. 1975. The effect of respiratory inhibitors on the physiology of host-parasite relations in *Dendrophthoae falcata* infection. *Indian Phytopath.* **28:** 289.

43. Kumar, N.C. and K.L. Mukherjee. 1973. Host-parasite relation in *Dendrophthoae falcata* infection. *Indian Phytopath.* **26:** 148.

44. Majumdar, P.K. and G.C. Sinha. 1972. Seasonal variation in the incidence of malformation in *Mangifera indica. Acta. Hortic.* **24:** 221.

45. Majumdar, P.K., G.C. Sinha and R.N. Singh. 1970. Effect of exogenous application of NAA on mango malformation. *Indian J. Hort.* **27:** 130.

46. Malo, S.E. and R.T.J. MacMillan. 1972. A disease of *Mangifera indica* in Florida similar to mango malformation. *Fla. St. Hort. Soc. Proc.* **85:** 264.

47. Manicom, B.Q. 1986. Factors affecting bacterial canker of mango caused by *Xanthomonas campestris* pv. *mangiferae-indicae. Ann. Appl. Biol.* **109:** 129.

48. McGuire, R.G. 1991. Concomitant decay reduction when mangoes are treated with heat to control infestation of Caribbean fruit flies. *Plant Dis.* **75:** 946.

49. Mishra, A.K. 1995. Control of bacterial canker of mango under suitable weather conditions. *Indian J. Mycol. Pl. Pathol.* **25:** 214.

50. Mishra, A.K. and Om Prakash. 1992. Bacterial canker of mango: Incidence and Control. *Indian Phytopath.* **45:** 172.

51. Narasimhan, M.J. 1954. Malformation of panicles in mango induced by a species of *Eriophyes. Curr. Sci.* **23:** 297.

52. Nariani, T.K. and M.L. Seth. 1962. Role of eriophyid mites in causing malformation disease in mango. *Indian Phytopath.* **14:** 231.

53. Nirvan, R.S. 1953. Bunchy top of young seedlings. *Sci. Cult.* **18:** 335.

54. Om Prakash and A.K. Mishra. 1997. Status and management of powdery mildew of mango. *Nat. Symp. Recent Advances in Diagnosis and Management of Important Plant Diseases.* IPS Zonal Meeting, Kanpur (India) 19-20 Dec. 1997.

55. Pandey, B.K. and Om Prakash. 1997. Control of stem-end rot disease of mango by hot water and fungicide treatment. *Nat. Sym. Recent Advances in Diagnosis and Management of Important Plant Diseases.* IPS Zonal Meeting, Kanpur. Dec. 19-20, 1997.

56. Pandey, G. and S. Ram. 1995. Presence of malformin-like activity in culture filtrates of *Fusarium moniliforme* var. *intermedium. Indian J. Hort.* **51:** 1.

57. Pathak, V.N. 1997. Post-harvest fruit pathology—Present status and future possibilities. *Indian Phytopath.* **50:** 161.

58. Pathak, V.N. 1998. Post-harvest fruit Diseases—Pathosis and Management. *J. Mycol. Pl. Pathol.* **28:** 87-113.

59. Pathak, V.N. and D.N. Srivastava. 1967. Mango losses due to Diplodia stem-end rot. *Trop. Agric.* **123:** 75.

60. Pathak, V.N. and D.N. Srivastava. 1967. Mode of infection and prevention of Diplodia stem-end rot of mango fruits. *Plant Dis. rep.* **51:** 744.

61. Pathak, V.N. and D.N. Srivastava. 1969. Epidemiology and prevention of Diplodia stem-end rot of mango fruits. *Phytopath. Z.* **65:** 164.

62. Pathak, V.N. and G.L. Khandelwal. 1969. Radiation and chemicals in the control of Diplodia stem-end rot of mango fruits. *Trop. Agric.* **125:** 99.

63. Pathak, V.N. and P.S. Shekhawat. 1976. Efficiency of some fungicides and hot water in control of anthracnose and Aspergillus rot of mango fruit. *Indian Phytopath.* **29:** 315.

64. Pathak, V.N., J.P. Goyal and H.C. Sharma. 1971. Screening trial for control of Diplodia and Rhizopus rots of mango fruits. *Plant·Dis. Rep.* **55:** 752.

65. Raina, K. and S. Ram. 1991 Occurrence of malformin-like substances in malformed mango seedlings. *Acta Hortic.* **291:** 272.

66. Ram, S. and L.D. Bisht. 1984. Occurrence of malformin-like substances in malformed panicles and control of floral malformation in mango. *Scientia Hortic.* **23:** 331.

67. Ramos, L.J., S.P. Lara, R.T. McMillan Jr. and K.R. Narayanan. 1991. Tip die-back of mango (*Mangifera indica*) caused by *Botyosphaeria ribis*. *Plant Dis.* **75**: 315.

68. Ramos, L. J., T.L. Davenport, R.T. McMillan Jr. and S. Pablo Lara. 1997. The resistance of mango (*Mangifera indica*) cultivars to tip die-back disease in Florida. *Plant Dis.* **81**: 509.

69. Raoof, M.A. and Om Prakash. 1983. Evaluation of some fixed oils for the control of black rot of mango fruits. *Indian J. Mycol. Pl. Pathol.* **13**: 348.

70. Reckhaus, P. 1987. Hendersonula die-back of mango in Niger. *Plant Dis.* **71**: 1045.

71. Sarkar, A. 1960. Leaf spot disease of *Mangifera indica* caused by *Pestalotia mangiferae*. *Lloydia* **23**: 1.

72. Sattar, A. and S.A. Mallick. 1939. Some studies on anthracnose of mango caused by *Glomerella cingulta* (*Colletotrichum gloeosporioides)* in the Punjab. *Indian J. Agric. Sci.* **9**: 511.

73. Schlosser, E. 1971. Mango malformation : Symptoms, occurrence and varietal susceptibility. *FAO Plant Prot. Bull.* **19**: 12.

74. Schlosser, E. 1971. Mango malformation: Incidence of bunchy top on mango seedlings in West Pakistan. *FAO Plant Prot. Bull.* **19**: 41.

75. Singh, K.K. and J.S. Jawanda. 1961. Malformation of mangoes. *Punjab Hort. J.* **1**: 18.

76. Singh, L.B., S.M. Singh and R.S. Nirvan. 1961. Studies on mango malformation: Review, symptoms, extent, intensity and cause. *Hort. Adv.* **5**: 197.

77. Singh, Z. and B.S. Dhillon. 1987. Occurrence of malformin-like substances in seedlings of mango (*Mangifera indica* L.). *J. Phytopath.* **120**: 245.

78. Sinha, P.P. and M.N. Hoda. 1988. Performance of mango varieties against bacterial canker. *Indian Phytopath.* **41**: 466.

79. Smoot, J.J. and R.H. Segall. 1963. Hot water as a postharvest control of mango anthracnose. *Plant Dis. Rep.* **47**: 739.

80. Sohi, H.S. 1983. Diseases of tropical and subtropical fruits and their control, pp. 73-86. In: *Recent Advances in Plant Pathology*. A. Husain, B.P. Singh, K. Singh and V.P. Agnihotri (eds). Print House, Lucknow (India).

81. Spalding, D.H. and W.F. Reeder. 1972. Postharvest disorders of mangoes as affected by fungicides and heat treatments. *Plant Dis. Rep.* **56**: 751.

82. Sridhar, R. 1973. Powdery mildew of mango and its control. *Indian Phytopath.* **26**: 361.

83. Srivastava, D.N. and J.C. Durgapal. 1965. Mode of infection and control of Diplodia stem-end rot of mango (*Mangifera indica*). *Indian J. Hort.* **22**: 27.

84. Suhag, L.S. and N. Mehta. 1982. Field evaluation of anti-powdery mildew chemicals on some fruits and vegetables. *Indian Phytopath.* **35**: 325.

85. Summanwar, A.S. and S.P. Raychaudhuri. 1968. The role of eriophyid mite (*Aceria mangiferae*) in the causation of mango malformation. *Indian Phytopath.* **21**: 463.

86. Summanwar, A.S., S.P. Raychaudhuri and S.C. Pathak. 1966. Association of the fungus *Fusarium moniliforme* Sheld. with the malformation in mango. *Indian Phytopath.* **19**: 227.

87. Sutton, B.C. 1980. *The Coelomycetes*. CMI, Kew, Surrey, England.

88. Tandon, I.N. and B.B. Singh. 1968. Control of mango anthracnose by fungicides. *Indian Phytopath.* **21**: 212.

89. Tandon, I.N. and B.B. Singh. 1968. Control of mango anthracnose by hot water treatment. *Indian Phytopath.* **21**: 331.

90. Tandon, R.N., U.S. Sisodia and K.S. Bilgrami. 1955. Pathological studies on *Pestalotia mangiferae*. *Proc. Indian Acad. Sci.* **42B**: 219.

91. Verma, A., V.C. Lele, S.P. Raychaudhuri and A. Ram. 1974. Mango malformation: A fungal disease. *Phytopath. Z.* **79**: 254.

92. Verma, K.S. and J. Singh. 1996. Perpetuation and management of *Macrophoma mangiferae* causing blight of mango. *Indian J. Mycol. Pl. Pathol.* **26**: 75.

93. Verma, K.S. and J. Singh. 1996. Pathology and epidemiology of Macrophoma blight of mango. *Indian J. Mycol. Pl. Pathol.* **26**: 98.

94. Verma, O.P. and R.D. Singh. 1970. Epidemiology of mango die-back caused by *Botryodiplodia theobromae* Pat. *Indian J. Agric. Sci.* **40**: 813.

95. Venugopal, Ram Kishun and T.B. Anilkumar. 1991. Cultural and biochemical variation in *Xanthomonas campestris* pv. *mangiferae-indicae. Indian J. Plant Pathol.* **9**: 27.

Banana Diseases

Banana (*Musa sapientum*) and plantain (*Musa paradisiaca*), both commonly called 'banana fruit', are one of the most important commercial fruit and vegetable crops grown all over the world in the tropical and subtropical areas. The cultivated forms yield much more food per hectare than even the potato. Combined world production of bananas and plantain in 1987 totalled 66 million metric tons which ranked them first among all fruits grown in the world (1987 *FAO Production Yearbook*). Although commercial plantations of banana are present extensively in the tropical regions of Americas, the fruit is grown over an area of 2,37,000 ha in India with an annual production of about 3,633,000 tonnes of edible ripe and unripe fruits (40). Banana plantations suffer from many serious diseases such as Fusarium wilt or Panama disease, Cercospora leaf spot or Sigatoka, Bacterial wilt or Moko disease, burrowing nematode infestation, Bunchy top (a nanavirus), anthracnose, and various other post-harvest fruit rots.

■ FUSARIUM WILT (PANAMA DISEASE)

The Fusarium wilt of banana was considered one of the six most destructive plant diseases in recorded history (1, 76). The disease was first reported from Australia in 1876 (cf. 62). Its presence was noticed in Hawaii in 1904 and in India in 1911. Since 1890s, the disease had been spreading in tropical America and the Caribbean Islands. In Jamaica the disease was first noticed in 1911. In 1912 there were 625 infected plants. The number rose to 2000 in 1920 and by 1939 over 4 million plants had been destroyed. By 1932, the Fusarium wilt had spread to almost all the countries that grow banana commercially. It is known in Sri Lanka since 1930 and was recorded in Myanmar, Thailand and Malaysia in 1925. Pakistan reported the disease in 1963. The disease continues to be the most widespread and important problem in the subtropics (64). In India, the disease is destructive in the southern states of Tamil Nadu, Kerala and Karnataka but is becoming common in other areas also. Upto 100%

incidence has been reported on a susceptible variety in certain parts of Karnataka. The economic loss is enormous in areas where the crop is grown for export such as the Caribbean Islands. The popular variety Gros Michel, mostly grown for export quality fruits, was most susceptible and had to be replaced with Cavendish bananas which were resistant. However, with the appearance of race 4 of the pathogen these bananas are also becoming susceptible in many countries (87, 90).

Fig. 19. Fusarium wilt of banana.
Courtesy: Dr. H.B. Singh.

Symptoms

The conspicuous symptoms of Fusarium wilt appear on at least 5 months old banana plants although 2-3 months old plants are also killed under highly favourable conditions for disease development. The earliest signs, mostly inconspicuous, are faint yellow streaks on the petiole of oldest, lowermost leaf. Two types of symptom development follow this stage. In the yellowing type, there is progressive yellowing of the old leaves and eventual collapse of the petiole. In the non-yellowing type (as in Gros Michel), the leaf petioles collapse without chlorosis of the leaf (83). Often all the leaves, except the youngest, collapse and the heart alone remains upright. Any new leaves that are formed are blotchy and yellow, often with wrinkling of the lamina (76, 88).

Fig. 20. Vascular discolouration in T.S. of pseudostem.
Courtesy: Dr. H.B. Singh

The pseudostem often shows a more or less conspicuous longitudinal splitting of the outer leaf sheath that forms its outer covering. But sometimes this symptom is not present. About 4-6 weeks after the appearance of streaks on the petiole, only the dead trunk of the pseudostem is left. These trunks remain standing for 1-2 months but ultimately they also fall down. The oldest plants in a mat are affected first. The young suckers rarely develop external symptoms. The suckers in an affected mat may eventually produce bunches of fruits that develop abnormally and ripen prematurely or irregularly. However, in Fusarium wilt the pathogen does not infect the fruits and does not cause discolouration of the fruits (88). These symptoms differentiate Fusarium wilt from bacterial wilt in which young suckers are also affected and fruits are discoloured. Discoloured vascular strands varying from light yellow to dark brown are the distinguishing internal symptoms of Fusarium wilt. Usually the discolouration appears first in the outer or oldest leaf sheath and extends up to the pseudostem. It is pronounced in the rhizome but is not common in the roots. However, roots of the diseased rhizomes are frequently blackened and decayed. Longitudinal sections through the diseased root bases show characteristic red strands passing into the stele of the rhizome.

The Causal Organism

The pathogen causing the Fusarium wilt or the Panama disease was first isolated from the host tissue collected in Cuba and was named *Fusarium cubense* in 1910. Full description of the fungus was published in 1913 and pathogenicity was conclusively proved in 1919. After reclassification of *Fusarium* the pathogen was named *Fusarium oxysporum* f. sp. *cubense* (E.F. Smith) Snyder and Hansen. Thirteen vegetative compatibility groups (VCGs) of the *forma specialis* have been described (59, 60, 63). Each VCG represents a genetically different subpopulation of the forma specialis. They have been grouped into at least four physiologic races.

The mycelium is mainly intracellular, being typically found in the xylem vessels, although in advanced stages of the disease intercellular hyphae may be observed in the cortex of roots and in the parenchymatous tissue in proximity to the site of infection. This mycelium produces microconidia in the host tissue. In nature, the macroconidia bearing bodies (sporodochia) appear at a late stage on the surface of petioles and leaves of the infected plants. These sporodochia emerge through the stomata on a globose mass of pseudoparenchymatous tissue measuring 26-30 μm in diameter. The conidiophores are verticillately branched and measure, on an average, 70 μm in length. The side branches are usually 1-celled and measure up to 14 μm. Conidia are borne at the apex of the main and lateral branches. The microconidia are 0– and 1-septate, ovate or somewhat elongated, and measure 5–7 $\times$ 2.5–3.0 μm. The macroconidia are pedicellate, sickle-shaped, 3-septate, and measure 22–36 $\times$ 4–5 μm. Chlamydospores are abundantly formed by hyphal and conidial cells. These spores are oval or spherical and usually in pairs.

Four physiologic races of the pathogen exist. These races are distinguished on the basis of their pathogenicity on different genotypes of banana and related hosts. Race 1 was the cause of devastation in Gros Michel. Race 2 is virulent on the cultivar Bluggoe and race 4 on Cavendish bananas. Race 3 is restricted to *Heliconia*, a close relative of banana. It has none or only a limited effect on banana.

Disease Cycle

Fusarium oxysporum f. sp. *cubense* is soil-borne through its thick-walled chlamydospores. These spores can remain dormant for several years after the host dies. Germination of chlamydospores occurs in response to host root exudates. The infection always occurs through injured lateral roots. There is no evidence of the fungus having the ability to attack living cells of the main root system of banana. Deep wounds to expose the xylem help in easy infection. Spore germination is stimulated in the vicinity of wounded root surface while it is inhibited by the intact root surface (71). Nematodes help in exposing the roots to infection (57). Further development and spread of most initial infections is usually stopped in the xylem by vascular occluding responses of the host, which include the formation of gels, tyloses, and the collapse of vessels (5, 6, 7). However, in susceptible cultivars some infections get established in the xylem and advance ahead of host defense mechanisms. After entry into the lateral roots the fungus proceeds to rhizomes where it develops extensively in the vascular tissue before passing up the vascular system into the pseudostem and the older leaf petioles.

The disease spreads mainly by contact of the root system of adjacent healthy plants with spores released by the diseased plants. The use of infected planting stock is another source of fresh introduction and spread. Floods help in local dispersal of the pathogen. The population of the fungus in soil increases considerably when the wilted banana plants collapse and declines shortly after their removal (67, 69).

Depth of soil penetration by the fungus, its survival and dispersal, and spread of the disease are influenced by soil conditions (81, 82). Depth of soil penetration is more in sandy loam than in silt loam (45). Survival and growth of the fungus are generally greater in acidic or light textured soils than in clayey or alkaline soils with high calcium content (83) Clay minerals retard the spread of the disease. Montorillonite clays have been associated with disease suppressive soils (84). Survival of the pathogen in field soil is best at 25 % saturation. With increase in soil moisture the survival ability is decreased. In general, the disease development is greatest when temperatures are optimal for host and pathogen growth. In Gros Michel, growth of the pathogen and disease development in the roots occurs at temperatures of 21°-27°C, but not at 34°C (8). Symptom development declines during cooler winter months (68). Saturated, poorly drained soils have been correlated with greater disease incidence (68), especially in Cavendish bananas (87). Low levels of zinc and high calcium : magnesium and potassium : magnesium ratios are associated with more severe disease. Low zinc reduces tylose formation while high Ca:Mg and K:Mg ratios adversely affect pectin formation in the host (61, 62).

In Taiwan, Hwang (35), studying the wilt of Cavendish bananas caused by race 4 of the pathogen, observed that long distance spread of the pathogen was by movement of water and infected suckers. The spread within the field was by irrigation water, pruning practices and planting of infected suckers. Mechanical wounding of roots helped infection. There was 10% disease at 17°C and 58% at 32°C. Rice rotation was very helpful in eliminating the pathogen which was mainly present in the top 20 cm soil. There is also a report that the fungus is very active in xylem at 34°C and is almost absent at 27°C.

Under the conditions in south India, the pathogen survives in soil supplied with farmyard manure for more than 4 months. In limed soil its survival was reduced to 2 months. In soils where soil moisture was maintained at 20 and 100% of water holding capacity the viability was retained beyond 4 months but in soils submerged under 5 cm depth of water the survival was reduced to only one month (66).

Flood fallowing and soil amendments have been extensively studied to find their suitability for eradication of the pathogen from soil and reclaiming the land for fresh plantations (56, 66, 68, 72, 79, 89). In flooded fields, the fungus is presumed to be eliminated due to the effect of toxins such as acetic acid and similar substances and due to low partial oxygen pressure to which the fungus is very sensitive (80). The carbon dioxide in submerged soil causes continued germination of conidia but prevents chlamydospore formation. Lack of these structures causes elimination of the fungus (53). Amendments of soil with carbohydrate sources and plant tissue reduces population of the fungus by stimulation of chlamydospore germination, lysis of germ tubes and conidia, and inhibition of further chlamydospore formation (72).

Management

Chemical control, use of organic amendments, flood fallowing, introduction of antibiotic organisms, and other less widely tested control measure have failed to give a practical solution to the problem of Panama disease. Host resistance is an attractive proposition but because most edible bananas are infertile, breeding for resistance is difficult. Using tissue culture methods, certain somaclonal variants of Giant Cavendish have been developed in Taiwan which have resistance to race 4 of the pathogen (36).

Meredith (45, 46) had reported that application of sodium nitrate and many mercuric salts to the soil checked the growth of the fungus. But the method is not practicable on a field scale and among standing plants. In large plantation areas, near big rivers, silting of the infested land by diverting silt-laden water from the river was found a successful method for reclamation of the land. The land is inundated under 2-5 feet of water for 6 mon (89). Application of this method also is limited by availability of water and size and topography of land. Substitution of flood fallowing method with wet rice culture, as tried successfully in Taiwan, appears to be a better approach. The land remains submerged under few inches of water for 3-4 months (35).

Attempts to control the disease by applying fungicides to the soil and by deep ploughing to 75 cm or by weed and brush fallowing for 3-5 years have not been successful. Corm injection of 2 per cent carbendazim plus 0.1 % Agallol or Aretan soil drench is claimed as an effective method of freeing the planting stock of the pathogen and protection of new plants (41).

Certain soil bacteria including actinomycetes capable of producing antibiotics musarin and monamycin and some lytic substances are found in the rhizosphere of resistant banana varieties (32, 70). Meredith (47, 48) had reported that addition of highly antagonistic organisms to soil around banana roots gives protection to plants against wilt. Inoculations with a strain of *Pseudomonas fluorescens* also provide biological control (78).

In general, sanitation by immediate removal of diseased plants with surrounding soil from the field, use of healthy rhizomes for planting, care during cultivation to avoid root injury, control of nematodes, rotation with rice and cultivation of resistant varieties where available constitute the management strategy against Panama disease. Certain varieties in south India are resistant to the wilt pathogen. These include Poovan, Moongil, Peyladen, Rajabale and Vamankeli (66). Cultivars Red Banana and Walha have also been found resistant in Karnataka.

■ CERCOSPORA LEAF SPOT OR SIGATOKA

This leaf spot disease of banana was first detected in Java in 1902. The name Sigatoka was given when it was found in epidemic form in the plains of Sigatoka in Fiji Island in 1913. The disease is destructive in Central America, Africa, Eastern Australia, Borneo, Java, Sumatra, Fiji, Malaysia, etc., but the economic importance varies with the region. In India, the disease occurs in Tamil Nadu, Kerala, Bihar and Bengal and mostly attacks the leaves. Although Sigatoka is not considered a fatal disease (64) in the subtropics, mortality of plants is reported from Bihar.

Sigatoka is the most important of all the leaf spots of banana (52, 53, 85). Significant losses are caused by destruction of the foliage and reduction of the functional leaf surface of the plant. It takes at least 12 healthy leaves on mature banana plants to carry the fruits to maturity. Destruction of most mature leaves by the leaf spot disease may leave only few functional leaves which may be insufficient to bring the fruits to maturity. If more than 4 leaves are destroyed on a plant, the fruits produced by them are mostly subjected to various types of rot. The fruits remain small, ripen unevenly, or they may fall or fail to ripen. Such fruits cannot tolerate low temperature in cold storage. They are also off-taste with discoloured flesh.

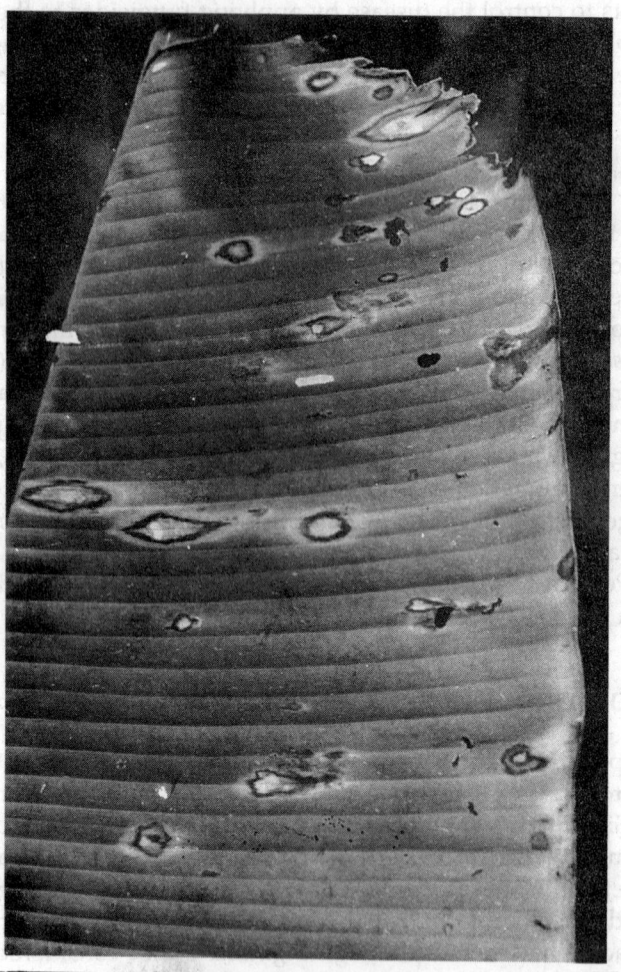

Fig. 20. Sigatoka leaf spot.
Courtesy: Dr. H.B. Singh

Symptoms

The early symptoms of Sigatoka appear on the third or fourth leaves from the top. The leaves that have opened about a month earlier show small, indistinct, longitudinal, light yellow spots parallel to the side veins. These early spots are 1-10 mm long and 0.5–1.0 mm wide. A few days later, the spots become 1-2 cm long, turn brown with light grey centres, and become readily visible. Upto this stage there is not much damage. But when the spots further increase in size, the tissue around them turns yellow and dies. Adjacent spots coalesce to form large dead areas on the leaf. In severe infections, the entire leaves die within a few days.

As a result of the loss of leaves the immature fruit bunches fail to fill out and ripen and may fall. If the fruit is nearing maturity at the time of heavy infection, the flesh ripens unevenly. Individual bananas appear undersized and angular in shape, their flesh develops a buff pinkish colour, and they store poorly.

The Causal Organism

The yellow Sigatoka is caused by the fungus *Cercospora musae* Zimm. but mostly the causal organism is referred to by its perfect stage, *Mycosphaerella musicola* Leach. There is another species, *Mycosphaerella fijiensis*, that causes black Sigatoka. The conidial stage of *M. musicola* develops while the spots are still light yellow. Conidia are produced on sporodochia on both sides of the leaf but are usually more abundant on the upper surface. Successive crops of conidia are produced by the same sporodochium during the brown spots stage of the disease. The conidia are slender, long, septate, and hyaline. They measure $40–80 \times 2.5–3.6$ μm. However, the conidial size is variously reported in different studies. Perithecia are formed during hot and humid weather on the brown spots as dark brown to black dot-like structures. They are more conspicuous in the central grey portion of the spots than on the margins. These structures are globose, ostiolate, and immersed. Their formation is initiated by fertilization of sexual hyphae by compatible spermatia formed in spermogonia. The perithecia measure $46.8–72.8$ (av. 61.8) μm in diameter. The asci in clusters are long, cylindrical, and $28–36 \times 8-11$ μm in size. Each ascus contains 8 hyaline, 2-celled ascospores which measure $14–18$ $(12–15) \times 3–4$ μm.

Disease Cycle

The pathogen can survive on dry infected leaves on the field soil. It is spread through conidia and ascospores. Conidia are formed in humid weather throughout the year but their release and germination depends on water or high humidity. They are dispersed by raindrop splashes and by wind. Ascospores are shot out violently through the ostiole in response to wetting of perithecia and are dispersed by air currents. They are responsible for long distance spread of the pathogen while conidia are generally the most important means of local spread. The infection by both type of spores produces the same type of spots and subsequent development of the disease.

Sigatoka spreads fast in a humid weather or periods of high rainfall at 23°–25°C. Little infection occurs at temperatures below 21°C even if humidity is optimum. In dry weather with high day temperature and little dew during night the disease fails to spread. Soils with poor drainage and low fertility favour the disease. Conditions which are conducive for increased humidity in the plantation are favourable for the disease. Thus, thick planting, presence of weeds and increased number of suckers promote disease development Sprinkler irrigation also promotes the disease.

Management

The Sigatoka disease is managed by a combination of practices involving quarantine, cultural methods and year-round chemical sprays.

Quarantine
The suckers brought from outside for planting should be clean. They should not have spotted or dry leaves adhering to them. Before planting, the suckers should be dipped in a fungicidal solution.

Cultural
The land for banana plantation should be well-levelled to permit drainage and prevent water accumulation around the plants. Periodical weeding is necessary to manage humidity. The diseased leaves should be picked up and destroyed.

Chemical
Before 1935, Burgundy mixture (2 %) with or without coconut oil had been used to give partial control of Sigatoka. Around 1935, Bordeaux mixture (5:5:50) was found better and until 1957 this fungicide was most commonly used to control Sigatoka. Later, zineb or copper oxychloride suspended in mineral oil or mineral oil alone (petroleum oil) were found to give better and cheaper control than Bordeaux mixture (86). Six to thirty lit. mineral oil was used per hectare. The interval between the

sprays varied from 10-12 days to 2-4 weeks depending on the severity of the disease. Mist sprayers were used for the spray. However, mineral oil (consisting largely of saturated hydrocarbons) is sometime phytotoxic, may reduce fruit yield, and may cause patchy and retarded ripening when directly applied to fruit bunches. Sometimes, other leaf spot diseases which could be controlled by Bordeaux mixture are enhanced by mineral oil sprays. In 1970, a combination of 250 g Benlate in 6 lit of oil, sprayed at 15 days interval was reported to give control of Sigatoka. Combination of maneb (2.5 kg), oil (18 lit), water (12 lit) and an emulsifier (200 ml) also gave good control. These days, chlorothalonil, mancozeb and other fungicides are routinely used in fungicide-oil-water emulsions for best all-round results. The strobilurin fungicide CGA 279202 at 70-90 g a.i./ha is reported to provide outstanding control of the black Sigatoka caused by *Mycosphaerella fijiensis*(43). It is a wide spectrum fungicide with high level of protective and curative activity.

Major banana varieties are susceptible to Sigatoka and at present there is no scope for control through resistant varieties.

■ POST-HARVEST DECAY OF BANANA FRUITS

Decay of bananas in the market and in the homes is a common occurrence throughout the tropical and subtropical regions of the world. The rot is fast spreading reducing the shelf life of the fruit to only a few days. The problem is complicated by the fact that bananas cannot be stored under refrigeration because it turns brown at very low temperatures. These rots are variously known as anthracnose rot, crown rot, stem-end or main stem rot, and black spots or black tip. Anthracnose rot is most common and destructive rot of bananas.

Anthracnose Fruit Rot

The anthracnose pathogen attacks the plant in all stages of its growth but significant losses are caused to fruits in storage. Two types of infection occur on fruits. In one type, inoculum is present in the plantation, infection occurs before the fruit bunches are harvested, and is followed by a quiescent period (latent infection) in which little, if any, fungus activity occurs for months; then a period of aggressive pathogen growth and decay occurs when fruits ripen, the rot increasing with ripening of the fruit. In the second type, there is no infection of the unripe fruit on the plants. Most infections and growth of the pathogen occur after the fruit is harvested or when it is in transit and storage, the infection occurring on wounds caused to skin of ripe fruits during handling. Both types of infection occur everywhere.

Minute, pin-point, black spots on the flowers and skin of green fruits are indicative of latent infection. Microscopic examination of these spots can reveal the penetrating germ tubes of the fungus. However, further development of the fungus stops for months. When the fruits ripen, the fungus becomes active and grows. These spots of different shapes can appear on any part of the fruit surface. Many spots enlarge and coalesce to form large black patches on the fruit surface, or the top and the tip of the fruit starts rotting. The internal flesh turns different shades of brown and is much sweeter than healthy fruits. The acervuli of the fungus may appear on these spots as pink masses. In storage rot, the infection occurs through bruises on the skin caused during transportation and handling. These infections very rapidly develop the black spots of large and small size. Infections can also occur on main stalk and finger stalks causing rot of these parts. In abaca, the anthracnose fungus causes leaf spots of seedlings.

Anthracnose rot is caused by *Colletotrichum musae* (Berk. and Curt.) von Arx (syn. *Gloeosporium musarum* Cke. and Mass.). Acervuli on fruits, stalks, petioles and occasionally on leaves are usually rounded or somewhat elongated and upto 400 μm in diameter. They are composed of epidermal or sub-epidermal, pale brown pseudoparenchyma which is sub-hyaline towards the conidiophore region. Setae are lacking. Conidiophores are cylindrical and tapering towards the apex, hyaline, septate, branched and subhyaline at the base. They are upto 30 μm long and 3-5 μm wide. Conidia are hyaline, aseptate, oval to elliptical or straight cylindrical with obtuse apices or flattened at the base and obtuse at the apex. They measure 11–17 × 3–6 (4.5) μm on the host and appear salmon pink in mass. The size of conidia is highly variable (97). Isolates studied in India have the conidial size of 11–16 × 4–7 μm. Appressoria are formed by germ tubes from conidia and also vegetative hyphae. They are navicular to ovate or irregularly lobed, medium to dark brown, and 6–12 (9–13) × 5–10 (9–11.5) μm in size.

Colletotrichum gloeosporioides also occurs on banana and causes same symptoms and effects. Both species can be present in the same material. *C. musae* can be differentiated from *C. gloeosporioides* by slightly longer and broader conidia and more rapid growth in culture at 24°C. The perfect stage of both species is *Glomerella cingulata*.

These fungi survive on banana plants and diseased crop debris. The spores are disseminated by rains, wind and insects. In areas where banana leaves are used for packing during transportation of fruit bunches chances of infection from lesions on leaves and petioles are high. The infected bunch and finger stalks also provide inoculum for infection of fruits during transit and storage. Humid weather and temperatures of 25°–30°C favour rapid growth of *C. musae*. There is no growth at 5°C or 36°C.

Optimum germination of conidia of *C. musae* occurs at 30°C. At this temperature the germ tubes also show faster growth and infection of ripe fruits also is maximum at this or higher temperatures (30°–35°C). Conidia can germinate at 80–100 % relative humidity with maximum at 100 % RH (54). The immature fruits are resistant (10). In nature, conidia germinate rapidly on fruits and cause "quiescent" or latent infection through formation of appressoria which remain inactive until the ripening of fruits (18, 55). During ripening of the fruit the appressoria form penetration hyphae that colonize the underlying tissues and lead to anthracnose lesions (55). Wounds on the fruit skin promote rot development (46) and increase losses through storage rot. Most of the table varieties of banana are susceptible to anthracnose.

Crown Rot and Stem-end Rot

In crown rot, the infection occurs through the cut ends of the main stalk and spreads into the crown and pedicel of fruits. Then the rot progresses to the fruits. The diseased portion turns black and dries. In main stalk rot (*Ceratocystis paradoxa*) the main stalk starts rotting. The tissues become soft and black and emit a pleasant smell. Black mycelium of the fungus is seen in the rotting area. This spreads to hands and fruit stalks and then to the stem-end of the fruit. Black spots are seen on the fruit surface. The flesh becomes soft and dark. The fruits show early premature ripening.

Many fungi are associated with these rots. In addition to the anthracnose fungus, others are *Botryodiplodia theobromae, Ceratocystis paradoxa, Fusarium roseum, Verticillium theobromae* and *Acremonium* sp. Diseased plant debris and diseased standing plants are the major sources of perennation of these fungi. Some of these fungi survive through chlamydospores and resistant conidia in soil. In the rot caused by *Ceratocystis paradoxa*, the pathogen is introduced in the plantation through infected rhizomes. The spores are dispersed by rain or wind. Usually it is the plantation soil adhering with fruit bunches that carries the inoculum and when fruits are injured they cause fruit rot. Infections in the plantations are through cut ends at the time of harvest. The infection of *Botryodiplodia theobromae* produces ethylene during the decay process which hastens the ripening of other fruits in the storage. Bananas are harvested when they are three fourth ripe and are exposed to ethylene gas to trigger the ripening process. The ripe fruits are much more susceptible to various rots than immature fruits. Thus, when ripening is hastened by fungal action the fungi that are latently present become active and the intensity of rot is increased.

The black spot, black tip, or cigar-end rot of bananas is caused by *Deightonella torulosa* (and also *C. gloeosporioides*) and is a field problem not very important as a post-harvest disease. *D. torulosa* is a common, weak parasite also reported on other tropical fruits. In this fruit rot, a black spot develops at the styler end which advances upward and becomes about 6 cm long. The spot is surrounded by a grey or light brown border. The spots increase in irregular manner and the fruits appear asymmetrical in shape.

Many other fungi have been isolated from decaying fruits in stores and market. These are *Curvularia lunata, Rhizopus stolonifer, Aspergillus niger, A. flavus, A. fumigatus* and *Geotrichum candidum*. The maximum development of most of these rots is at 25°C and 70-100 % relative humidity. They are generally absent at 10°C.

Management of Banana Decay

Management of banana fruit rots is not normally recommended through cold or high temperature treatments. Cultural practices such as choice of healthy rhizomes for planting and removal of decayed plant parts from the plantation are preliminary steps. Pre-harvest sprays of fungicides have been recommended but are only partially effective against some of the fungi listed above.

In the past, for the control of anthracnose, pre-harvest sprays of copper based fungicides (4 sprays during the period between fruit set and harvest) were recommended. These sprays left blue-stained spots on the fruit. Zineb, maneb and mancozeb are now commonly recommended for sprays in banana plantations for anthracnose control. Disinfection of the rhizomes against *Ceratocystis paradoxa* before planting is also done by dipping the rhizomes in 1% zineb or 0.3 % maneb. The pre-harvest sprays of these fungicides are also effective against *Deightonella torulosa* black spots.

Post-harvest chemical treatment of the fruits is usually done by dipping the fruit bunches in solutions of maneb, benomyl, thiabendazole and carbendazim. Maneb had found major application for post-harvest treatment to control crown rot caused by *F. roseum, V. theobromae* and *Gloeosporium musarum* (30). The cushions (cut hand stem) are dipped in 0.3 % maneb suspension (26). Benlate or thiabendazole (0.01–0.04 %) and Bavistin (MBC) at 500 ppm a.i. strength give good control of fruit rot when used as fruit dip (4, 58). Thiophanate-M (0.04 %) has also been found an effective fruit dip. However, complete elimination of rots is not achieved because of the multiplicity of organisms involved. Benzimidazole is reported to enhance fruit rot caused by *Curvularia lunata*. Strains of *Colletotrichum musae* resistant to benomyl (31), thiabendazole (23) and related fungicides are reported. For the control of Curvularia fruit rot, fruit dip in 2000 ppm solution of propionic acid, salicylic acid or sodium metabisulphite is reported to be effective by increasing the storage life of fruits upto 8 days. The antifungal antibiotics mycostatin, nystatin and pimaricin have also been used for the control of anthracnose fruit rot. The fruit bunches are dipped in 100 ppm solution (2, 49, 50, 51).

■ BACTERIAL WILT (MOKO DISEASE) OF BANANA

Bacterial wilt of banana has a long history of about 150 years. It was present in British Guyana in 1840-44 and was reported in Trinidad in the Moko plantain about the year 1890. Until recently, the disease was considered to be limited to parts of Central and South America and the West Indies (12, 53). It was believed that bacterial wilt was not present in any part of Asia. However, later it was recognized in countries like Malaysia and Sri Lanka, as well as Fiji, Hawaii, Ethiopia, Libya and many other countries making it a disease of worldwide occurrence (28). Although destructive in the Americas and spreading at other places, the disease has the potential to cause severe losses in the future (1). In India the disease was first reported in West Bengal in 1968 (19) and then in south Indian states of Tamil Nadu and Kerala (29, 77). However, except for reported losses of about 70 % in south India during 1977-78, there have been no reports of serious occurrence in India, probably because of varietal resistance and crop cultural practices.

Symptoms

Moko disease is a wilt disease which may be confused with Panama wilt of banana (*Fusarium oxysporum* f. sp. *cubense*). The symptoms start on rapidly growing young plants. The youngest three leaves turn pale green or yellow and collapse near the junction of lamina and petiole. Most leaves collapse within 3-7 days. The characteristic symptoms occur on young suckers (unlike Fusarium wilt) that have been cut back once and have begun regrowth. These are blackened, stunted, and may be twisted. If sucker leaves are present, these may turn yellow or become necrotic. Vascular discolouration in plants that have not produced fruit bunches is concentrated near the centre of the pseudostem, becoming less apparent on the periphery. A firm brown dry rot is found within fruits of infected plants. This is a distinguishing symptom of Moko disease. If the fruit symptoms are not present the disease may be Fusarium wilt (28).

The Causal Organism

Moko disease is caused by the bacterium *Ralstonia solanacearum*, earlier known as *Pseudomonas solanacearum*. Cells of the bacterium are rod-shaped,

Fig. 22. Moko disease. Internal brown rot of fruits.

0.5–0.7 × 1.5–2.5 μm in size and motile by 1-4 polar flagella. They are Gram-negative. The optimum temperature for growth is around 32°– 35°C, maximum 41°C and minimum 10°C. The bacterium has high tyrosinase activity on tyrosine supplemented medium and oxidizes mannitol, sorbitol, dulcitol, cellobios, maltose, lactose and trehalose. Arabinose, xylose and arginine are not characteristically utilized. In culture all strains of the bacterium produce mutants that are weakly virulent or avirulent and appear as small butyrous colonies with a distinct dark red centre on tetrazolium chloride agar. The virulent wild-type forms are irregularly round, fluidal white colonies with a light pink centre. The virulent type is best maintained by storage in distilled water at room temperature.

Ralstonia (Pseudomonas) solanacearum has following 5 races which differ in host range, geographic distribution, and the ability to survive under different environmental conditions (14).

Race 1: Solanaceous strain
Race 2: Musaceous strain
Race 3: Potato strain
Race 4: Ginger strain from the Philippines
Race 5: Mulberry strain from China

The species has also been classified into 5 biovars on the basis of their ability to oxidize certain hexose alcohols and disaccharides (33, 34).

Moko disease is mainly caused by Race 2 which affects *Musa* and *Heliconia*. The race comprises of several strains which differ in virulence and transmissibility. At least 8 strains have been identified which affect musaceous hosts: plantain, bananas and *Heliconia* spp. They cause symptoms ranging from stunting and distortion to rapid wilting. The tomato strain of the bacterium is also reported to cause wilt of certain seed-bearing *Musa* spp. (13). Race 1 of *R. solanacearum* is reported on certain species of *Musa*. Origin of race 2 of *R. solanacearum* was considered in Latin America. Banana strains of the pathogen could have evolved from a strain attacking heliconias, a common weed found near banana fields (11). In some areas, such as in India, evolution of the strains might be local and not introduced from outside.

Disease Cycle

The bacteria survive through infected rhizomes and also in soil for 6 months to 2 year. The spread is through use of infected rhizome, cutting machetes at the time of planting, and through insects which carry the bacteria from ooze on suckers and male flower bracts to healthy inflorescences and other parts of the plant. Entry into the host is mostly through wounds such as those caused during various cultural operations and during attack of insects and nematodes. The bacteria multiply rapidly in the xylem. Pathogenesis and virulence related factors are exopolysaccharides (EPS), lipopolysaccharides (LPS), auxins, ethylene and cell wall degrading enzymes.

Auxin balance of the plant is disturbed. The levels of indoleacetic acid (IAA) increase greatly in the susceptible banana plants (74). Highest accumulation of IAA is in the stem tissues. IAA is synthesized by the bacterium (75) and by the host and accumulates due to inhibition of the auxin degrading system (73). The production of ethylene by microorganisms and diseased plants is a well known phenomenon. The production of ethylene by the banana strain of *R. solanacearum* has been demonstrated (27). The premature ripening of fruit finger-ends is attributed to ethylene. Loss of virulence in the bacterium is generally accompanied by reduction in ethylene production.

Management

Resistant varieties

Mostly the banana varieties are susceptible to Moko disease. Varietal susceptibility differs with the bacterial strain used for inoculation or prevalent in the area. Indian banana varieties are generally resistant to the local strain. In south India the commercial varieties Monthan and Poovan were found resistant to the local strain of the bacterium (cf. 28).

Exclusion and eradication of the pathogen

The basis for Moko control is early detection of the infected plants and their rapid destruction along with the adjacent healthy looking plants (84). All pruning tools used in areas where Moko is present should be disinfected with a solution containing one part formaldehyde in 3 parts of water (15). The male flower buds should be removed to avoid infection by insect-transmitted bacteria after the emergence of female hands. To eradicate the infected plants and the adjacent healthy plants, herbicides such as 2,4-D have been used. After removal of the infected plants the area must remain fallow for 6-12 month. Exclusion should involve the use of disease-free suckers for planting. The suckers should be obtained from fields which have no trace of the disease.

Crop rotation

The pathogen survives well in the presence of susceptible host. In the absence of such hosts or in the presence of non-hosts the population of the pathogen declines. A long fallow period or flooding during the off season helps in eradication of the pathogen. In Honduras the use of sorghum or other gramineae in a 9-month rotation has been found useful in reducing the population of *R. solanacearum* in soil. In parts of Tamil Nadu (India) the farmers use a 3 yr rotation of banana, sugarcane, and rice (two crops) which is an effective wilt management practice.

Chemical control

The bacterial wilt cannot be controlled economically and effectively by chemical means. Various chemicals have been used as broadcast and spot treatments in planting holes but none has proved effective. Disinfection of planting holes with chloropicrin was found to eliminate *R. solanacearum* but is not cost effective.

Biological Control

Microbial antagonism is a promising strategy for bacterial wilt control (22, 91). In other wilt disease caused by *R. solanacearum*, cross protection by prior inoculation with non-pathogenic, avirulent, or incompatible forms of the bacterium or application of heat killed cells has been reported (37, 44). There are also reports of using bacteriocin-producing strains of *R. solanacearum* for the control of bacterial wilt of tomato, tobacco, potato and other hosts (20, 21, 22, 93). No work on these lines appears to have been done in bacterial wilt of banana. In south India, Anuratha and Gnanamanickam (3) used bacterization of banana plants with a strain of *Pseudomonas fluorescens*. The treatment increased survival percentage of inoculated plants and enhanced plant height and biomass. The antagonist produced a phenazin-like antibiotic.

■ BUNCHY TOP OF BANANA

Bunchy top is a destructive disease of banana in Asia, Australia, Egypt and Pacific Islands and is likely to cause severe losses in the future (1). Among banana diseases other than the two wilts, bunchy top is considered most destructive in the subtropics where it can have restricted distribution and can be managed by frequent roguing (64). It may not be a fatal disease but severely limits production. In India it is particularly destructive in the state of Kerala which contributes around 25% of total banana production in the country. It was first reported from Bengal in 1925 and then in Bihar in 1940, Assam in 1941 and Kerala in 1943. It is also prevalent in many other states of the country. In 1950, nearly 3000 sq. miles of banana plantations in Kerala were infected. Reports suggest an annual loss of Rs. 60 million (US $ 150,000) in Kerala alone. The disease was reported in Pakistan for the first time in 1993 (38).

Symptoms

The Symptoms of bunchy top may become apparent at any stage of plant growth, from very young plants to the fruit-bearing adult plants. In badly infected plants, the leaves are typically bunched together at the apex, forming a dense rosette, hence the name bunchy top. The young infected plants are usually stunted and show more erect leaves than normal.

The symptoms vary with the type of infection (84, 97). The primary infection refers to the infection which comes with the diseased suckers used for planting while the secondary infection is that which is caused during growth of the plant through the agency of insect vectors. The symptoms under these two conditions mainly differ in intensity. In acute

Fig. 23. Bunchy top of banana.

primary infection growth of the plant is slow, the plant does not reach a height of more than 2 feet, and never produces fruits. Young plants receiving secondary infection may show the same symptoms as the plants infected through suckers but with lesser intensity.

The first evidence of bunchy top is seen in the leaves. Green streaks appear on the secondary veins on the underside of the lamina and on the midrib and petioles. These streaks are about 0.75 mm wide and 1–1.25 cm long. At first, one to several streaks may be present and are generally followed by others in the same region. The abnormal markings vary from a series of dark green dots to a continuous dark green line. In some cases, symptoms may be evident in the newly emerged or emerging tightly rolled leaf as pale whitish streaks along the secondary veins. These markings are accompanied by a slight transverse wrinkling along the length of the compactly rolled lamina. On opening these leaves, dark green streaks along the secondary veins may be seen. Older leaves show the same symptoms and also dwarfing, marginal necrosis and curling. The leaves gradually become brittle and petioles are incompletely elongated.

In plants with secondary infection, the fruit bunches become choked and may split the stem. They are reduced in size and the fruits are unsaleable. The virus is systemic and is probably confined to the phloem. In transverse sections the vascular strands show suppression of fibrous sheath. In advanced stages of the disease, the root system presents a decayed appearance which is a secondary effect caused by invasion of the weakened roots by bacteria, fungi and nematodes.

The Causal Agent

The causal agent of bunchy top is a nanavirus known as Banana bunchy top virus (BBTV). It was considered a luteovirus in which the particles are isometric with ssRNA. However, in later studies it was reported as a non-geminated ssDNA virus. The particles are 60 nm in diameter.

The virus is transmitted by the banana aphid, *Pentalonia nigronervosa*, in a semi-persistent manner. It is not sap transmissible. All the suckers produced by a diseased plant carry the virus and they constitute the most important source of spread of the disease from plantation to plantation. The aphids cause the secondary spread during the growth of the plants. The rate of multiplication of aphids is maximum when the minimum temperature is about 18°–20°C with relative humidity ranging between 41 and 84%. All stages of the aphid acquire and transmit the virus but the nymphs are most efficient. To become infective, the aphid requires a minimum feeding period of 17 hours and transmits the virus to a susceptible plant by feeding on it for one and a half h or more. They retain the infective capacity for a period of 13 days after removal from the infected plants. These aphids occur around the base of pseudostem at soil level and for some inches below the soil surface. They are also found between the outer leaf sheaths and the pseudostems. In heavy infestations the aphids are also present in groups at the apex of the plants around the central leaf and on the base of petioles of older leaves. Presence of a strain of the virus producing no symptoms is causing its wider distribution (39).

Management

Phytosanitation is of paramount importance in the management of bunchy top of banana. The movement of infected planting material from place to place should be prohibited. No planting material should be obtained from areas where the disease is reported to occur. Since the secondary infection of healthy plants around a plantation where the disease is present is possible through aphids it is advisable that planting material should not be obtained even from the neighbourhood of the affected area.

As soon as the diseased plants are located in a plantation they should be dug out with roots and burnt. This step is effective in regions where the disease has restricted distribution. Spray of power kerosene or parathion is recommended to control the banana aphid. It had also been recommended that a wide banana-free zone should be created around an infested area. This may help in preventing the spread of the disease through insect vectors.

Immune or highly resistant varieties of *Musa* are not available. All the varieties yielding seedless fruits for eating are susceptible. In cultivar Veimama there is less disease and in Taiwan the local cultivar Lan-ya-chiao is reported to be resistant (99).

■ ROOT AND RHIZOME ROT OF BANANA

(Burrowing Nematode)

The disease caused by the burrowing nematode (*Radopholus similis*) in banana is known by several names such as black heart, banana decline and root and rhizome rot. It is fairly widespread in central and Latin America where 88 to 93 % loss in fruit yield has been reported. *R. similis* was first discovered feeding on the roots of banana plants in Fiji in 1893 (cf. 24). The nematode infects black pepper, coffee, coconut and sugarcane and causes serious yield losses. In south India the nematode has assumed economic importance as a serious pest of banana and many other plantation crops. Increased root population of the nematode directly reduces the yield of fruits and also affects other yield components such as number of hands, number of fingers and age to flowering (16).

Symptoms

The banana plants infected with *R. similis* first develop yellowing in the outer whorl of leaves. In 7-10 days yellowing extends to the inner whorl also. The leaves and bunches wither and drop leading to the death of the plant. The infected roots show reddish-brown cortical lesions. These lesions first appear as reddish, elongated flecks parallel to the root axis, then elongate as nematodes and their offspring feed on the host tissues. Distal portion of the root system is destroyed and the root mass is reduced (9). This results in poor anchorage of the plants which may topple down in high winds. In the parenchyma, cavities are formed by the destruction of cells. Lesions caused by the nematode create avenues for other organisms, especially those fungi which normally enter the plant only through wounds. According to Newhall (57) incidence of the Panama disease is almost doubled if the burrowing nematode is present. Root-knot nematodes have no such effect on the incidence of Panama disease. Loos (42) reported that the length of time for wilt-symptom expression was less in pots containing *R. similis* and the Panama wilt fungus than in those pots that contained the wilt fungus alone or in combination with root-knot nematode.

The attacked roots rot and have no laterals. Rotting tissues are also found on the rhizomes where lesions similar to those on roots are seen. The rotting is mainly due to the invasion of secondary bacteria and fungi which enter the root along with the nematodes. In advanced stages, rotting extends to the pseudostem. The damaged plants appear in isolated patches in the first crop. These patches coalesce to form larger patches. In about 5 yr the entire field becomes severely infested. The fruits on infected plants remain undersized. The application of fertilizers has no effect on plant growth.

Morphology and Life History

The genus *Radopholus* constitutes an important group of endoparasitic, migratory root nematodes inhabiting a wide variety of plants. Generally, these nematodes occur in tropical and subtropical regions of the world. The species *Radopholus similis* is considered indigenous to Australia and New Zealand because 9 of the 11 species of the genus are found in these countries (92). In India it appears that the nematode first established itself in Kerala as many varieties of banana infected by the nematode are cultivated there since long. It is possible that the introduction of suckers of Cavendish banana from abroad brought the nematode to this country. Taxonomy, biology, morphology and anatomy of the species were described by van Weerdt (94, 95, 96).

The adult males and females of *R. similis* are vermiform and measure 0.5–0.8 mm in length. In females the lip region is rounded and marked by three striae while in males the lip region is subspherical and striae may or may not be present. There are two ovaries (posterior and anterior) in the females. The cuticle is distinctly annulated. The spear of females is strong with well-developed basal knobs while in males the spear is very slender with tiny basal knobs. Males do not feed on the host.

Two distinct races of *R. similis* occur (9, 25). These are separated only by their ability to parasitize either citrus or banana or both. The black pepper yellows nematode may be a third race. The banana race infects banana but not citrus while the citrus race can infect both citrus and banana. The citrus race is restricted to Florida. Both races have a long list of other host plants. Only the banana race is reported in India. It attacks banana, coconut, arecanut, pepper and many other plant species. It causes spreading decline of tea in Sri Lanka.

All stages of the females can enter the roots at any point by puncturing the epidermal cells (38) and are found in root tissues as well as outside in soil. Occasionally, leaf bases near the soil line are also attacked by *R. similis* (98). Adult males lack the power to cause penetration of root cells. After entry, the nematodes migrate to the cortical region and move through the intercellular spaces laying eggs on the way at several points. Feeding on cells in the cortical region causes the breakdown of tissue resulting in tunnels and cavities. Often the eggs are seen in these cavities. However, the nematodes are not present in the disrupted tissue. They are present in intact tissue adjacent to the tunnels and cavities. Usually they migrate to fresh tissues in advance of the necrosis of cells or move out into the soil and re-enter the same or other roots at different points.

Fertilization is not essential for egg production and the females lay both, fertilized and unfertilized, eggs which measure 60–70 × 20–28 μm. The females lay 2-4 eggs per day for about 7-8 days. After embryonic development is complete the first molt occurs in the egg which hatches in the root tissue or in the soil. The second stage larvae in the tissues undergo 3 molts to become adult males or females. They move in the tissues forming cavities and tunnels and laying eggs. The larvae hatched outside in soil migrate to fresh roots to initiate fresh infections. Each individual female is capable of causing penetration, establishing infection and creating tunnels and cavities.

The burrowing nematode can survive in sandy loam soil for 6 months in the absence of the host roots. In India maximum number of the nematode are found in red loam soils of south India. The highest population occurs during May-June. The nematode is generally not found in the top 15 cm soil. The most common method of introduction of the nematode in new areas is the use of infected suckers and rhizomes for planting.

Management

In Latin America flood fallowing to control Panama disease was found effective in controlling the burrowing nematode also. After flooding for about 6 months, uninfested clean rhizomes are planted. This keeps the crop healthy for several years. In India also fallowing and flooding for 3 months after banana crop has been found effective in suppressing the nematode population in Tamil Nadu (65).

When clean rhizomes are not available, the discoloured portions on the rhizome should be removed with a sterilized knife. The cut should be deep enough to remove not only the discoloured tissue but some of the adjoining healthy looking tissue also. The rhizomes can then be dipped in Bordeaux mixture and DBCP paste applied on the cuts. Alternatively, the rhizome can be given hot water treatment at 53°–55°C for 20-25 min. In spite of these precautions some nematodes may survive in the rhizome and lethal populations may again build up within 2-3 year after planting (92).

Since the nematode is present in deeper layers of the soil fumigants are not very effective. Application of D-D, EDB or DBCP at the rate of 300 lit/ha at planting time results in increased yield for one year but the treatment is to be repeated the following year. These fumigants are now generally not recommended. Charles *et al.* (17) reported that application of carbofuran at the rate of 1 g a.i./plant at the time of planting and again after 3 months gives better control of the nematode than intercropping with different plant species. Low populations of the nematode in roots along with better vegetative and yield characters were observed in intercrop with *Crotalaria*.

■ REFERENCES

1. Agrios, G.N. 1988. *Plant Pathology*, 3rd Ed. p. 22, Academic Press.
2. Ananthanarayanan, A. and K. Seshadri. 1965. Mycostatin for prevention of anthracnose of banana in storage. *Indian Phytopath.* **18**: 367.
3. Anuratha, C.S. and S.S. Gnanamanickam. 1990. Biological control of bacterial wilt caused by *Pseudomonas solanacearum* in India with antagonistic bacteria. *Plant & Soil* **124**: 109.
4. Bailey, D.M., D.F. Cutts, L. Donegan, C.A. Phillips and R. Pope. 1970. The use of thiabendazole for the post-harvest treatment of bananas. *J. Food Technol.* **5**: 89.
5. Beckman, C.H. 1990. Host response to the pathogen. In: *Fusarium wilt of Banana.* (ed.) R.C. Ploetz. Academic Press.
6. Beckman, C.H. and S. Halmos. 1962. Relation of vascular occluding reaction in banana roots to pathogenicity of root-invading fungi. *Phytopathology* **52**: 893.
7. Beckman, C.H., M. E. Mace, S. Halmos and M.W. McGahan. 1961. Physical barriers associated with resistance to Fusarium wilt of banana. *Phytopathology* **51**: 507.
8. Beckman, C.H., S. Halmos and M.E. Mace. 1962. The interaction of host, pathogen and soil temperature in relation to susceptibility to Fusarium wilt of banana. *Phytopathology* **52**: 134.
9. Blake, C.D. 1972. Nematode diseases in banana plantations, pp. 245-266. In: *Economic Nematology.* (ed.) J.M. Webster. Academic Press, New York.
10. Brown, A.E. and T.R. Swinburne. 1980. The resistance of immature banana fruits to anthracnose (*Colletotrichum musae*). *Phytopath. Z.* **99**: 70-80.
11. Buddenhagen, I.W. 1960. Strains of *Pseudomonas solanacearum* in indigenous hosts in banana plantations of Costa Rica and their relationship to bacterial wilt of banana. *Phytopathology* **50**: 660.

12. Buddenhagen, I.W. 1961. Bacterial wilt of banana: History and known distribution. *Trop. Agric. (Trinidad)* **38**: 107.

13. Buddenhagen, I.W. 1962. Bacterial wilt of certain seed bearing *Musa* spp. caused by the tomato race of *Pseudomonas solanacearum*. *Phytopathology* **52**: 286.

14. Buddenhagen, I.W. and A. Kelman. 1964. Biological and physiological aspects of bacterial wilt caused by *Pseudomonas solanacearum*. *Annu. Rev. Phytopathol.* **2**: 203.

15. Buddenhagen, I.W. and L. Sequeira. 1958. Disinfectants and tool disinfection for prevention of spread of bacterial wilt of banana. *Plant Dis. Rep.* **42**: 1399.

16. Charles, J.S.K., T.S. Venkitesan, Y. Thomas and P.A. Varkey. 1985. Correlation of plant growth components to bunch weight in banana infected with burrowing nematode, *Radopholus similis*. *Indian J. Nematol.* **15**: 186.

17. Charles, J.S.K., T.S. Venkitesan and Y. Thomas. 1985. Comparative efficacy of antagonistic intercrops with carbofuran in control of burrowing nematode, *Radopholus similis*, in the banana cultivar "Nendran". *Indian J. Nematol.* **15**: 241.

18. Chakravarty, T. 1957. Anthracnose of banana (*Gloeosporium musarum*) with special reference to latent infection in storage. *Trans. Brit. Mycol. Soc.* **40**: 337.

19. Chattopadhyay, S.B. and N. Mukhopadhyay. 1968. Moko disease of banana: A new record. *FAO Plant Prot. Bull.* **16**: 57.

20. Chen, W.Y. and E. Echandi. 1984. Effect of avirulent bacteriocin producing strains of *Pseudomonas solanacearum* on the control of bacterial wilt of tomato. *Plant Pathology* **33**: 245.

21. Cuppels, D.A., R.S. Hanson and A. Kelman. 1978. Isolation and characterization of a bacteriocin produced by *Pseudomonas solanacearum*. *J. Gen. Mycol.* **109**: 295.

22. Daheb, M.K. and M.K. Goorani. 1969. Antagonism among strains of *Pseudomonas solanacearum*. *Phytopathology* **59**: 1005.

23. de Lapeyre de Bellaire, T. and C. Dubois. 1997. Distribution of thiabendazole resistant *Colleotrichum musae* isolates from Guadeloupe banana plantations. *Plant Dis.* **81**: 1378.

24. Du Charme, E.P. 1968. Burrowing nematode decline of citrus. A Review, pp. 20-37. In: *Tropical Nematology*. (Eds.) G.C. Smart and V.G. Perry. Univ. Florida Press.

25. Du Charme, E.P. and W. Birchfield. 1956. Physiologic race of the burrowing nematode. *Phytopathology* **46**: 615.

26. Eckert, J.W. and N.F. Sommer. 1967. Control of post-harvest diseases of fruits and vegetables by post-harvest treatments. *Annu. Rev. Phytopathol.* **5**: 391.

27. Freebairn, H.T. and I.W. Buddenhagen. 1964. Ethylene production by *Pseudomonas solanacearum*. *Nature* **202**: 313.

28. Gnanamanickam, S.S. and C.S. Anuratha. 1992. Moko disease of banana, pp. 283-299. In: *Plant Diseases of International Importance*. Vol. III. *Diseases of Fruit Crops*. Eds. J. Kumar *et al.* Prentice-Hall, New Jersey.

29. Gnanamanickam, S.S., T.S. Lokeshwari and K.R. Nandini. 1979. Bacterial wilt of banana in Southern India. *Plant Dis. Rep.* **63**: 525.

30. Greene, G.L. and R.D. Goos. 1963. Fungi associated with crown rot of boxed bananas. *Phytopathology* **53**: 271.

31. Griffee, P.J. 1973. Resistance to benomyl and related fungicides in *Colletotrichum musae*. *Trans. Brit. Mycol. Soc.* **60**: 433.

32. Harper, J.L. 1950. Studies on the resistance of certain varieties of banana to Panama disease. I. Internal factor for resistance and antibiotic. II. The rhizosphere. *Plant Soil* **2**: 374.

33. Hayward, A.C. 1964. Characterization of *Pseudomonas solanacearum*. *J. Appl. Bacteriol.* **27**: 265.

34. He, L.Y., L. Sequeira and A. Kelman. 1983. Characteristics of strains of *Pseudomonas solanacearum* from China. *Plant Dis.* **67**: 1357.

35. Hwang, S.C. 1985. Ecology and control of Fusarium wilt of banana. *Plant Prot. Bull. Taiwan* **27**: 233.

36. Hwang, S.C. 1990. Somaclonal resistance in Cavendish banana to Fusarium wilt, p. 146. In: *Fusarium Wilt of Banana*. (Ed.) R.C. Ploetz. APS Press.

37. Kempe, J. and L. Sequeira. 1983. Biological control of bacterial wilt of potatoes: Attempts to induce resistance by treating tubers with bacteria. *Plant Dis.* **67**: 499.

38. Khalid, S., M.H. Soomro and R.H. Stover. 1993. First report of banana bunchy top virus in Pakistan. *Plant Dis.* **77**: 101.

39. Kiritani, K. and Su-Hong Ji. 1999. Papaya ringspot, banana bunchy top, and citrus greening in the Asia and Pacific Region: occurrence and control strategy. *Japan Agric. Res. Quarterly* **33**: 23.

40. Krishnamurthy, M.M. 1993. Nematode problem of banana, pp. 180-187. In: *Handbook of Economic Nematology*. Eds. K. Sitaramaiah and R.S. Singh. Cosmo Publications, New Delhi (India).

41. Lakshamanan, P. and S. Mohan. 1989. A novel method to control Panama disease of banana incited by *Fusarium oxysporum* f. sp. *cubense*. *Indian J. Mycol. Pl. Pathol.* **19**: 93.

42. Loos, C.A. 1959. Symptom expression of Fusarium wilt of the Gros Michel banana in the presence of *Radopholus similis* and *Meloidogyne incognita acrita*. *Proc. Helm. Soc. Wash.* **26**: 103.

43. Margot, P., F. Huggenberger, J. Amrein and B. Weiss. 1998. CGA 279202: a new broad spectrum strobilurin fungicide. *Proc. Intern. Crop Prot. Conf.*, Brighton, Nov. 1998. Vol. **2**: 375.

44. McLaughlin, R. J. and L. Sequeira. 1988. Evaluation of an avirulent strain of *Pseudomonas solanacearum* for biological control of bacterial wilt of potato. *Amer. Potato J.* **65**: 255.

45. Meredith, C.H. 1941. The growth of *Fusarium oxysporum cubense* in soil. *Phytopathology* **31**: 91.

46. Meredith, C.H. 1943. The effect of soil and chemical mixtures on the growth of *Fusarium oxysporum cubense*. *Phytopathology* **33**: 398.

47. Meredith, C.H. 1944. The antagonism of soil organisms to *Fusarium oxysporum cubense*. *Phytopathology* **34**: 426.

48. Meredith, C.H. 1946. Soil actinomycetes applied to banana plants in the field. *Phytopathology* **36**: 983.

49. Meredith, D.S. 1960. Studies on *Gloeosporium musarum* causing storage rots of Jamaican bananas. I. Anthracnose and its control. *Ann. Appl. Biol.* **48**: 279.

50. Meredith, D.S. 1960. Studies on *Gloeosporium musarum* causing storage rots of Jamaican bananas II. Some factors influencing anthracnose development. *Ann. Appl. Biol.* **48**: 518.

51. Meredith, D.S. 1961. Chemical control of transport and storage diseases of banana. *Trop. Agric.* (London) **38**: 205.

52. Meredith, D.S. 1970. Banana leaf disease (Sigatoka) caused by *Mycosphaerella musicola*. *CMI Phytopath. Pap.* **11**: 147 pp.

53. Meredith, D.S. 1970. Major diseases of banana: past and present status. *Rev. Plant Pathol.* **149**: 539.

54. Mishra, A.P. and T.B. Singh. 1962. Effect of temperature and humidity on the development of banana anthracnose. *Indian Phytopath.* **15**: 11.

55. Muirhead, I.F. and B.J. Deverall. 1981. Role of appressoria in latent infection of banana fruit by *Colletotrichum musae*. *Physiol. Plant Pathol.* **19**: 77.

56. Newcombe, M. 1960. Some effects of water and anaerobic conditions on *Fusarium oxysporum* f. sp. *cubense* in soil. *Trans. Brit. Mycol. Soc.* **43**: 51.

57. Newhall, A.G. 1958. The incidence of Panama disease of banana in the presence of root knot and burrowing nematodes (*Meloidogyne* and *Radopholus*). *Plant Dis. Rep.* **42**: 853.

58. Ogawa, J.M. , H.J. Su, Y.P. Tsai and T.M. Lee. 1969. Post-harvest decay development affected by systemic activity of benomyl in bananas, oranges and pine apples. *Phytopathology* **59**: 1043 (abstr.).

59. Ploetz, R.C. 1990. Vegetative compatibility in *Fusarium oxysporum* f. sp. *cubense*: Classifying previously non-characterized strains. *Acta Hort.* **275**: 699.

60. Ploetz, R.C. 1990. Population biology of *Fusarium oxysporum* f. sp. *cubense*. In: *Fusarium Wilt of Banana*. Ed. R.C. Ploetz. APS Press, St. Paul, Minn.

61. Ploetz, R.C. (ed.). 1990. *Fusarium Wilt of Banana*. APS Press, St. Paul.

62. Ploetz., R.C. 1992. Fusarium wilt of banana (Panama disease), pp. 270-282. In: *Plant Diseases of International Importance*. Vol. III. *Diseases of Fruit Crops*. (eds.) J. Kumar, *et al*. Prentice Hall, New Jersey.

63. Ploetz, R.C. and J.C. Correll. 1988. Vegetative compatibility among races of *Fusarium oxysporum* f. sp. *cubense*. *Plant Dis.* **72**: 325.

64. Ploetz, R.C. and V. Galan-Sauco. 1998. Banana diseases in the subtropics: a review of their importance, distribution and management. *Acta Hortic.* **490**: 263.

65. Rajendran, G., T.C. Nayanathan and V. Sivagami. 1979: *Indian J. Nematol.* **9**: 54.

66. Ramakrishnan, T.S. and S. Damodaran. 1956. Observations on the wilt disease of banana. *Proc. Indian Acad. Sci.* **43B**: 21.

67. Rishbeth, J. 1955. Fusarium wilt of banana in Jamaica. I. Some observations on the epidemiology of the disease. *Ann. Bot.* **19**: 293.

68. Rishbeth, J. 1957. Fusarium wilt of banana in Jamaica. III. Attempted control. *Ann. Bot.* **21**: 597.

69. Rishbeth, J. 1960. Factors affecting the incidence of banana wilt (Panama disease). *Emp. J. Exp. Agric.* **28**: 109.

70. Rombouts, J.E. 1953. The microorganisms in the rhizosphere of banana plants in relation to susceptibility and resistance to Panama disease. *Plant Soil* **4**: 276.

71. Sequeira, L. 1958. Role of root injury in Panama disease infections. *Nature* **182**: 309.

72. Sequeira, L. 1962. Influence of organic amendments on survival of *Fusarium oxysporum* f. sp. *cubense* in soil. *Phytopathology* **52**: 976.

73. Sequeira, L. 1964. Inhibition of indoleacetic acid oxidase in tobacco plants infected by *Pseudomonas solanacearum*. *Phytopathology* **54**: 1078.

74. Sequeira, L. and A. Kelman. 1962. The accumulation of growth substances in plants infected by *Pseudomonas solanacearum*. *Phytopathology* **52**: 439.

75. Sequeira, L. and P.H. Williams. 1964. Synthesis of indoleacetic acid by *Pseudomonas solanacearum*. *Phytopathology* **54**: 1240.

76. Simmonds, N.W. 1966. *Bananas*, 2nd ed., p. 512. Longman, London.

77. Sivamani, E. and S.S. Gnanamanickam. 1987. Occurrence of Moko wilt in "poovan" banana in Puduottai district of Tamil Nadu. *Indian Phytopath.* **40**: 233.

78. Sivamani, E. and S.S. Gnanamanickam. 1988. Biological control of *Fusarium oxysporum* f. sp. *cubense* in banana by inoculation with *Pseudomonas fluorescens*. *Plant Soil* **107**: 3.

79. Stover, R.H. 1954. Flood fallowing for eradication of *Fusarium oxysporum* f. sp. *cubense*. II. Some factors involved in fungus survival. *Soil Sci.* **77**: 401.

80. Stover, R.H. 1955. Flood fallowing for eradication of *Fusarium oxysporum* f. sp. *cubense*. III. Effect of oxygen on fungus survival. *Soil Sci.* **80**: 397.

81. Stover, R.H. 1956. Studies on Fusarium wilt of banana. I. The behavior of *Fusarium oxysporum* f. sp. *cubense* in different soils. *Can. J. Bot.* **34**: 927.

82. Stover, R.H. 1958. Studies on Fusarium wilt of banana. II. Some factors influencing survival and saprophytic multiplication of *Fusarium oxysporum* f. sp. *cubense*. *Can. J. Bot.* **36**: 311.

83. Stover, R.H. 1962. *Fusarium wilt of banana and other* Musa *species*. Longman, London.

84. Stover, R.H. 1972. *Banana, Plantain and Abaca Diseases*. CMI, Kew , Surrey, England.

85. Stover, R.H. 1980. Sigatoka leaf spots of banana and plantain. *Plant Dis.* **64**: 750.

86. Stover, R.H. and J.D. Dikson. 1968. Leaf spot of banana caused by *Mycosphaerella musicola*: action of oil on life cycle of the pathogen. *Can. J. Bot.* **46**: 1495.

87. Stover, R.H. and S.E. Malo. 1972. The occurrence of fusarial wilt in normally resistant Dwarf Cavendish banana. *Plant Dis. Rep.* **56**: 1000.

88. Stover, R.H. and N.W. Simmonds. 1987. *Banana*, 3rd Ed. Longman, London.

89. Stover, R.H. *et al.* 1961. Studies on Fusarium wilt of banana. VII. Field control. *Can. J. Bot.* **39**: 197.

90. Su, H.T., S.C. Chang and W.H. Ko. 1986. Fusarium wilt of Cavendish bananas in Taiwan. *Plant Dis.* **70**: 814.

91. Tanaka, H. and H. Maeda. 1988. Biological control. *Bact. Wilt News* **4**: 5.

92. Thorne, G. 1961. *Principles of Nematology*. McGraw Hill.

93. Tsai, J.W., S.T. Hsu and L.C. Chang. 1985. Bacteriocin producing strains of *Pseudomonas solanacearum* and their effect on development of bacterial wilt of tomato. *Plant Prot. Bull. Taiwan* **27**: 267.

94. van Weerdt, L.G. 1957. Studies on the biology of *Radopholus similis* (Cobb 1893) Thorne 1949. Part I. *Plant Dis. Rep.* **41**: 832.

95. van Weerdt, L.G. 1958. Studies on the biology of *Radopholus similis*. Part II. *Nematologica* **3**: 184.

96. van Weerdt, L.G. 1960. Studies on the biology of *Radopholus similis*. Part III. *Nematologica* **5**: 43.

97. Wardlaw, C.W. 1961. *Banana diseases, including plantain and abaca*. John Wiley and Sons, Inc., New York 98.

98. Wehunt, E.J. and D. J. Edwards. 1968. *Radopholus similis* and other nematode species on banana, pp. 1-19. In: *Tropical Nematology*. (Eds.) G.C. Smart and V.G. Perry. Univ. Florida Press.

99. Yang, I.L. 1970. Studies on the varietal resistance to bunchy top disease in banana. *J. Taiwan Agr. Res.* **19**: 57.

⤶ 5

Diseases of Grapevines and Grapes

■ DOWNY MILDEW OF GRAPEVINES

Downy mildew is the most destructive fungal disease of grapevines and is one of the most widely studied plant diseases. Prior to 1870, this disease was considered endemic to North America where it appears to have co-evolved with wild grapes and then shifted to cultivated grapes. Since 1875 its epidemics were reported in France where it was introduced through grape cultivars imported from the USA as rootstocks resistant to grape phylloxera (*Daktulo vinifolii*). This aphid pest of grapevines was also introduced into Europe from North America before 1863. In western Europe, particularly France, the downy mildew caused heavy losses to the wine industry during 1870s-1880s (1). It was in the search of chemical control of this disease that Prof. Millardet of the Bordeaux University made the accidental discovery of Bordeaux mixture in 1885. The disease had reached England in 1894, Brazil in 1893, South Africa in 1907, and Australia in 1916 (29). At present the downy mildew is a major disease of grapes in grapevine-growing areas of the world (1) and has been recorded from 91 countries from temperate zone to the tropics. In India, the disease is known to occur in Maharashtra.

Most of the economic loss is due to cluster destruction and loss of foliage resulting in the loss of photosynthetic area. When conditions of temperature, leaf wetness duration and relative humidity and foliage canopy are favourable, crop losses vary from 50-100%. The damage is greater following early infection. In epidemic years, defoliation of vines prevents maturation of fruits and canes and exposes fruits to sun scald.

Symptoms

Symptoms of the disease appear on all aerial and tender parts of the vines (49). They are more pronounced on leaves, young shoots, and immature berries which are most susceptible (87-89). On the upper surface of leaves irregular light yellow spots are seen. On the opposite surface, below these spots, downy growth of the fungus may be present. Later, due to necrosis, these leaf spots turn dark brown. The growth on the lower surface becomes dirty grey. A number of spots coalesce to form large necrotic patches. The infection of aged leaves results in a mosaic of small, angular, yellow to reddish brown lesions limited by veinlets. The affected leaves fall prematurely. When the leaves with suspected infection are moistened, enclosed in a polyethylene bag and incubated overnight at 20°–25°C in darkness, white cottony growth is produced on active infected tissue.

The diseased shoots remain stunted and may be swollen due to hypertrophy. Necrosis may also occur. The shoots are rarely infected and only when about 10-15 cm long. The nodes are more susceptible than internodes (49). The infected leaves, shoots and tendrils become covered with whitish growth of the fungus. The flowers and young berries are also infected. The flower clusters are highly susceptible. The infected inflorescence first turns oily yellowish-brown and may develop the downy growth during periods of high humidity at night. Infection of peduncles results in death of the entire cluster. The young berries are most susceptible from setting until 5-6 mm in diameter. The berries stop growing, turn hard, bluish green and then brown. They wither and fall down. After the growth of the fungus the entire bunch of berries is destroyed. Sometimes, only a part of the bunch is affected. When the infection of berries occurs in early stages of their development fungal growth is visible on the surface. However, when the fruits are half grown and infection occurs at this stage the fungal growth is confined to the inside of the berries. This results in light green or brown colour of fruits. Normally, the fully grown or maturing berries do not contact fresh infection, probably because the stomata turn non-functional (49) but they are also lost if the supporting pedicels and cluster stem, which remain susceptible, are diseased.

The Causal Organism

The downy mildew of grapevines (*Vitis vinifera*) is caused by *Plasmopara viticola* (B. & C.) Berl. and de T. The intercellular mycelium of the fungus consists of coenocytic, thin-walled, hyaline hyphae with granular protoplasm. These hyphae are 1-2 to 60 μm wide according to the size and shape of intercellular spaces through which they pass. Haustoria are spherical or pear-shaped and 4-10 μm in diameter.

Sporangiophores arise from hyphae congregated in the substomatal space. One to twenty sporangiophores emerge from a single stoma. Sometimes they may emerge directly through the cuticle. On young berries they come out through lenticels. These spore-bearing structures are mostly produced during the night when conditions of high humidity are present. They are 300-500 (mostly 140-250) × 7-9 μm in size. The branching of the sporangiphores is almost at right angles to the main axis and at regular intervals. The first branches arise from the apex. From the lower branches secondary branches are also produced. From the apex of each branch 2-3 sterigmata arise and bear sporangia.

The sporangia are hyaline, thin-walled, oval or lemon-shaped, papillate, and measure 15-30 × 11-18 μm. Their germination is either by production of zoospores formed inside or outside in a vesicle or through germ tubes, depending on humidity. In indirect germination, each sporangium produces 1-6 zoospores. The zoospores are pear-shaped and 7-9 × 4-5 μm in size. They have two apical flagella which may be 30 μm long. On the host, the zoospores encyst and germinate by producing a flexuous aseptate germ-tube, 50-80 μm long. The oospores are produced mostly in tissues adjacent to the midrib. They are thick- and wrinkled- walled, pigmented, and 25-36 μm in diameter. Antheridia are rarely seen. The oospores germinate by producing a germ-tube of variable length, and 2-3 μm in diameter. The germ-tube bears an apical sporangium, 35-40 × 25 μm in size. This sporangium produces 8-20 zoospores. Strains and ecotypes of the pathogen with varying virulence to *Vitis* and related genera have been reported (68, 77).

Disease Cycle

Plasmopara viticola is a biotroph. In areas with a short and mild winter where grapevines are evergreen the fungus can survive in its active phase between bud scales and in diseased leaves that remain on the vines until the next bud burst. It can also survive in buds even in those areas where there is complete leaf fall due to prolonged and severe winter (cf. 29). The wild species of *Vitis*, where present, also harbour the fungus. However, the main sources of survival are oospores formed in the leaf and shoot tissue around the end of the growing season when conditions are no longer suitable for asexual reproduction. The fallen leaves contain enormous numbers of oospores. More than 200 oospores/sq. mm have been reported (cf. 29).

Some scientists believe that oospore formation occurs best at 14° – 18°C while others feel that temperature is not as important as rainfall. After formation, the oospores undergo a temperature-dependent dormancy period. They can survive in leaf debris for at least 2 years. Generally, the oospores germinate more abundantly and in a shorter time early in the season when they have been subjected to frequent rains and mild temperatures. The risk of downy mildew is highest under these conditions. Rainy season with heavy rains during the oospore formation period hastens their maturity, hence, there is more disease in the next crop season (98). The date of optimum oospore maturation can be used for predicting disease severity in spring. Under controlled conditions oospores mature at alternating weekly temperatures of 10° and 5°C (72). Low temperature reduces the dormancy period. On maturity, oospores germinate over a period of 2-3 months in late winter or in spring. Minimum temperatures for germination range from 7° to 13°C and optimum 20° – 25°C. At 21°C the germination process takes 0.5 – 2 hours compared to 0.2-6 hours at 10°C. For good germination the soil must be saturated. High light intensity may retard or prevent germination.

Primary infection is caused by the soil-borne inoculum. The sporangia and/or zoospores are splashed or blown to the wet vine surfaces. Zoospores are released by sporangia within 30 min at 20°C. The primary infection starts on leaves near the ground level. The zoospores swim towards the stomata. Width of stomata and possibly exudates from stomata direct the movement. The zoospore swarming may continue for 3-5 hours. Within 20-30 min of reaching the stomata, the zoospores become motionless and encyst by absorbing their flagella and getting enclosed in a membrane. Under favourable conditions they germinate by a germ tube that grows through stomatal aperture into the substomatal cavity. Many germ tubes may penetrate a single stoma.

In the substomatal cavity, a vesicle is formed. This develops a short hypha at its distal end. When this hypha touches a host cell, a primary haustorium is formed. This occurs within 3.5 hours after the sporangium has come in contact with the susceptible host surface at 23° C. For 12-15 hours there is no development although the size of primary haustorium increases. Within 24 hours of sporangium deposition, hyphae have grown between the cells and produce additional haustoria. After 36 hours, the host cells are noticeably plasmolysed. Hypertrophy of cortical cells may cause a slight swelling of infected stem. Mycelial growth is more abundant in young tissues. The growth ceases at 30°C although the mycelium can remain viable at 42°–43°C for 12 days.

The first crop of sporangia is then produced from these primary infections. Sporulation (sporangia formation) requires continuous RH ranging from 95 to 100% or saturation. It also requires a minimum of 4 hours of darkness. Moist, dark conditions following a period of light favour maximum sporulation. Under continuous light or darkness and high humidity there is little or no sporulation (10, 47). Sporulation occurs at temperatures ranging from 9° to 34°C. The optimum temperatures have been cited as 18° – 24°C and 18° – 28°C. The maximum sporangial production is reported to occur between 23° and 28° C. Under favourable conditions sporulation can continue for at least several successive nights (sometimes over a period of 2 months) on the same lesion. Sporulation is also favoured by high water content of the infected tissue.

The sporangial dehiscence occurs when the cross wall of the callus between sporangium body and stalk is dissolved in water. Thus, moist air is essential for dissemination of sporangia. The sporangia are detached by movement of leaves and are wind-borne to new sites to cause secondary infection. Water also disperses these sporangia. The viability of sporangia decreases with increasing temperature and decreasing RH. The attached sporangia survive longer than detached sporangia. Various reports cite sporangium survival for 4-14 days at temperatures of 10°–20°C. At 30°C sporangia survive only for 6 hours. Sunlight adversely affects sporangial viability. In India, where temperatures reach more than 30°C, sporangia harvested during day hours do not germinate while those harvested during night from 2 a.m. to 6 a.m. germinate within one hour (87, 88). The optimum temperature for germination of sporangia is 10° – 16°C while the minimum is 5°C. Under conditions of optimum temperature and humidity, germination occurs within an hour. Sporangium germination and the liberation of zoospores occurs only in water at temperatures ranging from 2° – 9°C to 28° – 30°C with optimum between 15° and 23°C.

The germ tube from the zoospore enters the host through the stomata and lenticels. The incubation period varies from 7 or 8 days to 20 days depending on susceptibility of the host variety, air temperature and relative humidity. As temperature increases from 5°C to 26°C, the incubation period decreases. At 100% RH it is 11-12 days. At 28°C the incubation period is 5-6 days at 100% RH and 8 days at 70% RH.

Management

Sanitation is very important in the management of downy mildew of grapevines. The fallen leaves and twigs should be collected and burnt. The microclimatic conditions under the canopies being closely related to infection, inoculum production, and spread of the disease, vines should be planted with proper spacing and should be trained in such a manner that the leaves are not near the soil surface. These precautions permit free circulation of air and ensure low humidity.

The disease has been controlled with well-timed application of copper and dithiocarbamate fungicides. Bordeaux mixture (4: 4: 50). copper oxychloride such as Blitox-50 (0.3%), zineb, maneb and mancozeb (0.2%) and captan (0.2 – 0.5%) have been successfully used against the disease. Bordeaux mixture in new formulations is still used extensively in French vineyards for management of grape downy mildew. Five sprays are generally recommended at the following stages: (i) immediately after pruning, (ii) 3-4 weeks after pruning, (iii) before the buds open, (iv) when berries have formed, and (v) during growth of shoots. Sprays should be stopped 2 weeks before harvest. Some workers have found the antifungal antibiotic Aureofungin effective in controlling the disease. The above schedule of spray is suited to areas where the downy mildew is of known regular occurrence. It may be costly and unproductive in areas where the disease is of irregular occurrence such as in the arid zones. For this, forecasting based on weather parameters and plant growth conditions is recommended so that chemicals are used only when required (50, 98).

Since the mid-1970s the mixture of metalaxyl (Ridomil) with either copper fungicides or mancozeb or folpet has been found to give effective control of the disease when applied before or after the infection. Most mixtures containing metalaxyl persist for 2-3 weeks. Metalaxyl inhibits the formation of secondary haustoria and the growth of mycelium inside host tissue, and stops lesion formation and sporangial formation through action of its volatiles (105-107). Because of resistance development in the pathogen against metalaxyl, the latter is mostly recommended in combination with protectant fungicides such as mancozeb. Dimethomorph (CME 151) has better protective and curative activity than metalaxyl-mancozeb combination (108). It inhibits sporulation when used at 25 mg/L before infection or up to 9 days after infection.

Phosphorus (phosphonic) acid or its derivative phosethyl-Al is considered a true systemic downy mildew fungicide and an alternative to metalaxyl (110). Applied at 1.2 g/L up to 12 days after infection, the fungicide reduces the incidence and severity of downy mildew through reduced production of inoculum of *P. viticola*. The fungicide is translocated from leaves via the xylem and to some extent via the phloem to extremeties of shoots and to clusters at least until berry set. It provides protection to actively growing foliage and clusters for 14-21 days. In plant tissue, phosethyl-Al rapidly breaks down to phosphorus acid and carbon dioxide. The acid is the active component and acts directly on the fungus. Its indirect effects on host metabolism are secondary (27).

A new downy mildew fungicide, RPA 407213, is reported to have high level of protective and curative effect against *P. viticola*. Alone or in combination with fosetyl-Al or cymoxanil it has been found highly effective against established infections.

Azoxystrobin, a new strobilurin wide spectrum systemic fungicide, has preventive and curative effect against the grape downy mildew and grape black rot fungi (12). On young plants, the preventive efficacy of azoxystrobin at 250 g a.i./ha applied 10 days before inoculation was equal to that of mancozeb at 2800 g a.i./ha. Its efficacy at the same dose was higher than that of mancozeb when applied 7 days before inoculation. The mixture of azoxystrobin (187 g a.i.) with cymoxanil (120 g a.i.) was also more effective than a mixture of mancozeb (1400 g a.i.) and cymoxanil (120 g a.i.). When sprayed at 10 days intervals the bunch protection was almost perfect.

The use of systemic fungicides is handicapped by the phenomenon of development of resistance in the pathogen against the fungicide (11, 42, 91). The resistant races of *P. viticola* have been reported for several phenylamide fungicides including metalaxyl, ofurace, milfuram and cyprofuram. The level of resistance varies (46). A high level of resistance and a high degree of fitness is a disturbing feature of metalaxyl resistance in fungi (18). When metalaxyl-mancozeb combinations are used the resistant isolates persist but their frequency decreases if fungicide use in the field is confined to mancozeb or stopped completely. Metalaxyl resistant strains show resistance to other phenylamide systemic fungicides also. The aggressiveness of metalaxyl-resistant isolates may be related to heterokaryosis in *P. viticola*. The combination of protectant and systemic fungicides has synergistic effect (43). In metalaxyl-mancozeb combination, the protectant component may increase the permeability of cell membrane to increase the concentration of the site-specific inhibitor (metalaxyl) reaching the site of action in the fungus. Furthermore, since metalaxyl is usually fungistatic, the activity of sublethal concentration of the protectant component may make their combined effect more fungistatic. Control of metalaxyl-resistant isolates has also been achieved by using other types of systemic fungicides such as phosethyl-Al (76).

As a part of the strategy to combat the problem of resistance development, the fungicide industry has come up with certain decisions (25, 26, 91). These include (i) the fungicides should be made available only in pre-packed mixtures with protectant fungicides, (ii) only 2 to 4 applications per season should be recommended, and (iii) post-infection use of mixtures would not be recommended. The pre-packed mixtures have several advantages such as (i) the protectant component should control the resistant strains, (ii) the dose of the systemic fungicide may be reduced due to additive or synergistic factor, (iii) reduced concentration of the systemic fungicide should reduce the selection pressure for resistance, and (iv) most importantly, the use of pre-packed mixture is an enforceable strategy.

In Germany, some scientists have reported that leaves sprayed with water extract of composted horse manure-straw-soil mixture develop resistance to infection of *P. viticola*. The extract has no direct effect against the pathogen (104). Subsequently, it was reported (79) that the disease regulation is the result of biological control through enhanced microbial activity on the leaf surface. Acqueous extracts of fermented products of many types of materials have since been used to control downy mildew, powdery mildew, and other fungal diseases of grapevines. *Fusarium proliferatum* has been identified as a biocontrol agent against *P. viticola*. Post-infection application of microconidial suspension of the fungus on leaf discs reduced sporulation of *P. viticola* by 97 % (34). In vineyards, weekly application of the conidial suspension reduces the incidence of downy mildew. The hyphae of the mycoparasite coil around and penetrate the sporangiophores of *P. viticola*. Induction of resistance to downy mildew by treatments with *Pseudomonas syringae* pv. *syringae* or salicylic acid is also reported.

The sources of resistance to downy mildew are present in many species of *Vitis* and these have been used for developing resistant cultivars. Under Indian conditions cultivars Amber Queen, Champion, and Digraset have been found resistant to downy mildew while cultivars Athens, Buckland, Sweet Water, Essabela, Goethe, James, Malaga, Khalili, Westfield and Solanis x Riperia are moderately susceptible (22). In another report, the varieties found resistant in south India were Amber Queen, Cardinal, Champa, Champion, Dogridge, and Red Sultana. Some promising resistant mutants have also been induced by irradiating the seeds.

■ POWDERY MILDEW OF GRAPEVINES

The known history of powdery mildew of grapevine dates back to 1834 when the fungus causing the disease was first described as *Erysiphe necator* in the eastern part of North America. However, it was not considered important until it was reported from England in 1847 and by 1850 it had spread to France and other major grape-growing areas of Europe where it caused considerable loss to grapevine growers and the wine industry. Today, powdery mildew can be found in most grape-growing areas of the world, including the tropics (13, 47). It appears in almost epidemic form in all the vineyards in India when the conditions are favourable for its development. The disease is much more serious than the downy mildew of grapes and is more dangerous than other powdery mildews of different crops. It is also known to occur in a mild or severe form in North and South America, Europe, parts of Africa, and in Australia. Losses in the yield of fruits may be up to 40-60 %. In addition to loss of yield, the fungus infection makes the grapes unsuitable for wine making. Berries infected by the fungus tend to be higher in acid than healthy berries. This is not considered desirable by the brewers. The fungus itself produces an off-flavour in wine made from infected grapes. The infected berries tend to crack, thus, providing entry to other pathogens and saprophytes (13).

Symptoms

The disease attacks the vines at any stage of their growth. All the aerial parts of the plant are attacked. As in other powdery mildews, the characteristic symptom of the disease is the appearance of white, powdery patches on affected parts. Cluster and berry infections usually appear first. Cluster infection before or shortly after the bloom results in poor fruit set and considerable crop loss. Young fruits (berries), just after bloom, show whitish mycelial growth on the surface. Berries are susceptible to infection until their sugar content reaches about 8 %, although established infections continue to produce spores until berries contain 15 % sugar. When the infection of berries occurs before they attain full size, the epidermal cells are killed and the growth of epidermis is prevented. As the internal pulp continues to grow, the skin cracks. Such berries either dry up or rot. If the attack occurs when fruits are nearing maturity or beginning to ripen, they fail to colour properly, become irregular in form and only few of them ripen, remaining undersized with a blotchy surface. Often the infected berries develop a net-like pattern of scar tissues.

Leaf lesions appear late and do not cause much damage. On young leaves, small whitish patches appear on the upper, or sometimes on the lower, surface of the leaf. These patches grow in size and finally coalesce to cover large areas on the lamina. Similar floury patches are formed on the stem, tendrils, and flowers (64). The powdery growth gradually turns grey and finally dark coloured. Malformation and discolouration of the affected leaves are common symptoms. The stems turn brown in colour. The diseased vines have a wilted appearance and remain stunted in growth. Necrosis of the penetrated epidermal cells and even of adjacent cells is a characteristic reaction of resistance in some varieties of the host (112).

The Causal Organism

The disease is caused by *Uncinula necator* (Schw.) Bur. (anamorph *Oidium tuckeri* Berk.). The mycelium is entirely superficial on the attacked parts to which it adheres by means of bilobate or multilobate appressoria. The hyphae are hyaline, slender, septate, branched and 4-5 μm in diameter. They turn darker in colour when the formation of conidia is over. The conidiophores arising perpendicular to the creeping hyphae on the host surface are simple, multiseptate and erect. They are attached to the mycelial hyphae by a cylindrical foot cell measuring 24-40 μm. Cells of the conidiophores are generally wider than those of the mycelial hyphae, measuring 6.2-7.5 μm. They bear a chain of 3-4 conidia under field conditions. Under static humid conditions the chains may contain 8-10 conidia. These hyaline conidia are oval in shape and measure 25-30 x 15-17 or 27- 47 x 14-21 μm. The oldest conidium is at the distal end of the chain. The conidia germinate by a short germ tube terminated by a bilobed or multilobed appressorium.

Cleistothecia (sexual fruiting bodies) of the fungus have been found in North America (65), Europe, Russia, Peru, and Australia (109). Under the climatic conditions prevailing in most vine growing areas of India, the perfect stage of the fungus is not found. The fungus is heterothallic and most populations consist of two mutually exclusive mating types (41, 83). A small percentage of isolates have the capacity to form cleistothecia in protracted association with isolates that initially appear to be of an incompatible mating type. When the mating types are present, the cleistothecia can form on all infected tissues during the later part of the growing season. They are found embedded in the superficial mycelium on the leaves, or on shoots, chiefly at the nodes or in buds among the scales and hairs. They are hyaline and spherical when young but turn yellow due to the accumulation of a yellow lipid in the ascocarp (39). When the outer cells of the ascocarp darken the mature cleistothecium turns black, almost round with a flattened top, and measures 75-100 μm in diameter. The peridium is covered with 8-25 septate appendages which appear inserted in the equatorial region of the ascocarp. The appendages are coiled (uncinate) at the distal end and brown at the base. They are 1-6 times as long as the diameter of the cleistothecium (47). When the cleistothecium is immature it retains functional connections with the hyphal mass. Mature cleistothecium develops a basal concavity and the connections with the hyphal mass die (60). Each perithecium contains 2-6, sometimes more, ovate to subglobose asci measuring 48- 60 x 37-45 μm. In each ascus there are 4-6 ascospores which are oval to ellipsoid in shape and hyaline. The ascospores are low in water content and their measurements are affected by the mounting medium used. Unmounted ascospores measure 23-28 x 14-16 (mean 26 x 15) μm. In lactophenol mount, they measure 18-29 x 10-15 (22 x 12) μm (66). These ascospore germinate by a short germ tube which terminates into a multilobed appressorium. Multiple germ tubes may also arise from the ascospore (40, 65).

Disease Cycle

The conidia can resist desiccation but it is not known how long they can remain viable. The fungus perennates through hyphae inside dormant vegetative buds (13, 63). Where cleistothecia are formed they serve as additional source of perennation (20, 21, 38, 65). In tropical climate the survival is mainly through mycelium and conidia on green tissues remaining on the vines (13). Environmental factors are very much responsible for the formation of cleistothecia (80). Rains disperse the cleistothecia to the bark of the vines where they are retained between leaf fall and bud break in the next season (20, 21). Density of cleistothecia is higher on fallen leaves than on the bark but their viability is more on the bark than on fallen leaves on soil (21). After the infection from ascospores or from the hibernating mycelium in the host buds (63, 75) enormous numbers of conidia are produced to carry on the conidia-to-conidia life cycle. The conidia on the leaf surface start germination within 90 min and in about 20 hours about 52 % conidia have germinated. Almost all the germ tubes coming in contact with a hard surface form appressoria (56) and in 117 hours after conidial germination on leaves, colonies of *U. necator* with profuse conidiophores and conidia are seen (56). Temperature appears to be the major determinant of fungal development. It determines the extent of asexual (conidial) reproduction of *U. necator* (17) and fluctuating temperatures decide the rate of conidial formation, germination and colony development (113, 114). In the temperate climate of California (USA) the early warm spring climate usually precedes severe epidemic of powdery mildew (74). Rapid germination of conidia, infection, and development of the pathogen occurs at temperatures of 21°- 30° C (24), although the fungus can grow at temperatures from 5° to 30°C. The minimum temperature for germination of conidia is 6°C and for infection and growth 7°C. The optimum for germination is 25°C. Conidial germination ceases at 3°C and 35°C and at 40°C conidia are killed (24). At 25°C conidia germinate in about 5 hours and time from inoculation to sporulation is 5 days whereas at 23°C and 30°C it is 6 days. Mildew colonies are killed after exposure to 36°C for 10 hours or to 39°C for 6 hours.

Free water causes poor and abnormal germination of conidia. Rainfall is detrimental because it removes conidia and disrupts the mycelium. Atmospheric RH of 40-100 % is sufficient for germination of conidia and infection although germination can occur even at less than 20 per cent RH. Humidity has a greater effect on sporulation than on germination. Low diffuse light favours development of powdery mildew. In bright sunlight conidial germination is inhibited. In areas where cleistothecia play a major role in survival of the fungus and in primary infection, it is reported that cleistothecia in the bark of the vine discharge ascospores when rainfall occurs between bud burst and bloom. Ascospore discharge requires free water. Temperatures within the range of 10°and 25°C have little effect on ascospore release but a temperature of 4°C or lower can suppress ascospore discharge. Ascospores germinate equally well in free water and in saturated atmosphere. Germination declines rapidly as humidity decreases. Appressoria are not formed by ascospore germ tubes at below 10°C and at above 31°C.

It was reported that climatic conditions during October and November in India are ideal for development of the disease. Warm, dry weather with just enough humidity is very favourable. However, after a microclimatic study in Maharashtra, Chavan, *et al.* (15) have reported more rapid spore production and disease development during December and January, when the weather is cool and humid, than in November and February. According to them temperature in the range of 12.2°- 30.1°C and relative humidity greater than 57.4 % favours sporulation of *U. necator*. At temperatures below 8.8°C and above 34°C and RH below 47.4 % the rate of multiplication is zero. The disease development is retarded in sunshine. In south India, the disease incidence is reported to be significantly influenced by relative humidity and maximum temperature whereas the influence of minimum temperature, rainfall, and total rainy days was not significant (4). Increase of RH by 1% increased the disease incidence by 2.4 %. Increase of temperature by 1°C decreased the disease by 4.4 %. Maximum temperature in the range of 27°- 31°C along with RH up to 91 % favoured disease incidence while maximum temperature range of 31°-34°C inhibited the development of powdery mildew. In Punjab, mycelial growth and production of conidia was found to be very rapid at 25°C followed by 20° and 30°C. At 35°and 40°C there is no powdery mildew development (56). Although infection can occur at zero % RH extremely low humidity adversely affects growth of the pathogen on the host surface.

Management

Due to the widespread occurrence of powdery mildew in grapevine-growing areas of the world, regulatory measures (quarantine) are of no value except where introduction of fungicide-resistant strains is feared. Clean cultivation of vines is an important part of disease management in grapevine orchards. Pruning after shedding of leaves, thinning out and cutting back of laterals and removal and destruction of all diseased parts constitute clean cultivation. Excessive nitrogen fertilization tends to promote succulent growth which is associated with increased incidence of powdery mildew (6).

The control of powdery mildew in commercial orchards is generally based on the use of fungicides. Fungicidal control measures should start in the early stages of vine development and repeated at 7- 21 days interval depending on the type of fungicide being used. Dusting of vines with sulphur (300 mesh) had been an effective control method in the past and is still most extensively used chemical control measure (13). The first dusting should be done when new shoots are 7-15 cm long, second during or just before blossoming. A third application can be made 40-50 days later. Sulphur is also applied as a wettable powder. In dry climates sulphur dust is preferred whereas in areas with plenty of rainfall wettable powder or flowable formulations are recommended for their retention qualities. Much of the fungicidal activity of sulphur is due to its vapour action which is dependent on particle size. As the particle size increases fungicidal activity decreases. Therefore, amorphous sulphur is best for fungicidal activity. The optimum temperature for sulphur activity is 25°- 30°C. Above 30°C there is the risk of phytotoxicity. The activity of sulphur is reduced in humid air compared to dry air (13). Because sulphur has poor retention qualities, application schedules of 7-10 days are usually required. Pre-bud swell or dormant stage spray of lime sulphur delays the development of epidemics (38).

In addition to sulphur dust (15 kg/ha), Sulfex (0.25%). Karathane (0.05%), Calixin (0.05%), Topsin-M (0.1%), Thiovit (0.25%), all have been found effective against powdery mildew of grapevines (9, 14). However, the triazole fungicides have proved more effective. Generally, Bayleton (triadimefon) is considered a most effective triazole fungicide against grape powdery mildew fungus. Triadimefon at the rate of 40 g/100 lit. or penconazole or cyprconazole at the rate of 40 ml/100 litre sprayed in mid-March, last week of April, and first week of May are highly effective against the disease. Balamuralikrishnan and Jeyrajan (5) found fenarimol (Rubigan 12 EC), used in 7 sprays at 0.05%, at 10 d interval between 15 cm cane length and 60 cm cane length highly effective against powdery mildew. The fungicide was highly compatible with wide-spectrum protectants such as mancozeb and its residue persisted for 14 days after the last spray but at a very low level. The dissipation rate was high during summer. Bharadwaj (7) compared efficacy of 7 fungicides against *U. necator* under dry temperate conditions of Himachal Pradesh (India) and found triadimefon (Bayleton), flutriazole (Impact), or hexaconazole (Anival) more effective than carbendazim (Bavistin), dinocap (Karathane) and sulphur (Sulfex). Pearson *et al.* (67) have recommended single application vapour-action treatment with triadimefon, triadimenol (Bayton), etaconazole (Vangard), flusilazole, mycobutanil and penconazole. They used 0.2 to 4.0 g a.i. per vine between the bloom and two weeks after the shatter. The treatment gave effective control of the disease. Cyprodinil and related anilopyrimidine fungicides are more recent compounds being used against powdery mildews. One advantage with the sterol- inhibiting triazole fungicides is that the interval between sprays can be increased and, thus, the number of sprays decreased. Dinocap (Karathane) is used at 10-14 days interval while the triazoles can be used at 14-21 days interval. The strobilurin fungicide compound, CGA 279202, is reported to give excellent control of grape powdery mildew at 6.75-7.5 g a.i/100 L. Spray schedules developed on the basis of weather station information and some model such as Gubler-Thomas model save cost of unnecessary sprays (19).

Although organic fungicides including the systemic ones listed above are commercially used (61, 66), their use is second to sulphur. The organic fungicides have solved the problem of phytotoxicity encountered with sulphur but they have the problem of resistance development in the pathogen. Isolates of *U. necator* resistant to benomyl, triadimefon, (Bayleton), myclobutanil and fenarimol are reported (45, 61, 62, 101, 114) including in India. Resistance to benomyl became widespread after 3-4 years of its use in vineyards (62). Triadimefon resistance in *U. necator* is most likely to manifest itself under high disease pressure which is in part a function of temperature (113). Reduced sensitivity of some isolates of *U. necator* to triadimefon in India was reported by Thind *et al.* (101). Cross resistance to sterol biosynthesis inhibitor fungicides is common in grape powdery mildew (114). To overcome the problem of resistance development, integration of sulphur and triazole (DMI) fungicides has been adopted in many countries. In Australia, DMI fungicides are applied before or around flowering and sulphur is applied after flowering (111). In Germany and some other European countries one application of sulphur followed by 3 applications of Bayleton or penconazole have given complete control of powdery mildew.

In attempts to encourage the alternative methods of control of powdery mildew, non-conventional chemicals, plant products and many approaches to biological control are in advanced stage of experimentation. Mineral oil and plant oils have been used against grape powdery mildew with positive results (59). Petroleum (mineral) oil applied as emulsion (1% v/v) in water has provided moderate protection, excellent pre-lesion and post-lesion curative action and is antisporulant. Plant oils have also showed significant action in pre-lesion treatment and as antisporulants in treatments applied to established lesions. These treatments are not effective against the downy mildew in vineyards. The mineral oils were as effective in vineyards as myclobutalin. Azami *et al.* (3) have reported that a rape oil derivative gives good control of grape powdery mildew. Repeated sprays of the rape oil derivative at the rates of 2.0 and 5.0 ml (formulated product) per litre prevented the foliar symptoms as effectively as either wettable sulphur 2 g/L (formulated product) or fenarimol 0.2 ml/L (formulated product). However, in heavily diseased vines the fungicides were better than the oil derivative.

Spray of 1 per cent sodium bicarbonate 3 times during the growing season, commencing after the appearance of first symptoms gives good control of powdery mildew. A single spray of 0.1 M solution of phosphate is reported to induce a systemic protection against powdery mildew and suppress lesions on diseased foliage (70). The efficacy of these inducers of systemic acquired resistance (SAR) is increased by the addition of 0.1 % sulphur.

In biological control, suppression of grapevine powdery mildew by the mycoparasite *Ampelomyces quisqualis* Ces. (syn. *Cicinnobolus cesatii* De

Bary) is reported (33). It is a naturally occurring pycnidial hyperparasite which is wholly internal within the mycelium, conidiophores, conidia, and ascocarps of *U. necator*. It requires free water on leaf surface to infect the mildew colonies. In *Sphaerotheca fuliginea* this mycoparasite is reported to germinate within 24 h of inoculation and the germ tubes form appressorium-like structures at the point of contact with powdery mildew. The hyperparasitic hyphae penetrate from cell to cell through septal pores of the host hyphae and continue growing during degeneration of the infected cells. After 7-10 days the parasite produces pycnidia with mature conidia in host hyphae and conidiophores. When pycnidia are wet by rain, conidia are exuded in a cirrhus which dissolves and conidia are dispersed by raindrop splashes. However, in nature, populations of *A. quisqualis* lag several weeks behind populations of powdery mildew fungi in their development and its colonies are seen late in the season allowing the mildew to develop to damaging levels (33). Sprays of concentrated conidial suspensions earlier in the season have given partial control of the powdery mildew of grapevines (33), particularly in wet seasons. In a recent report, English *et al.* (32) have found 45 % reduction in powdery mildew invasion of leaf area by applying *Orthotydeus lambi* mites to the leaves. The pathogen could develop cleistothecia only on leaves not inhabited by the mites. Application of myclobutanil and strobilurin fungicide azoxystrobin did not greatly affect abundance of the mites but mancozeb and wettable sulphur greatly reduced their population. Induced resistance by application of *Pseudomonas syringae* pv. *syringae* or salicylic acid is also reported.

Acqueous extracts of composts have been used with positive results. Populations of *Bacillus, Pseudomonas, Serratia, Penicillium,* and *Trichoderma,* all are enhanced on the host surface and they provide biological regulation of the plant pathogens (73). Fermented cowdung, various neem products, sodium bicarbonate, a formulated rape oil product, and a silica preparation were tried in Germany (79). Neem products gave reasonable control of powdery mildew but the treatment was very costly. The silica preparation and sodium bicarbonate have given acceptable control. These are inexpensive and environmentally safe.

The species of the genus *Vitis* and cultivars within the species differ in susceptibility to powdery mildew. Some varieties of grapes possess a certain degree of resistance to mildew in India (23, 84). Under dry temperate conditions of grape growing areas of Himachal Pradesh, the local varieties were found resistant compared to other cultivars like Black Prince, Isabella, Sultana Red, Champion, Cardinal, Perlett, etc., grown in the country. Cultivars Chholtu White and Chholtu Red are immune while Skibba White, Skibba Red and Bhatta Ribba White are resistant with 0.1- 5.0 % disease severity (8).

■ BUNCH ROT OF GRAPES

Bunch rot of grapes caused by *Botrytis cinerea* is an economically important disease in many viticultural regions of the world (58). From the economic point of view only infections in flowers and berries are of importance in terms of lowering quality and quantity of yield. Heavy damages to flower bunches are reported during unusually wet and cool conditions and a prolonged flowering period. Under these conditions upto 30 % of the bunches can be destroyed. Juices of Botrytis-infected grapes lose much of their fruity components and tend to age soon after fermentation. These traits are undesirable for wine making. The molded berries are often colonized by fruit flies and acetic acid bacteria. They are prone to secondary infection of such pathogens as *Aspergillus* sp., *Penicillium* sp. and *Trichothecium roseum*. These fungi produce a bitter taste, an undesirable off-flavour, and mycotoxins. The losses are heavier if the stalks of berries are infected and the whole bunches eventually drop to the ground.

Symptoms

Infection can occur on young as well as relatively older leaves which show irregularly shaped necrotic spots. In humid weather these spots may rapidly enlarge and coalesce. Infected flowers do not show any visible symptom. Microscopic observation of the floral parts reveals necrosis of stamens and growth of the fungus on the style and stigma. During the period of flowering, the infected stamens dehisce, and the solitary ovary can often be seen covered with tufts of sporulating mycelia.

The most prominent symptom of the disease is found on the berries. The infected berries become dark coloured and show the typical "grey mould" symptom, i.e., greyish, hairy mycelial growth all over the fruit surface. Often the fungus can be seen growing along the cracks or splits on the berries. Tufts of conidiophores and conidial bunches protrude from stoma and peristomatal cracks on the skin of the berry. In severe infections all the berries in a bunch are involved and totally lost.

The Causal Organism

Botrytis cinerea Pers.: Fr. The generic name is derived from the Greek word *botrys* meaning "a bunch of grapes". The conidia are formed in aggregations looking like a bunch of grapes. The perfect stage of the fungus is *Sclerotinia fuckeliana* in Discomycetes of Ascomycotina. The fungus was first described by de Bary. It has historical importance because studies on physiology of parasitism were initiated by de Bary with this fungus. The mycelium consists of 8-16 μm wide, septate, brown hyphae. The young hyphae are thin and hyaline.

Conidiophores and conidia are produced free. Conidiophores are lighter-brown than hyphae with hyaline apical portion. They are septate, 8-24 μm wide and irregularly branched at the apical portion. The tip of conidiophores and their branches is slightly enlarged and bears minute sterigmata. Conidia formed on these sterigmata are hyaline, aseptate, oval, globose, or short cylindric, and borne in clusters. On the host the conidia measure 4-24 x 4-18 μm. They are dry and easily wind-blown. They readily germinate in water. The fungus produces sclerotia which appear as sporodochial layers on many hosts. In addition to functioning as a major source of perennation of the fungus, these structures produce spermatia which initiate the formation of sexual fruit bodies. *Botrytis cinerea* has a very wide host range in regions with cool climate or during winters in sub-tropical regions. It attacks apples also causing blossom blight as well as post-harvest fruit rot. It is a serious pathogen of many field and vegetable crops.

Disease Cycle

The major sources of primary inoculum are sclerotia of the fungus present on dormant vines and on fallen berries. In temperate climate the mycelium can also survive through the winter and initiate primary infection in new growth in the spring. Mainly the infection is by conidia germinating on and infecting susceptible tissue or from mycelium growing from infected tissue that has fallen onto healthy tissue. In flower infections, latent infection may also occur (54). The path of floral infection is through stigma and style and then into the styler end of the ovary. The dead floral parts remain adhering to developing berries and later the mycelium and conidia on them cause infection of berries when their skin starts softening (58). Conidia are agents of secondary spread.

Sclerotia may be directly infective by producing hyphae or they serve as sources of conidia by sporogenous germination. Some sclerotia show carpogenous germination and form apothecia for ascospore production. Sclerotia survive better in dry soil at 20°-25°C. The optimum temperature for sclerotial germination followed by infection is also 20°-25°C. The fungus requires cool (18°-23°C), damp weather for best growth, sporulation, spore release and germination, and establishment of infection. Fungal growth outside the cuticle is enhanced by different nutrient sources such as pollen, debris of floral parts, or exudates from berries (48). The maximum primary infection of flowers during bloom from germinating sclerotia occurs when the daily air temperature during spring reaches 20°C. The optimum temperature-moisture conditions for infection are 20°C and saturation (leaf wetness) for 16 h. Germinating spores seldom penetrate actively growing tissue directly. Apart from cultivar susceptibility, climatic factors and microclimate depending on canopy management (78), infection of grape clusters is increased by insect injury, especially the grape berry moth (*Lobesia botrana* and *Eupoecilia ambiguella*). Larvae of the first generation attack the flowers, causing losses only in yield whereas those of the second and later generations damage green, ripening berries predisposing the fruit to attack of *B. cinerea* (35, 36). Losses may increase from 5% in non-infested clusters to 90% in clusters infested with the moth. Dense canopies are conducive to rapid spread of infection in flowers and bunches. High nitrogen treatment, especially application of ammonium nitrate predisposes grapevines to infection of *B. cinerea* and increases disease severity (71).

Management

Sanitation and canopy management are non-chemical approaches to the control of bunch rot. Removal of debris and shoots likely to be containing sclerotia should be removed. Humidity among the vines must be reduced by proper training and removal of weeds (44). The soil-borne inoculum of *B. cinerea* can be reduced by soil solarization (52). Leaf removal or green pruning is a new viticultural practice involving partial defoliation of basal portions of shoots near the clusters, performed shortly after bloom (30, 31, 90) or done 2-3 times during the season (71). This method of canopy management has been studied intensively as a technique to control Botrytis bunch rot. Wind speed through grapevine canopies increases markedly after the leaf removal (30), and the development of *B. cinerea* decreases inversely with wind speed (102). As much as 69 % disease reduction has been reported.

B. cinerea quickly develops resistance to fungicides. Therefore, so far no highly effective and durable fungicide spray schedule has been found. During the late 1950s and until 1968 captan, captafol, folpet, chlorothalonil and thiram were commonly used in vineyards to control bunch rot but the control ranged between 20 and 50%. *In vitro*, chlorothalonil and folpet fail to inhibit mycelial growth of the fungus (71). After 1968, the release of benzimidazole systemic fungicides (benomyl, thiophanate methyl and carbendazim) were found to give consistently better control. However, around 1972 resistance to MBC was reported from many countries. Natural variants of *B. cinerea* can also be resistant to benzimidazoles. However, thiphanate methyl (Topsin M) is still recommended since in absence of resistance in the pathogen it gives excellent control.

Since 1976, the new group of dicarboximide fungicides such as iprodione (Rovral) and vinclozolin (Ornalin, Ronilan, or Vorlon) have exhibited good efficacy against Botrytis diseases. In the initial years no resistance to these fungicides was reported but since 1980 resistance has been reported from many countries (51, 55, 103). Despite abundant presence of resistant strains, the efficacy of dicarboximides has not broken down as quickly as benzimidazoles. Vinclozolin is preferred over iprodione for prolonged protection. However, the economics of this treatment is not very favourable. At present, benzimidazoles, dicarboximides, dichloran, cupric hydroxide and mancozeb are the common choice of fungicides available for bunch rot control. Most of the dicarboximide-resistant strains exhibit resistance to benzimidazoles. It is, therefore, essential to use suitable alternating protectant and systemic fungicides in a schedule such as in the control of apple scab. Since the increased numbers of sprays enhance the cost of grape production, non-chemical methods stated above become especially significant. The use of neem oil, horticultural oils, silicates, phosphates, bicarbonates and sulphur dioxide is among the new approaches to reduce grey mold or bunch rot.

In developed viticulture regions, table grapes are periodically fumigated with sulphur dioxide during refrigerated storage to prevent post-harvest decay caused by many fungi. The typical commercial practice is to fumigate with 5000-10000 ppm sulphur dioxide for 20-30 min immediately after the harvested grapes enter the storage followed by weekly fumigation with 1000-2000 ppm sulphur dioxides for 30 min. In this way grapes can be kept in cold storage for 5 months. *B. cinerea* is killed at a much lower dose of 100 ppm for 20-30 min (82). Sulphur dioxide causes bleaching of some fruits but the colour returns after the fruits are exposed to oxygen. Acetalaldehyde vapours are also effective as fumigants for post-harvest treatment of grape bunches against bunch rot (2).

Biological control of Botrytis grey mold of grapes by using *Trichoderma harzianum* was reported by Elad (28).

■ BLACK ROT OF GRAPES

Black rot was described in USA as early as 1850. In 1911, Reddick (69) had published a bulletin incorporating the information on the disease available at that time. In India the disease was first reported in 1932 by J.C. Luthra. It is a destructive disease of grapes in areas where it occurs every year, particularly in warm humid regions. If proper preventive measures are not adopted during conditions favourable for the disease there may be complete loss. In dry climate the disease is not severe. The damage is through either direct rotting of berries or through blight of blossom clusters. Leaf spots are not so injurious. Necrotic lesions on stems weaken the vines but are not responsible for their death.

Symptoms

During the active growing season of the vines the infection occurs on new shoots. Numerous scattered, sometimes in groups, circular and red necrotic spots are formed on the leaves, especially on the narrower part of the lamina. The spots develop usually between the veins and are most apparent on the upper surface of the leaf. Later when the spots are about 2-6 mm or more in diameter, the main area of the spots turns brown to greyish-tan while the margins appear as a black line. With the increase in size the spots lose their circular outline. Black, dot-like structures (pycnidia) are formed on the upper side of the spots in a ring near the outer edge of the brown area. On the shoots, tendrils, petioles and flower stalks and on leaf veins, the lesions are purple or black, sometimes depressed and elongated, and bear scattered pynidia. With growth of the shoots the bark splits in the lesioned area.

Generally, the berries show symptoms when they are half grown, although sometimes younger fruits may also show symptoms. These spots are at first whitish and 1-2 cm in diameter. Soon, a rapidly widening brown ring with a black margin develops around these spots. The central area of these lesions remains flat or becomes depressed and when the spots are about 6 cm in diameter dark pycnidia develop in the centre. The whole berry soon rots and shrinks, and becomes coal black as the surface becomes covered with the black pycnidia.

The Causal Organism

Black rot of the grapes is caused by the fungus *Guignardia bidwellii* (Ellis) V.& R. of Loculoascomycetes in Ascomycotina. The fungi of this group form small perithecioid pseudothecia with small spherical locules. In *Guignardia* the mycelium is at first hyaline, much branched, and septate. With age it becomes brown in the host as well as in cultures. The sexual and asexual stages are present at the same time. In the sexual stage the fungus develops perithecium-like structures in which clusters of 8-spored, clavate and thick-walled asci arise from base of the locule. They measure 62 - 80 x 9 - 12 μm. Ascospores are aseptate, hyaline, obovoid and measure 12 - 17 x 5 - 7 μm. Paraphyses are absent. Each hyaline ascospore, just at maturity, forms a septum dividing it into two quite unequal parts.

In the asexual stage, which is of *Phoma* or *Phyllosticta*-type, ellipsoidal, hyaline conidia are produced in pycnidia formed on all parts of the host. These pycnidia have a more definite outline than perithecia and their wall has fewer cell layers. With formation of pycnidia the growth of the fungus in cultures looks black. These pycnidia measure 80-100 μm in diameter. Inside the pycnidia, conidia are formed on short conidiophores.

The conidia are hyaline, spherical to ovoid, and measure 8-11 x 4-6 μm. Towards the end of the season, groups of pycnidia form pseudosclerotia which have a thick wall made up of irregular cells. These sclerotial structures either remain sterile or get converted into pseudothecia of the sexual stage. Pycnidia also contain microconidia which do not germinate and function as spermatia to initiate the formation of pseudothecia or perithecioid structures.

Disease Cycle

The fungus survives mainly through perithecia present on mummified fruits fallen on the orchard floor. Conidia can also survive on the host. Thus, acospores and conidia both can initiate primary infection. The release of ascospores and conidia can take place only when the perithecia and pycnidia get thoroughly wet. While the ascospores are shot out forcibly and then carried by air currents, the conidia are exuded in a viscid mass from which they can be washed down or splashed away by rain. Ascospores are discharged continually throughout the spring and summer.

The primary infection, whether from ascospores or conidia, takes place on young, rapidly growing leaves and on fruit pedicels. The spores adhere to the host surface with the help of their sticky wall. The ascospores take 36-48 hours for germination. They produce a single germ tube that can cause direct penetration of the host epidermis. It produces an appressorium but no haustoria are formed. The invading hyphae are inter- and intra-cellular. The incubation period varies from 8 to 25 days. In the ensuing spots, pycnidia are formed which provide conidia for secondary infections. The conidia also germinate slowly, taking about 10-12 hours even in favourable weather. Warm and humid weather favours rapid spread of the disease. For infection, short duration rains or 2-3 days of mist after rains is ideal (85).

Management

Sanitation is important because the fallen rotting berries contain the source of primary inoculum. The yearly pruning and training of the vines generally removes the source of inoculum on the vines. Mummified berries buried in soil do not permit ascospore formation. Therefore, ploughing in early spring has been recommended as a sanitary measure. High humidity among the vines favours spore germination and infection. Canopy management to reduce humidity and encourage flow of air and permit exposure of berries to sunshine are essential.

The management of black rot is, however, primarily achieved through fungicide sprays (86). The fungicides used for downy mildew and powdery mildew control, generally help in the control of black rot also. In USA, protectant fungicides were applied every 7-10 days of new growth and upto 10-14 applications were required. With the development of systems for prediction of infection periods and efficacy of sterol biosynthesis inhibiting fungicides more effective schedules with lesser number of sprays after infection are now recommended (37). The protectant and systemic fungicides that have been in use are ferbam, captan, folpet, copper oxychloride, benomyl, and triadimephon (Bayleton 50 WP at the rate of 140 g a.i./ha). Sprays of ferbam or benomyl just before bloom, immediately after bloom, and 10-14 days later give good control of the disease. The antitranspirant compound, *gao-zhi-mo*, used as spray for control of many diseases in China, is reported to give good control of black rot also. Azoxystrobin, the strobilurin fungicide, used at 250 g a.i./ha has shown excellent preventive and curative properties. At 7 days after inoculation its efficacy was comparable with hexaconazole at 2 g a.i./ha (12).

■ ANTHRACNOSE OF GRAPEVINES

Anthracnose is also a serious disease in vineyards in many areas of the world. This disease was first reported in France in 1839 but now it is found in many countries. In India the disease is present throughout the country and is a limiting factor in the southern parts. Where it is present, vines and fruits of susceptible varieties are badly damaged. The disease causes damage to leaves, stems, and berries. It is especially serious on the new sprouts during the rainy season. In north Indian conditions the disease appears only during the rainy season. Due to restricted growth of the diseased shoots there is less fruit yield. Infection of the berries leads to rotting and this may reduce the yield by 50 %. The berries escape infection if the crop matures before the onset of rains. Vineyards in South America and South Africa have reported severe losses. During 1950-51, about 90% of the vineyards in Chile had reported losses varying from 80 to 100 %. Among various foliar diseases of grapevines in India, anthracnose has the longest spell spread over the period from June to October (92, 93, 95). The leaves show higher disease intensity than the canes.

Symptoms

All green parts of the plant can be attacked. The new shoots and fruits are particularly susceptible. On young leaves the symptoms appear as round or irregularly shaped red or dark brown spots. Later, the central portion of the spots turns grey and the margin remains brown. The central portion may separate and drop leaving a hole in the lamina. When there are too many lesions they remain small and cause the leaves to die off within 6 weeks. When they are isolated and few they rapidly increase in size, coalesce with each other forming large necrotic patches, often involving the entire lamina surface. The leaves ultimately turn into crumpled brown masses.

Symptoms are more specific on the shoots and tendrils. Large number of small, scattered, light brown spots appear on these parts. They are at first round but later become elliptical and depressed. Tissue on the margins is slightly raised and darker in colour. The central portion becomes ash-coloured. Petioles and leaf veins may show similar lesions. In warm and wet weather pinkish spore masses develop in the central area of the spots. Growth of severely infected shoots stops and the leaves are small and light green in colour. Infection of petioles and leaf veins causes curling of the leaves but they do not shed.

On the berries the disease appears as dark red spots. A single spot may increase in size to cover half the fruit surface. Generally, these spots are 6 mm in diameter. There is no softening of the flesh. However, the tissue below the spotted surface dries and turns hard. The adjacent healthy tissue continues to grow. This causes deformity of the berry. Sometimes, the skin of berry splits. Fruit stalk also shows the spots. If the infected shoots are not annually pruned, anthracnose can be seen on 2-3 years old shoots also.

The Causal Organism

Anthracnose of grapes is caused by the fungus *Sphaceloma ampelinum* de Bary (Melonconiaceae of Coelomycetes in Deuteromycotina) which is the conidial stage of the ascomycetous fungus *Elsinoe ampelina* Shear. The conidial stage has also been known as *Gloeosporium ampelophagum*.

The mycelium consists of hyaline to pale-brown septate hyphae. The conidia are formed in acervuli which appear as pink conidial masses on the spots. The acervuli are often confluent. They are composed of pale brown polyhedral cells closely packed by mutual pressure leaving no intercellular spaces. Conidiophores are 1-2 septate, cylindrical, unbranched, hyaline or pale brown and are formed from the upper cells of the acervulus. Conidia are aseptate, hyaline, oval, and measure 5–6 × 2–3 μm. In mass they look pinkish.

In the sexual stage (*Elsinoe*) perithecia (pseudothecia) are formed in old cankers on the shoots or in fallen berries. They are small and inconspicuous. Globose asci are scattered in the fruit body. Ascospores are hyaline, 3-septate, and 15–16 × 4.5 µm in size. The pseudothecia have no ostiole and ascospores are released by softening and disintegration of. the pseudothecial walls. There appears to be not much of a role of this stage in the epidemiology of grape anthracnose.

Cultural and pathogenic variations in the species have been reported (16). In studies conducted by Suhag *et al.* (96), 167 isolates from different grape cultivars were grouped into three broad subgroups (SA1, SA2, and SA3). The disease reaction on 20 commercial cultivars was studied. The isolates showed variation in production and germinability of conidia *in vitro*. The nature and intensity of leaf symptoms varied with subgroups and grape cultivars. Three cultivars, Angur Kaian, Karachi Gulabi and Nigara were resistant to two and some to only one subgroup. Thus, there is a possibility of existence of pathotypes in the species. The subgroup SA1 was most virulent and took less time for initiating necrosis than the other two subgroups on the susceptible cultivar Thomson Seedless. The culture filtrates behaved in a similar manner suggesting production of toxic metabolites capable of causing necrosis (97).

Disease Cycle

The pathogen primarily survives as mycelium in cankers on the shoots. In the next growth season this mycelium produces conidia for primary infection of new growth. It may also form pseudothecia and ascospores as primary inoculum. The spores on green tissue germinate in the presence of moisture (dew) and cause infection. The optimum temperature for germination is 30°C (95). Germination is by a germ tube and more on the lower surface than on the upper surface of the leaf. The germ tubes elongate till 12 hours after germination on the leaf surface. The tip bulges when the germ tube approaches stomata indicating appressorium formation and entry through stomata. The age of the leaf has no effect on the infection (57). The hyphae do not grow deep but grow in the cortical tissue. Under favourable conditions acervuli and conidia are formed which cause secondary infections and spread the disease. Spores are not produced on the berries. The infecton is generally limited to the cold season.

Management

Chemical control measures adopted for downy mildew and black rot control generally manage anthracnose also. The fungicides used against the grape anthracnose include benomyl, carbendazim, Difolatan, captan, thiophanate methyl, chlorothalonil, mixture of thiophanate methyl and chlorothalonil, ziram, difenoconazole, metalaxyl + folpet, dichlofluanid (Euparen), fluazinam (Shirlan), etc. (53, 99). Mixture of thiophanate-M and chlorothalonil sprays at 10-13 days interval until buds reach 5 cm length has given a high degree of protection (99). Ziram, dichlofluanid, fluazinam and captan applied from bud burst to flowering stage give good control without the risk of residue exceeding the safe limit (53). In Haryana (India), plants sprayed with 10% ferrous sulphate + 0.2% Difolatan showed lowest disease incidence on the leaves (81). Carbendazim sprays have been found very useful against anthracnose. However, the fungus develops resistance when this systemic fungicide is used excessively (100). Sanitary precautions need emphasis. Annual pruning operation and canopy management should ensure that all the shoots bearing cankers are removed (94). Resistance to anthracnose exists in some varieties of grape. In India, the cultivar Anabe-shahi is susceptible while Bangalore Blue, Isabella, Golden Muscat, Himrod, Schuyler White Beauty Seedless, Niabel, St.Valior, Large White, Golden Queen, and Muscat are resistant (96). The variety Champion is moderately resistant. The resistance in the variety Bangalore Blue has been attributed to a high level of organic acids in the leaves.

■ REFERENCES

1. Agrios, G.N. 1988. *Plant Pathology*, 3rd Ed. Academic Press.
2. Avissar, I. and E. Pesis. 1991. The control of postharvest decay in table grapes using acetaldehyde vapors. *Ann. Appl. Biol.* **118**: 229.
3. Azam, M.G.N., G.M. Gurr and P.A. Magarey. 1998. Efficacy of a compound based on canola oil as a fungicide for control of grapevine powdery mildew caused by *Uncinula necator. Aust. Plant Pathol.* **27**: 116.
4. Balamuralikrishnan, M. and R. Jeyrajan. 1997. Effect of weather factors on powdery mildew incidence in grapevine. *J. Mycol. Pl. Pathol.* **27**: 225.
5. Balamuralikrishnan, M. and R. Jeyrajan. 1998. Effect of fenarimol sprays on powdery mildew incidence in grapevine. *J. Mycol. Pl. Pathol.* **28**: 181.
6. Bavaresco, L. and R. Eibach. 1987. Investigations on the influence of N fertilizer on resistance to powdery mildew (*Oidium tuckeri*), downy mildew (*Plasmopara viticola*) and on phytoalexin synthesis in different grapevine varieties. *Vitis* **26**: 192.
7. Bharadwaj, L.N. 1997. Field evaluation of fungicides against powdery mildew of grapes under dry temperate conditions. *Plant Dis. Res.* **12**: 58.
8. Bharadwaj, L.N., I.M. Sharma and V.K. Gupta. 1997. Field evaluation of grape cultivars to powdery mildew under dry temperate conditions of Himachal Pradesh. *J. Mycol. Pl. Pathol.* **27**: 341.
9. Bhujbal, B.G., B.P. Patil and N.S. Madhane. 1982. Comparative efficacy of some fungicides against the powdery mildew of grapes. *Pesticides* **16**: 17.
10. Brooks, P.J. 1979. Effect of light on sporulation of *Plasmopara viticola. N.Z.J. Bot.* **17**: 135.
11. Bruin, G.C.A. and L.V. Edgington. 1981. Adaptive resistance in Peronosporales to metalaxyl. *Can. J. Plant Pathol.* **3**: 201.
12. Bugaret, Y., P. Sauris and M. Clerjeau. 1998. Downy mildew and black rot of grapevines: efficacy of azoxystrobin in the vineyard. *Phytoma* **504**: 69.
13. Bulit, J. and R. Lefon. 1978. Powdery mildew of vine. In: *The powdery Mildews*. D.M. Spencer, (Ed.) Academic Press.

14. Chandrasekhara Rao, K. 1989. Comparative efficacy of fungicides for the control of grape powdery mildew (*Uncinula necator*). *Indian J. Mycol. Pl. Pathol.* **19**: 200.

15. Chavan, S.B., S.V. Khandge, M.C. Varshney and J.D. Patil. 1995. Influence of weather parameters on conidia formation in powdery mildew of grape. *Indian Phytopath.* **48**: 40.

16. Cheema, S.S., S.P. Kapoor, J.S. Chohan and R. Jeyrajan. 1978. Studies on the cultural and pathogenic variations of *Sphaceloma ampelinum*, the causal organism of the anthracnose disease of grape. *Indian Phytopath.* **31**: 163.

17. Chellemi, D.O. and J.J. Marois. 1991. Sporulation of *Uncinula necator* on grape leaves as influenced by temperature and cultivar. *Phytopathology.* **81**: 197.

18. Cohen, Y. and M.D. Coffey. 1986. Systemic fungicides and the control of Oomycetes. *Annu. Rev. Phytopathol.* **24**: 311.

19. Correiar, B.R. 1999. Use of automated weather stations and the Gubler-Thomas model for control of powdery mildew (*Uncinula necator*) in California grapevines. *Summa Phytopathologica* **25**: 70.

20. Cortesi, P., D.M. Gadoury, R.C. Seem and R.C. Pearson. 1995. Distribution and retention of cleistothecia of *Uncinula necator* on the bark of grapevine. *Plant Dis.* **79**: 15.

21. Cortesi, P., M. Bisiachm, M. Ricciolini and D.M. Gadoury. 1997. Cleistothecia of *Uncinula necator*—additional source of inoculum in Italian vineyards. *Plant Dis.* **81**: 922.

22. Datar, V.V. 1986. Sources of resistance in grapevine against downy mildew caused by *Plasmopara viticola*. *Indian Phytopath.* **39**: 120.

23. Datar, V.V., D.S. Mukadam and K.B. Deshpande. 1979. Reaction of some grape varieties to powdery mildew. *Indian Phytopath.* **32**: 302.

24. Delp, C.J. 1954. Effects of temperature and humidity on the grape powdery mildew fungus. *Phytopathology* **44**: 615.

25. Delp, C.J. 1980. Coping with resistance to plant disease control agents. *Plant Dis.* **64**: 652.

26. Delp, C.J. 1984. Industries' response to fungicide resistance. *Crop Protection* **3**: 3.

27. Dercks, W. and L.L. Creasy. 1989. Influence of phosetyl-Al on phytoalexin accumulation in *Plasmopara viticola*-grapevine interaction. *Physiol. Mol. Plant Pathol.* **34**: 203.

28. Elad, Y. 1994. Biological control of grape grey mold by *Trichoderma harzianum*. *Crop Prot.* **13**: 35.

29. Emmett, R.W., T.J. Wicks and P.A. Magarey. 1992. Downy mildew of grapes, pp. 90-128. In: *Plant Diseases of International Importance.* Vol. III. *Diseases of Fruit Crops.* J. Kumar, H.S. Chaube, U.S. Singh and A.N. Mukhopadhyaya. (eds.). Prentice Hall, New Jersey.

30. English, J.T., C.S. Thomas, J.J. Marios and W.D. Gubler. 1989. Microclimate of grapevine canopies associated with leaf removal and control of Botrytis bunch rot. *Phytopathology* **79**: 395.

31. English, J.T., M.L. Kaps, J.F. Moore, J. Hill and M. Nakova. 1991. Leaf removal for control of Botrytis bunch rot of wine grapes in the Mid-western United States. *Plant Dis.* **77**: 1224.

32. English, L.G., A.P. Norton, D.M. Gadoury, R.C. Seem and W.F. Wilcox. 1999. Control of powdery mildew in wild and cultivated grapes by a tydeid mite. *Biological Control* **14**: 97.

33. Falk, S.P., D.M. Gadoury, R.C. Pearson and R.C. Seem. 1995. Partial control of grape powdery mildew by the mycoparasite *Ampelomyces quisqualis*. *Plant Dis.* **79**: 483.

34. Falk, S.P., R.C. Pearson, D.M. Gadoury, R.C. Seem and A. Sztejnberg. 1996. *Fusarium proliferatum* as a biocontrol against grape downy mildew. *Phytopathology* **86**: 1010.

35. Fermaud, M. and R. Le Menn. 1989. Association of *Botrytis cinerea* with grape berry moth larvae. *Phytopathology* **79**: 651.

36. Fermaud, M. and A. Giboulot. 1992. Influence of *Lobesia botrana* larvae on field severity of Botrytis rot of grape berries. *Plant Dis.* **76**: 404.

37. Funt, R.C., M.A. Ellis and L.V. Madden. 1990. Economic analysis of protectant and disease-forecast-based fungicide spray programs for control of apple scab and grape black rot in Ohio. *Plant Dis.* **74**: 638.

38. Gadoury, D.M. and R.C. Pearson. 1988. The use of dormant spray to control grape powdery mildew. *Phytopathology* **78**: 1198.

39. Gadoury, D.M. and R.C. Pearson. 1990. Ascocarp dehiscence and ascospore discharge in *Uncinula necator*. *Phytopathology* **80**: 393.

40. Gadoury, D.M. and R.C. Pearson. 1990. Germination of ascospores and infection of *Vitis* by *Uncinula necator*. *Phytopathology* **80**: 1198.

41. Gadoury, D.M. and R.C. Pearson. 1991. Heterothallism and pathogenic specialization in *Uncinula necator*. *Phytopathology* **81**: 393

42. Genet, J.L. and O. Vincent. 1999. Sensitivity of European *Plasmapara viticola* populations to cymoxanil. *Pesticide Science* **55**: 129.

43. Gisi, U., H. Binder and E. Rimbach. 1985. Synergistic interaction of fungicides with different modes of action. *Trans. Brit. Mycol. Soc.* **85**: 229.

44. Gubler, W.D., J.J. Marois, A.M. Bledsoe and L.J. Bettigs. 1987. Control of Botrytis bunch rot of grape with canopy management. *Plant Dis.* **71**: 599.

45. Gubler, W.D., H.L. Ypema, D.G. Quinmette and L.J. Bettiga. 1996. Occurrence and development of resistance in *Uncinula necator* to triadimefon, myclobutanil and fenarimol in California grapevines. *Plant Dis.* **80**: 902.

46. Harzog, J. and H. Schuepp. 1985. Three types of sensitivity to metalaxyl in *Plasmopara viticola*. *Phytopath. Z.* **114**: 90.

47. Kapoor, J.N. 1967. *Uncinula necator*. CMI Description of Pathogenic Fungi and Bacteria, No. 160.

48. Kosuge, T. and W.B. Hewitt. 1964. Exudates of grape berries and their effect on germination of conidia of *Botrytis cinerea*. *Phytopathology* **54**: 167.

49. Lafon, R. and J. Bulit. 1981. Downy mildew of the vine. In: *The Downy Mildews*. D.M. Spencer, (ed.). Academic Press.

50. Lalancette, N., L.V. Madden and M.A. Ellis. 1987. A model for predicting the sporulation of *Plasmopara viticola* based on temperature and duration of high relative humidity. *Phytopathology* **77**: 1699.

51. Latorre, B.A., V. Flores, A.M. Sara and R. Roco. 1994. Dicarboximide resistant isolates of *Botrytis cinerea* from table grapes in Chile: Survey and characterization. *Plant Dis.* **78**: 990.

52. Lopez-Herrera, C.J., B.Verdu-Valiente and J.M. Melero-Vera. 1994. Eradication of primary inoculum of *Botrytis cinerea* by soil solarization. *Plant Dis.* **78**: 594.

53. Magarey, R.D., R.V. Emmett and G.S. Roberts. 1995. Residues of four fungicides applied for control of grapevine anthracnose caused by *Sphaceloma ampelinum* in bud burst—flowering spray program. *Australian Plant Pathol.* **24**: 209.

54. McClelland, W.D. and W.B. Hewitt. 1973. Early Botrytis rot of grapes, time of infection, and latency of *Botrytis cinerea* Pers. in *Vitis vinifera*. *Phytopathology.* **63**: 1151.

55. Moorman, G.W., R.J. Lease and R.J. Vali. 1994. Bioassay for dicarboximide resistance in *Botrytis cinerea*. *Plant Dis.* **78**: 890.

56. Munshi, G.D. and T. Singh. 1994. Conidia to conidia cycle of *Uncinula necator* on grapes. *Indian Phytopath.* **47**: 195.

57. Naik, S.T., M.N.L. Shastry and S. Lingaraju. 1987. Host-parasite interaction of *Sphaceloma ampelinum*. *Indian Phytopath* **40**: 547.

58. Nair, N.G. and G.K. Hill. 1992. Bunch rot of grapes caused by *Botrytis cinerea*, pp. 147-169. In: *Plant Diseases of International Importance*. Vol. III. *Diseases of Fruit Crops*. J.

Kumar, H.S. Chaube, U.S. Singh and A.N. Mukhopadhyay (eds.). Prentice Hall, New Jersey.

59. Northover, J. and K.E. Schneider. 1996. Physical modes of action of petroleum and plant oils on powdery and downy mildews of grapevines. *Plant Dis.* **80**: 544.

60. Pady, S.M. and J. Subbayya. 1970. Spore release in *Uncinula necator*. *Phytopathology* **60**: 1702.

61. Pearson, R.C. 1986. Fungicides for disease control in grapes: advances in development. In: *Fungicide Chemistry, advances and practical applications*. Green and Spilker (eds.). Amer. Chemical Society Symposium. Ser. 304.

62. Pearson, R.C. and E.F.Taschenberg. 1980. Benomyl-resistant strains of *Uncinula necator* on grapes. *Plant Dis.* **64**: 677.

63. Pearson, R.C. and W. Gartel. 1985. Occurrence of hyphae of *Uncinula necator* in buds of grapevine. *Plant Dis.* **69**: 149.

64. Pearson, R.C. and A.C. Goheen (eds.). 1988. *Compendium of Grape Diseases*. Amer. Phytopathol. Soc. Press, St. Paul, Minn.

65. Pearson, R.C. and D.M. Gadoury. 1987. Cleistothecia, the source of primary inoculum for grape powdery mildew in New York. *Phytopathology* **77**: 1509.

66. Pearson, R.C. and D.M. Gadoury. 1992. Powdery mildew of grape, pp. 129-146. In: *Plant Diseases of International Importance*, Vol. III. *Diseases of Fruit Crops*. J. Kumar *et al.* (eds.). Prentice-Hall, New Jersey.

67. Pearson, R.C., D.G. Riegel and D.M. Gadoury. 1994. Control of powdery mildew in vineyards using single application vapor action treatments of triazole fungicides. *Plant Dis.* **78**: 164.

68. Rafaila, C., V. Sevcenco and Z. David. 1968. Contributions to the biology of *Plasmopara viticola. Phytopath. Z.* **63**: 328.

69. Reddick, D. 1911. The black rot of grapes. *N.Y. Agric. Exp. Sta. (Ithaca) Bull.* **293**: 289.

70. Reuveni, R. and M. Reuveni. 1998. Foliar fertilizer therapy: a concept in integrated pest management. *Crop Protection* **17**: 111.

71. R'-Houma, A., M. Cherif and A. Boubaker. 1998. Effect of nitrogen fertilization, green pruning and fungicide treatments on Botrytis bunch rot of grapes. *J. Plant Pathol.* **80**: 115.

72. Ronzan-Tran, M.S.C. and C. Clerjeau. 1988. Techniques for formation, maturation and germination of *Plasmopara viticola* oospores under controlled conditions. *Plant Dis.* **72**: 938.

73. Sackenheim, R., H.C. Weltzien and W.K. Kast. 1994. Effects of microflora composition in the phyllosphere on biological regulation of grapevine fungal diseases. *Vitis* **33**: 235.

74. Sall, M.A. 1980. Epidemiology of grape powdery mildew. A Review. *Phytopathology* **70**: 338.

75. Sall, M.A. and J. Wyrsinski. 1982. Perennation of powdery mildew in buds of grapevine. *Plant Dis.* **66**: 678.

76. Samoucha,Y. and Y. Cohen. 1985. Efficacy of fosetyl-Al in controlling metalaxyl-resistant strains of *Phytophthora infestans* and *Pseudoperonospora cubensis*. *Phytopathology* **75**: 1384.

77. Santelli,V. 1957. A physiologic form of *Plasmopara viticola* found for the first time. *Phytopathology* **47**: 3.

78. Savage, S.D. and M.A. Sall. 1984. Botrytis bunch rot of grapes: Influence of trellis type and canopy microclimate. *Phytopathology* **74**: 65.

79. Schlosser, E. 1994. Alternative control of powdery mildew on grape vine. Abstr. in *46th. Inte. Symp. Plant Prot.* Gent, Belgium, May 3, 1994.

80. Schnathorst, W.C. 1965. Environmental relationships in the powdery mildews. *Annu. Rev. Phytopathol.* **3**: 343.

81. Singhrot, R.S., J.P. Singh and L.S. Suhag. 1982. Effects of anthracnose disease on productiveness of Thompson Seedless grape. *Indian J. Mycol. Pl. Pathol.* **12**: 13.

82. Smilanick, J.L., J.M. Harvey, P.L. Hartsell *et al.* 1990. Influence of sulphur dioxide fumigation dose on residues and control of postharvest decay of grapes. *Plant Dis.* **74**: 418.

83. Smith, C.G. 1970. Production of powdery mildew cleistocarps in a controlled environment. *Trans. Brit. Mycol. Soc.* **55**: 355.

84. Sohi, H.S. and T.S. Sridhar. 1972. Notes on relative resistance and susceptibility of grape varieties to powdery mildew (*Uncinula necator*). *Indian J. Agric. Sci.* **42**: 641.

85. Spotts, R.A. 1977. Effect of leaf wetness duration and temperature on the infectivity of *Guignardia bidwellii* on grape leaves. *Phytopathology* **67**: 1378.

86. Spotts, R.A. 1977. Chemical eradication of grape black rot caused by *Guignardia bidwellii*. *Plant Dis. Rep.* **61**: 125.

87. Srinivasan, N. and R. Jeyrajan. 1976. Viability of *Plasmopara viticola* sporangia produced at different times in a diurnal cycle. *Curr. Sci.* **45**: 106.

88. Srinivasan, N. and R. Jeyrajan. 1976. Grape downy mildew in India. I. Foliar, floral and fruit infections. *Vitis* **15**: 133.

89. Srinivasan, N. and R. Jeyrajan. 1983. Influence of positions in grapevine flowers and berries on growth of downy mildew (*Plasmopara viticola*). *Madras Agric. J.* **70**: 557.

90. Stapleton, J.J. and R.S. Grant. 1992. Leaf removal for non-chemical control of the summer bunch rot complex of wine grapes in the San Joaquin Valley. *Plant Dis.* **76**: 205.

91. Staub, T. and D. Sozzi. 1984. Fungicide resistance: A. continuing challenge. *Plant Dis.* **68**: 1026.

92. Suhag, L.S. and J.C. Duhan. 1982. Studies on four pathogenic fungi on grapevine in North India. *Indian Phytopath.* **35**: 344.

93. Suhag, L.S. and R.K. Grover. 1977. Epidemiology of grapevine anthracnose caused by *Sphaceloma ampelinum* in North India. *Indian Phytopath.* **30**: 460.

94. Suhag, L.S. and B.S. Daulta. 1981. A note on the incidence and distribution of grapevine anthracnose under different systems of training. *Indian J. Mycol. Pl. Pathol.* **11**: 108.

95. Suhag, L.S., J.C. Kaushik and J.C. Duhan. 1982. Etiology and epidemiology of fungal foliar diseases of grapevine. *Indian J. Mycol. Pl. Pathol.* **12**: 191.

96. Suhag, L.S., R.K. Grover and J.C. Kaushik. 1982. Variability in *Sphaceloma ampelinum* and the performance of grapevine cultivars against the pathogen. *Indian Phytopath.* **35**: 526.

97. Suhag, L.S., R.K. Grover and J.C. Kaushik. 1984. Production of toxic metabolites by *Sphceloma ampelinum* causing grapevine anthracnose. *Indian Phytopath.* **37**: 679.

98. Sung, C.T.M. 1990. Simulation of the date of maturity of *Plasmopara viticola* oospores to predict the severity of primary infections in grapevines. *Plant Dis.* **74**: 120.

99. Thind, S.K., P.K. Monga and N. Kaur. 1995. Efficacy of agrochemicals on grape anthracnose. *Recent Horticulture* **2**: 40.

100. Thind, T.S., C. Mohan and S. Kumar. 1994. Observations on field isolates of *Gloeosporium ampelophagum* with reduced sensitivity to carbendazim in Punjab. *Indian J. Mycol. Pl. Pathol.* **24**: 46.

101. Thind, T.S., M. Clerjeau, S.S. Sokhi, C. Mohan and F. Jailloux. 1998. Observations on reduced sensitivity of *Uncinula necator* to triadimefon in India. *Indian Phytopath.* **51**: 97.

102. Thomas, C.S., J.J. Marois and J.T. English. 1988. The effects of wind speed, temperature, and relative humidity on development of aerial mycelium and conidia of *Botrytis cinerea* on grapes. *Phytopathology* **78**: 260.

103. Vali, R.J. and G.W. Moorman. 1992. Influence of selected fungicide regimes on frequency of dicarboximide-resistant and dicarboximide-sensitive strains of *Botrytis cinerea*. *Plant Dis.* **76**: 919.

104. Weltzien, H.C. and N. Ketterer. 1986. Control of downy mildew *Plasmopara viticola* on grapevine leaves through water extract of composted organic wastes. *Jour. Phytopathol.* **116:** 186..

105. Wicks, T.J. 1980. The control of *Plasmopara viticola* by fungicides applied after infection. *Aust. Plant Pathol.* **9:** 2.

106. Wicks, T.J. and T.C. Lee, 1982. Effect of fungicide volatiles on sporangia production of *Plasmopara viticola. Plant Dis.* **66:** 195.

107. Wicks, T.J. and T.C. Lee. 1982. Evaluation of fungicides applied after infection for control of *Plasmopara viticola* on grapevine. *Plant Dis.* **66:** 839.

108. Wicks, T.J. and B. Hall. 1990. Efficacy of demethomorph (CME 151) against downy mildew of grapevines. *Plant Dis.* **74:** 114.

109. Wicks, T.J. and P. Magarey. 1985. First report of *Uncinula necator* cleistothecia on grapevine in Australia. *Plant Dis.* **69:** 727.

110. Wicks, T.J., P. Magarey, M.F. Wachtel and A.W. Frensham. 1991. Effect of post-infection application of phosphorus (phosphonic) acid on the incidence and sporulation of *Plasmopara viticola* on grapevines. *Plant Dis.* **75:** 40.

111. Wicks, T.J., R. Emmett and C.R. Anderson. 1997. Integration of DMI fungicides and sulphur for the control of powdery mildew. *Austr. and N.Z. Wine Industr. J.* **12:** 280.

112. Yarwood, C.E. 1978. History and taxonomy of the powdery mildews. In: *The Powdery Mildews*, D.M. Spencer (ed.). Academic Press.

113. Ypema, H.L. and W.D. Gubler. 1997. Long term effect of temperature and triadimefon on proliferation of *Uncinula necator*: Implications for fungicide resistance and disease risk assessment. *Plant Dis.* **81:** 1187.

114. Ypema, H.L., M. Ypema and W.D. Gubler. 1997. Sensitivity of *Uncinula necator* to benomyl, triadimefon, myclobutanil and fenarimol in California. *Plant Dis.* **81:** 293.

6

Guava Diseases

- ## GUAVA ANTHRACNOSE AND FRUIT ROTS

Anthracnose is a common disease of guava, sometimes causing significant losses in the north-western part of Uttar Pradesh (India) where it was first investigated in 1951. It occurs in other parts of western India also. Although the disease causes die-back of twigs, the major damage is to the fruits which are destroyed when immature or when ripe and during transportation and storage. The anthracnose affected fruits are not fit for consumption.

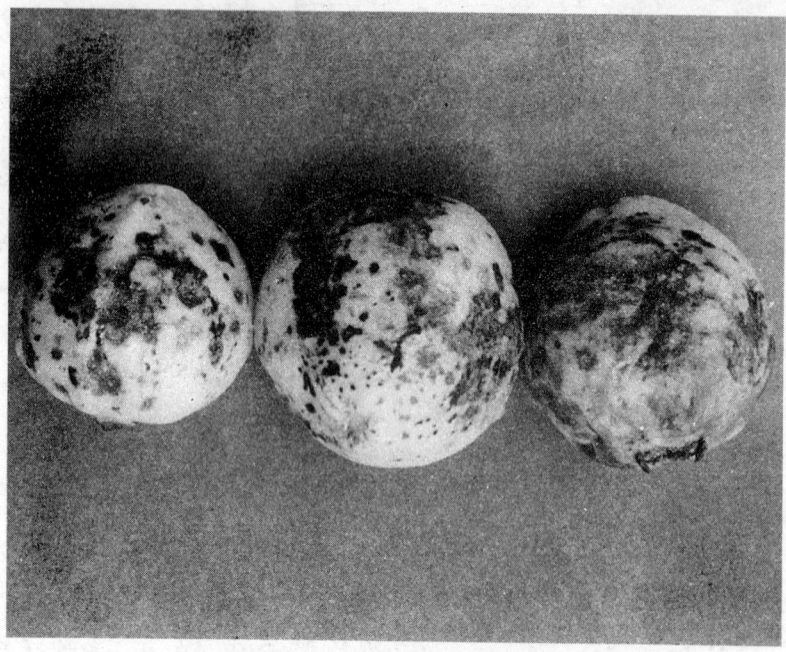

Fig. 24. Anthracnose of guava fruits.

Symptoms

Anthracnose of fruits appears during the rainy season on ripening fruits as rough blisters which coalesce to form large necrotic areas of 5-6 mm diameter. With increase in size these blisters cover a major portion of the fruit. The infected immature and ripe fruits turn brown, shrink and get mummified. These mummies hanging on the trees or fallen on the ground are most conspicuous symptom of the presence of anthracnose in the orchard. The fruits carrying incipient infection start rotting after harvest and during storage. In the post-harvest decay phase the blisters may be upto 10-20 mm in diameter. Anthracnose also affects the buds and the flowers and the immature fruits get infection from the flowers. In some cases the infected part of immature or ripe fruit turns hard and corky and shows cracks in the skin. Die-back is an additional symptom which is seen on new growth during the rainy season (30).

The Causal Organism

Anthracnose of guava is caused by *Colletotrichum psidii* (*Gloeosporium psidii*). However, most scientists now refer it as a strain of *Colletotrichum gloeosporioides* with its perfect stage *Glomerella cingulata* (26). The branched, septate mycelium is intracellular and light grey in colour. The acervuli of the fungus develop on the affected tissue. They contain short, aseptate, yellowish and smooth conidiophores and long, needle-like, dark brown or black, septate setae. Each conidiophore bears a terminal conidium. The conidia are sickle-shaped, 1-celled, hyaline, and measure $11\text{-}24 \times 4.0\text{-}5.5$ μm in size. They germinate by a germ tube which forms appressorium on the host surface. Perithecial stage of the fungus on guava has not been described.

Disease Development

The pathogen survives in or on the mummified fruits present on the trees and on the orchard floor. It can also survive in infected tissue of the branches and twigs on the tree. The conidia formed during warm and wet weather on these sources of primary inoculum cause fresh infections during the rainy season. Inoculum for the secondary spread is provided by acervuli resulting from these infections. The conidia are dispersed by wind, raindrop splashes, and also insects. Wounds caused by insects or driving rains facilitate infection. The optimum temperature for infection of immature and ripe fruits is 30°C. Maximum and minimum temperatures for the infection are 35°C and 10°-15°C, respectively. A high relative humidity of 95% is best for spread of the disease (30). Potassium deficiency promotes anthracnose (16).

Management

Orchard and tree sanitation is an important part of management strategy against guava anthracnose. Removal and destruction of the affected twigs and mummies present on the tree and on the orchard floor should be carried out immediately after the rainy season. In chemical control, Difolatan (0.2%), zineb (0.2%), Bordeaux mixture (3:3:50) and copper oxychloride or cuprous oxide (0.2-0.3%) sprayed at 10 days interval starting before the fruit set and continued during the rainy season, when weather permits, gives satisfactory control of the disease. Four to five sprays are sufficient. These fungicide sprays are effective against anthracnose as well as other serious foliage and fruit diseases such as scab and algal leaf and fruit spots. For storage, the fruits should be picked only from such treated trees. Fumigation of the fruits with sodium metabisulphite or bleaching powder gives highly satisfactory control of post-harvest anthracnose rot of fruits (26).

Fig. 25. Phytophthora rot of guava fruits.

In addition to the anthracnose fungus, post-harvest fruit rots of guava are also caused by *Botryodiplodia (Lasiodiplodia) theobromae. Phomopsis psidii, Pestalotiopsis versicolor, Aspergillus awamori* and *Rhiozopus arrhizus* which are common cause of decay of other fruits also (18, 19, 28). A fruit rot of guava caused by *Phytophthora nicotianae* var. *parasitica* is common in the wet areas of many Indian states during the rainy season. The loss is estimated to be 20-36 kg of fruits per tree (27). Circular brownish spots are formed on the green fruits at the blossom end. On the immature fruits dirty white fungal growth can cover the entire fruit surface which usually shows dry rot and mummification. However, if the fruits are reaching the stage of ripening there may be rotting of fruits on the tree otherwise the infection further develops in storage.

Benzimidazole fungicides (benomyl, carbendazim, thiabendazole) have been found effective as fruit dip for control of post-harvest decay of guava fruits (1, 2, 10, 13, 15). According to Majumdar and Pathak (15) 5-min dip of fruits in 0.1% carbendazim completely controls the Phomopsis rot and 0.1% benomyl is most effective against the anthracnose rot (25, 32). Botryodiplodia rot and Pestalotiopsis rot are effectively controlled by 0.1% imazalil when used as pre-inoculation dip and benomyl is most effective when used as post-inoculation dip. Rhizopus rot is best controlled by Cuprasol-50 (88% copper oxychloride) and 0.01% Planofix (*L*-naphthyl acetic acid). Arya, *et al.* (1) reported control of Aspergillus rot of guava by 2-min dip in 1250 ppm Bavistin or Saprol. Gamma radiation and hot water treatments of guava fruits are also reported (9, 13).

Biological control of the post-harvest decay of guava fruits has been attempted. *Trichoderma* spp. are effective biocontrol agents against *Lasiodiplodia theobromae, C. gloeosporioides, Pestalotiopsis versicolor, Phomopsis psidii* and *Rhizopus arrhizus* (14).

GREY BLIGHT, LEAF SPOTS AND SCAB

This disease appears on leaves and fruits and causes damage similar to that caused by anthracnose. Reddish brown spots of different shape and size appear on the leaves. With increase in their size the central portion turns grey or ash-coloured. The margin remains dark brown. In the central grey portion large number of dark, pin-point structures (acervuli) are formed. On green, immature fruits deep scaby spots appear which distort the fruit. These scabs may be 2-4 mm in diameter. The fruits fail to attain full size.

The disease is caused by *Pestalotia psidii* Pat. (20). Guba (8), in his monograph of the genus, had not included this species as a recognized taxon. *Pestalotia podocarpi* Denis and *Pestalotia disseminata* Thuem. with which *P. psidii* was mentioned as synonym are reported on guava. The branched septate mycelium is brown in colour. The acervuli are at first immersed and then erumpent and are composed of thick-walled, closely packed polyhedral cells. The conidiogenous area has thinner-walled cells. The conidiophores are short, simple, or irregularly branched, septate and hyaline. The conidia are 5-celled (4-septate), fusiform, straight or curved. The basal and apical cell is hyaline while the three middle cells are dark brown. The pointed end of the apical cell bears three hyaline setulae (appendages) which may be branched. These conidia measure 20-24 × 5.3 μm. The presence of 5 cells in the conidium suggests that this species may be called *Pestalotiopsis psidii* in the same manner as *Pestalotia mangiferae* (grey blight of mango) is now named *Pestalotiopsis mangiferae* on the basis of 5-celled conidia. Like *Colletotrichum* this fungus also survives in fallen leaves and fruits in the orchard. It is also present in the leaves which remain attached to the tree. Conidia produced on these sources serve as primary inoculum. The disease is managed by sanitary precautions and the spray fungicides listed for anthracnose disease (11).

■ WILT DISEASE OF GUAVA

In northern and eastern parts of India (in U.P. and West Bengal), wilt is the most destructive disease of guava. The disease was first observed in U.P. in 1935 and is now present in approximately 51,200 sq. km area of the state (22). In the central and eastern parts of the state the disease is estimated to cause an annual tree mortality of 5-10%. It appears to be a complex of fungal invasion, insect attack and nutritional disorder. Guava wilt disease is reported from other countries also.

Symptoms

The symptoms of fungal wilt caused by *Fusarium* appear during the rainy season (July-August). The affected plants are lustreless with brown leaves. These leaves remain attached to the tree for some time. The bark of the trunk is discoloured. Sometimes, the symptoms appear on one side of the grown-up tree or on some branches while the other side or other branches remain healthy with normal green leaves and normal coloured bark of the trunk on that side. In severe cases, the branches and young twigs start drying one after another. In few weeks the entire tree is dead. In the roots of the tree the xylem is blackened and fungal mycelium can be seen in the vessels in a cross section of the root. In a guava wilt disease reported from South Africa, the symptoms include an abrupt wilting of the foliage that begins on the upper branches and spreads to the whole tree. The leaves become chlorotic and, eventually, most of them absciss.

Fig. 26. Guava wilt.
Courtesy: Dr. H.B. Singh

In severe cases the leaves shrivel and die on the tree which gives a scorched appearance.

The Causal Organism

Das Gupta and Rai (4) and Prasad *et al.* (22) attributed the guava wilt disease in U.P. to infection of *Fusarium oxysporum* f. sp. *psidii*. In West Bengal, Chattopadhyaya and Bhattacharya (3) attributed the wilt to *Fusarium solani* under high soil moisture conditions and to *Macrophomina*

Fig. 27. Guava wilt. Internal discolouration of stem.

Courtesy: Dr. H.B. Singh

phaseolina in dry soils. Dwivedi (5) also has attributed the wilt to *Macrophomina phaseolina*. In Taiwan, a similar disease existing since 1923 is reported to be caused by *Myxosporium psidii* (12). In South Africa (24) the guava wilt disease (GWD) is reported to be caused by *Penicillium vermoesenii*. The pathogenicity of *Myxosporium* and *Penicillium* has been proved in the respective countries. The typical vascular wilt is, however, caused by the pathogen reported from U.P. (17, 22). Pathogenicity of *F. oxysporum* was proved by Prasad *et al.* (22). Edward (6, 7) further studied the pathogenesis of the fungus on guava and its variability.

The microconidia of the fungus are oval to ellipsoidal, 1-2 celled, hyaline, and measure 5.8-9.5×3.5-4.5 μm. The macroconidia are elongated, curved, with both ends thin, hyaline and 3- or more-septate. They measure 25.2-42×2.8-4.2 μm. The spherical chlamydospores, formed by hyphal and conidial cells, are terminal or intercalary, thick-walled, and 10-15.5 μm in diameter. Several physiologic variants of the fungus were reported by Edward (6).

Disease Cycle

The fungus survives in roots of dead trees and through chlamydospores formed by the saprophytic growth in roots. The infection is through root hair or through wounds or natural openings (7). Although the symptoms are visible during the rainy season, infection takes place much earlier and the fungus mycelium ramifies in the xylem vessels. When these vessels are blocked by fungus structures and the breakdown products of the host tissue, the passage of nutrients and water is checked. Leaves of the affected trees are deficient in nitrogen and zinc (29). The disease is more common in alkaline soils with moderate soil moisture.

Management

Recovery of the affected trees has been claimed by some scientists (29) who have used an integrated approach involving (i) pruning to remove branches 30-40 cm from discoloured conducting vessels in March, (ii) drenching of soil around the pruned trees with 10-15 litres of 0.2% carbendazim solution per tree in March, June, and September, and (iii) spray of 0.05% (a.i.) Metacystox and 0.3% zinc sulphate twice, i.e., March and September. Use of nitrate nitrogen for fertilization is useful against the disease. Application of lime or gypsum to soil (1 kg per tree) or margosa seed cake (10 kg per tree) after exposing the roots to a depth of 15 cm is also reported to check the disease. While selecting seedlings for planting in the orchard care should be taken to select only plants with normal looking leaves and bark. When the trees show advanced stage of wilting they should be cut down and all vestiges of roots removed from the soil. The same site should not be used for new planting. Resistant varieties of guava include Banarsi (Andhra strain), Dholak Sind, Nasik, White Guava No. 6299, Supreme, Clone 32-12 and Lucknow-49 (27).

■ ZINC DEFICIENCY IN GUAVA

A destructive zinc deficiency disease of guava in Pushkar valley of Rajasthan (India) was reported in 1954 (31). During a survey in 14 orchards

Fig. 28. Zinc deficiency symptom on guava leaves.

not a single plant was found healthy. It was estimated that growers were losing about Rs. 650 thousand annually. The disease was often tagged with Fusarium wilt by some workers (29). However, it is more of a decline type of disease than a typical vascular wilt.

The disease is characterized by yellowing of leaf lamina surface enclosed by veins, reduction in size of leaves, stunted growth of trees and die-back of twigs. The leaves are soft. The diseased parts do not bear flowers and if some flowers form fruits the latter dry and crack. The badly affected trees do not bear any fruit. During investigations of the disease no fungus, bacterium or virus was found in the host tissue. When 20 young diseased plants were brought from the area to Delhi and planted they started showing recovery. Spray of zinc sulphate with lime enabled the trees to recover and form healthy normal leaves.

The control of the disorder is, thus, based on zinc sulphate sprays. The spray contains 450 g zinc sulphate and 65 g of slaked lime in 72 lit water (23). Three sprays given at 2 months interval give complete recovery of the trees. Although application of 113-140 g zinc sulphate to soil per tree also gives effective control, the recovery is slow and zinc ions may interfere with the uptake of other nutrients from soil.

■ REFERENCS

1. Arya, A., D.K. Dwivedi *et al*. 1981. Chemical control of Aspergillus rot of guava. *Indian Phytopath*. **34**: 359.
2. Bhargava, S.N. and A.P. Singh. 1974. Thiabendazole storage of guava fruits. *Indian Phytopath*. **27**: 613.
3. Chattopadhyay, S.B. and S.K. Bhattacharya. 1968. *Indian J. Agric. Sci*. **38**: 65.
4. Das Gupta, S.N. and J.N. Rai. 1947. Wilt disease of guava, *Psidium guajava*. *Curr. Sci*. **16**: 257.
5. Dwivedi, S.K. 1990. Guava wilt incited by *Macrophomina phaseolina*. *National Acad. Sci. India, Lett*. **13**: 301.
6. Edward, J.C. 1960a. Variation in the guava wilt pathogen, *Fusarium oxysporum* f. *psidii*. *Indian Phytopath*. **13**: 30.
7. Edward, J.C. 1960b. Penetration and establishment of guava wilt pathogen, *Fusarium oxysporum* f. *psidii* in guava root. *Indian Phytopath*. **13**: 168.
8. Guba, E.F. 1961. *Monograph of Monochaetia and Pestalotia*. Harvard Univ. Press, Cambridge, Mass., U.S.A.
9. Gupta, J.P. and M.S. Chatrath. 1973. Gamma radiation for the control of postharvest fruit rot of guava (*Psidium guajava*). *Indian Phytopath*. **26**: 506.
10. Gupta, J.P., M.S. Chatrath and A.M. Khan. 1973. Chemical control of fruit rot of guava caused by *Colletotrichum gloeosporioides*. *Indian Phytopath*. **26**: 650.
11. Kaushik, C.D. 1972. Parasitism and control of *Pestalotia psidii* causing cancerous disease of ripe guava fruits. *Indian Phytopath*. **25**: 61.
12. Leu, L.S., C.W. Kao, C.C. Wang, W.J. Liang and S.P.Y. Hsieh. 1979. *Myxosporium* wilt of guava and its control. *Plant Dis. Rep*. **63**: 1075.
13. Majumdar, V.L. and V.N. Pathak. 1991. Effect of hot water treatment on postharvest diseases of guava (*Psidium gaujava*) fruits. *Acta Botanica Indica* **19**: 79.
14. Majumdar, V.L. and V.N. Pathak. 1995. Biological control of postharvest diseases of guava fruits by *Trichoderma* spp. *Acta Botanica Indica* **23**: 263.
15. Majumdar, V.L. and V.N. Pathak. 1997. Control of fruit rots of guava by chemical fungicides. *J. Mycol. Pl. Pathol*. **27**: 17.
16. Midha, S.K. and J.S. Chohan. 1970. Role of potassium in the pathogenesis of *Colletotrichum gloeosporioides* in guava fruits. *Indian Phytopath*. **23**: 716.
17. Pandey, R.R. and R.S. Dwivedi. 1985. *Fusarium oxysporum* f. sp. *psidii* as a pathogen causing wilt of guava in Varanasi district, India. *Phytopath. Z*. **114**: 243.
18. Patel, K.D. and V.N. Pathak. 1995. Development of Botryodiplodia rot of guava fruits in relation to temperature and humidity. *Indian Phytopath*. **48**: 86.
19. Patel, K.D. and V.N. Pathak. 1996. Temperature and relative humidity in relation to development of Rhizopus rot of guava fruit. *Indian J. Mycol. Pl. Pathol*. **26**: 9.

20. Patel, M.K., M.N. Kamat and G.M. Hingorani. 1950. *Pestalotia psidii* Pat. on guava. *Indian Phytopath.* **3**: 165.
21. Prasad, N., R.L. Mathur and I.S. Chattri. 1966. *Indian J. Agric. Sci.* **36**: 201.
22. Prasad, N., P.R. Mehta and S.B. Lal. 1952. Fusarium wilt of guava in Uttar Pradesh (India). *Nature* **169**: 753.
23. Raychaudhuri, S.P., T.K. Nariani and H.C. Joshi. 1961. Deficiency disease of guava in Rajasthan and its control. *Indian Phytopath.* **14**: 134.
24. Schoeman, M.H., E. Benade and M.J. Wingfield. 1997. The symptoms and cause of guava wilt in South Africa. *J. Phytopathol.* **145**: 37.
25. Singh, A.P. and S.N. Bhargava. 1977. Benlate as an effective postharvest fungicide for guava fruits. *Indian J. Hort.* **34**: 309.
26. Singh, J.P. and S.K. Sharma. 1982. Controlling anthracnose of guava caused by *Glomerella cingulata* by fumigation. *Indian Phytopath.* **35**: 273.
27. Sohi, H.S. 1983. Diseases of tropical and subtropical fruits and their control, pp. 81-82. In: *Recent Advances in Plant Pathology.* A. Husain, B.P. Singh, Kishan Singh and V.P. Agnihotri (eds.). Print House, Lucknow (India).
28. Srivastava, M.P. and R.N. Tandon. 1969. Postharvest diseases of guava in India. *Plant Dis. Rep.* **53**: 206.
29. Suhag, L.S. and A.P. Khera. 1986. Studies on the variation in nutritional level of wilted, regenerated and healthy trees of guava cultivar Banarasi Surkh. *Indian Phytopath.* **39**: 90.
30. Tandon, I.N. and B.B. Singh. 1969. Studies on anthracnose of guava and its control. *Indian Phytopath.* **22**: 322.
31. Vasudeva, R.S. and S.P. Raychaudhuri. 1954. Guava disease in Pushkar Valley and its control. *Indian Phytopath.* **7**: 78.
32. Wills, R.B.H., B.I. Brown and K. J. Scott. 1983. Control of ripe rot of guava by heated benomyl and guazatine dips. *Aust. J. Exp. Agric. Animal. Husb.* **22**: 437.

7

Diseases of Papaya

■ STEM AND FOOT ROT

Also known as collar rot or root rot, this is the most serious fungal disease of papaya (*Carica papaya*). Under conditions favourable for its development it can annihilate the entire plantation within one season making the soil unfit for replanting (47). It is a widespread disease in India, Sri Lanka, Hawaii and South Africa. In India, it usually appears during the rainy season (July-August). The severity of the disease depends upon the intensity of rainfall coupled with high temperature.

Symptoms

The earliest symptom of the disease is the appearance of water-soaked patches at the base of the stem. These patches expand and ultimately girdle the stem base. Due to rotting, the tissues turn dark brown or black. Simultaneously, the terminal leaves begin to droop and wilt, becoming yellow and falling prematurely. If fruits are formed they also drop. Due to disintegration of parenchymatous tissues at the base of the stem the entire plant topples down and dies. If the bark is removed, the internal tissues appear dry, brown and give honeycomb appearance. Rotting may spread above and below on the stem down to the roots. The roots deteriorate and may be destroyed.

The typical stem rot is most common in 2-3 years old plants. However, the younger plants have also been found dying due to early infection. Damping off of papaya seedlings in the nurseries is also fairly common and is caused by the same organisms. Seedlings raised in such nurseries carry the disease when transplanted in the main field and under favourable environments develop symptoms of stem rot. Often, the plants do not die quickly but linger on for some time.

The above described symptoms are for the stem and foot rot of papaya caused by *Pythium*. Additional symptoms for the disease caused by *Phytophthora* are given in the next section.

The Causal Organism

Several species of *Pythium* are reported to cause stem rot of papaya but in India and in Hawaii *P. aphanidermatum* is mainly responsible for stem rot as well as for damping off in nurseries (29, 41, 43, 46). In the Hawaii Islands, *Phytophthora parasitica* (34), subsequently confirmed as *P. palmivora* (45), is also reported to be normally associated with the disease (26).

The mycelium of *P. aphanidermatum* consists of intracellular, much branched hyphae. The entire thallus is full of oospores, oogonia, and antheridia within the host tissue and on the surface. The hyphae are 2.5-8 μm thick, hyaline, and coenocytic. However, in old cultures of the fungus and in hyphae present in old tissues irregular septation may be seen.

The fungus is soil inhabitant. The papaya residue left in the soil harbours the fungus mostly in its oospore form (47). Sugars, present in the host residue, help growth of the fungus and production of oospores. Optimum temperature for the disease development is 36° (47). Abundance of moisture around the base of the stem is conducive to disease development and its spread.

Management

The stem rot of papaya can be avoided if the plants are grown on well-drained land where water logging can be avoided. If any plant is badly diseased it should be carefully uprooted and destroyed. The same pit should not be used for replanting. Sometimes the grown-up diseased plants have been cured by removing the affected tissues and applying some fungicidal paste. Thorough drenching of the exposed soil around the stem base and also spraying the stem with 6 : 6 : 50. Bordeaux mixture or 0.2% captan considerably reduces the incidence of stem rot. The damping-off phase of the disease is effectively checked by soil treatment (drenching) with 0.2% captan before and during emergence (43) or soil fumigation with vapam (44). Seed treatment with organo-mercurials or Thiram or 0.25% Difolatan improves the emergence. Combination of seed and soil treatment is recommended for best results.

■ PHYTOPHTHORA ROOT ROT OF PAPAYA

A number of Phytophthora diseases of papaya have been reported from Hawaii, Australia, Philippines, Sri Lanka, Malaysia, Brazil, Spain and Taiwan. Phytophthora root rot reported from Hawaii, Australia, Malaysia and Taiwan is considered a serious disease causing as much as 20% or more mortality in papaya plantations. In addition to direct mortality of plants, the disease causes fruit rot and replant problem also (17, 26).

Symptoms

Root and stem infections produce symptoms similar to those caused by *Pythium aphanidermatum*. Initially, the fungus attacks the lateral roots and then proceeds to tap root in poorly drained soil during the rainy season. The whole root system decays. The leaves turn yellow, dry and hang down. Finally, only a tuft of small leaves is left at the top. Infection of the main stem can occur at any point. It causes water-soaked areas and fall of the leaves and the fruits. The destruction of the bark and underlying parenchyma is similar to that caused by *P. aphanidermatum*. The infected fruits show water-soaked lesions with a milky latex. On mature fruits, whitish fungal growth appears. The diseased fruits fall prematurely or shrivel, turn dark brown, and become mummified.

The Causal Organism

The *Phytophthora* associated with root and fruit rot in Hawaii and Australia was initially identified as *P. parasitica* (34) but later it was confirmed to be *Phytophthora palmivora* (16, 26, 45). The species has a very wide host range that includes papaya, citrus and various palms. The mycelium of the fungus is inter- and intracellular, aseptate but forms septa in old hyphae. The hyphae are large, upto 7 μm thick, and often swollen at regular intervals. The papaya isolate of *P. palmivora* produces sporangia on compact sympodial sporangiophores. The sporangia are 55×22 μm in size with a length/diameter ratio of 1.6 (16). The range of measurements on different hosts has been reported as $38\text{-}72 \times 33\text{-}42$ μm. Light is highly stimulatory to sporangia production on artificial media but not on green papaya fruits (3). The sporangia germinate in nutrient solution by producing germ tubes but in water they germinate by producing zoospores (4). Aggregations of swimming zoospores in soil water, usually at low temperature around 16° C, occur when concentration of zoospores is high (27). The zoospores are large and 8-10 μm in diameter when encysted.

   P. palmivora is heterothallic. The mating types are distributed in nature (49). For induction of sexual reproduction hormones produced by the opposite mating type play the crucial role (24). When different isolates are grown in mixed cultures oospores are formed. The antheridium is amphigynous. Although the presence of mating types of the same species was considered necessary for oospore production, reports from south India suggest possible inter-specific hybridization in plantations, where several species of *Phytophthora* parasitize closely growing different plant species, thus leading to the common occurrence of oospores (see Phytophthora diseases of citrus). On papaya also, several species of

Phytophthora such as *P. parasitica, P. cinnamomi* and *P. capsici* are reported (26). Oospores are spherical and measure 35-45 μm in diameter with 4 μm thick wall. The average diameter of oospores of papaya isolates is 23 μm (26). The oospores germinate to produce secondary sporangia (2).

P. *palmivora* produces thick- and thin-walled chlamydospores on artificial substrates (21) as well as in host tissues. These spores measure, on an average, 39 μm in diameter. In water, these spores germinate by short germ tubes which bear apical sporangia.

Disease Cycle

In papaya plantations the pathogen survives through chlamydospores, sporangia, zoospores, and oospores (when formed). More than 85% colonies recovered from a soil, known to contain the fungus, originate from chlamydospores, a few from sporangia, and rarely from oospores (26, 28). The diseased plant residue containing these structures in soil serves as an important source of primary inoculum for root infections (36). The propagules can become air-borne when the soil is splashed by falling raindrops. These propagules cause infection of the stem and the fruits. The secondary spread is caused by sporangia and zoospores formed on the primary lesions and dispersed by wind and raindrop splashes.

Rain and wind are the two most important factors in epidemiology of Phytophthora disease of papaya (or other crops). Release of sporangia and their projection into air currents is helped by rains and the wind disperses the inoculum when it reaches the air (20). Thus, the wind-blown rains are ideal for rapid spread of the disease among the plants.

Root rot is most serious during the rainy season, especially in poorly drained land (45). In water-logged conditions, defence mechanism of the roots against invasion is weakened and high mobility of zoospores under such conditions contributes greatly to the severity of the disease. Chemotactic response of zoospores is related to their mobility. Mobile zoospores are attracted to roots more easily than the non-mobile zoospores (22).

Temperature is another important determinant of disease severity. It affects the growth and sporulation of the pathogen. The optimum temperature for mycelial growth of *P. palmivora* is 30° C, the maximum 36° C and minimum 12° C (17). The optimum temperature for sporangial formation is 25° C, the maximum 35° C and minimum 15° C (3).

Management

Field sanitation is important in the management of the Phytophthora disease of papaya. Removal of the infected fruits from the trees and the orchard floor and destruction of badly affected trees reduces the soil-borne inoculum and, thus, less damage to seedling roots in the next planting as well as less chances of aerial infection of standing trees and fruits.

A virgin soil planting method (23, 25) was found very effective, inexpensive and non-hazardous in Hawaii. The seeds are planted in virgin soil in 10×35 cm holes. Since the roots become resistant when the plants are 3 mon of age, the plants growing in small islands of virgin soil grow vigorously and ward off infection.

Dry weather sprays of tribasic copper sulphate and mancozeb have been found to give protection to fruits (18) and stems. Systemic fungicides such as metalaxyl and fosetyl-Al are effective against *Phytophthora* but there are no reports of their use in papaya plantations.

■ ALGAL FRUIT SPOTS

Algal fruit and leaf spots (red rust) are very common in the wet and warm areas growing papaya. Mostly, the fruits are affected although spots, as described in mango, may also develop on papaya leaves. The spots appear as greenish-brown, reddish yellow or almost black areas of about 1 mm size. The margins of these spots are indefinite. They are dendritic and slightly raised above the host surface. The fruits in all stages of development may be affected. As the fruits grow the old spots often split the skin. These cracks facilitate the entry of secondary rot causing organisms. These spots are caused by the alga *Cephaleuros virescence*. The alga attacks many fruit trees including guava. The fruiting bodies of the alga are rusty in colour and appearance. Regular sprays of copper based fungicides at monthly interval have been recommended for control of the alga.

■ POST-HARVEST FRUIT ROTS

Post-harvest decay of papaya fruits is caused by many fungi. These include species of *Fusarium*, *Rhizopus stolonifer*, *Colletotrichum gloeosporioides*, *Colletotrichum (Gloeosporium) papayae*, *Phytophthora palmivora* and *Phytophthora parasitica*, *Lasiodiplodia (Botryodiplodia) theobromae*, *Macrophomina phaseolina*, *Ascochyta caricae*, *Phomopsis* sp. and *Alternaria citri*. Species of *Phytophthora*, *Colletotrichum* and *Alternaria citri* are the most common cause of papaya fruit rot.

The anthracnose rot (*Colletotrichum* spp.) is a major post-harvest disease of papaya throughout the tropical and subtropical areas of the world(6).

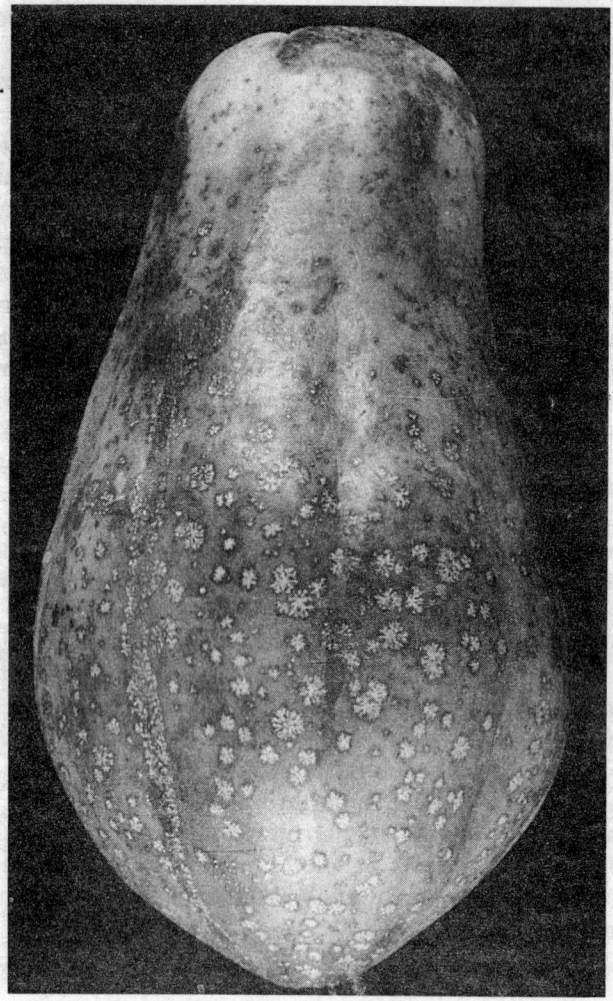

Fig. 29. Algal spots on papaya fruit.

It generally appears on half-ripe fruits but under favourable conditions fully ripe fruits are equally affected. Immature green fruits are not affected. Early season infection of fruits is correlated with post-harvest anthracnose rot of fruits (19). Small, circular, water-soaked dark spots appear on the fruit surface. As the fruits mature, these spots enlarge rapidly forming the characteristic sunken, circular lesions. These lesions may be 20-30 mm to few cm in size. Eventually, many spots coalesce forming large sunken, brown, rotting patches. Their surface becomes zonate and acervuli of the fungus appear in concentric rings. At high temperatures and in high humidity these spots progress very rapidly. The maximum rot occurs at 30°C.

The infection of fruits generally occurs through bruises and wounds, especially after harvest (42). However, spray trials (1) have suggested that the infection occurs in the field while the fruits are still attached to the plant. Ultrastructural evidence obtained from detached green fruits inoculated in the laboratory showed that *C. gloeosporioides* penetrates directly through the cuticle (10). The enzyme cutinase is involved in the process (14, 15). The cutinase deficient mutants of the pathogen are non-pathogenic to papaya fruits (13). The fungus initially establishes itself on attached immature fruits by penetrating the presumably intact cuticle of the fruit directly. The fruits could be infected at all stages of maturity but the infection remains latent until after the fruit reaches the critical stage of growth when the symptoms in the form of lesions appear whether the fruits were attached or detached (12).

Colletotrichum is effectively controlled by biweekly pre-harvest protective field sprays of fungicides. Copper based fungicides are satisfactory. Fungicides recommended for Phytophthora diseases of papaya and for anthracnose of other fruit trees can also be used. Post-harvest fruit treatments include hot water treatment at 48°-49° C for 20 min; fumigation of fruits with isothiocyanate and coating of fruits with groundnut (peanut) oil. The oil coating effectively controls the rot caused by *Alternaria, Colletotrichum* and *Aspergillus* (35). Forced hot air (48.5° C for 3-4 h) combined with thiabendazole (4 g a.i./L) fruit dip or with hot water immersion (49° C, 20 min) is reported to protect the papaya fruits from most rots such as those caused by *Lasiodiplodia theobromae, C. gloeosporioides, Mycosphaerella* sp. and *Phomopsis* sp. (33).

■ PAPAYA MOSAIC (RINGSPOT)

The papaya mosaic in India was first reported in 1948 from Maharashtra (7). Plants contact the disease at all stages of growth but are seriously affected when they are about a year old. The initial symptoms of papaya mosaic vary considerably according to age of the host, time of inoculation, season, and finally, under natural and artificial conditions of infection. The disease causes significant yield losses in papaya and cucurbits. In a recent survey (48) the highest severity of this disease was recorded in the state of Bihar (60.6%), followed by Uttar Pradesh (57.4%) and Haryana (48.8%). The loss is hundred per cent if the plants are infected in early stage of their growth.

The disease appears as profuse mottling and puckering of leaves, especially the young ones. Within 30-40 days of appearance of the first

Fig. 30. Papaya ring spot. Leaf deformity.
Courtesy: Dr. H.B. Singh

symptoms the individual plants show degeneration and a marked reduction in growth. The first infection of the plant occurs at the top leaves, the lower mature leaves showing no abnormality. However, subsequent leaves produced after the primary infection invariably show the symptoms. In the plants allowed to remain standing after infection, the symptoms appear every year with greater vigour after the summer is over. Other more conspicuous symptoms are reduced size of leaves, chlorotic and malformed appearance and defoliation of older leaves leaving only a tuft of small leaves at the top. The leaves are often modified into tendril-like structures (shoestrings). The number of lobes per leaf is increased and all these lobes are not normal but thin and distorted structures. In some leaves only half of the lamina is found to undergo these malformations while the other half remains normal.

Often, conspicuous dark green spots and elongated streaks, appearing like water-soaked areas, are formed on the petiole and stem. This symptom does not occur in many cases (31). Generally, the fruits remain much smaller than normal and are deformed. On some fruits large mosaic patches are also formed. In severe infections the plants have only a few small chlorotic or tendril-like leaves and fail to flower.

The typical papaya mosaic, described only from the USA and Venezuela, is caused by Papaya mosaic virus, a potexvirus. It is sap transmissible and no vectors are known. The so-called papaya mosaic in India is precisely the Papaya ringspot virus (PRSV). It is a definitive member of the virus genus potyvirus. Its transmission in nature is by aphids. In papaya sap, the ringspot virus has thermal inactivation point at 54°- 56°C and dilution end point at 1 : 1000. *In vitro* longevity at room temperature is 8 hours. Isolates at various places in India are reported to have thermal inactivation point at 55°C and dilution end point at 1:1000 to 1:10000 with *in vitro* longevity of 26 hours at room temperature. The particles of the virus are flexuous filaments measuring 800 × 12 nm (763 for Indian isolates). The virus is present in all aerial parts of the affected plants including flowers and pollen but is not seed-borne. Concentration of the virus in floral parts is low (30). The papaya ringspot virus is serologically related to watermelon mosaic virus although the latter does not infect the papaya. General properties of the virus are similar to those of other potyviruses.

In addition to members of the family Caricaceae (papaya) the ringspot virus can infect many cucurbit hosts such as *Cucurbita pepo*, *C. maxima*, *C. vulgaris*, *Cucumis sativus*, *Luffa acutangula* and *Trichosanthes anguina* (8). Some investigators have failed to obtain infection of *C. melo* and *C. sativus* (31). However, this host range is related to the pathotype of the virus. Since 1974, two pathotypes have been distinguished. The pathotype P is exclusively or predominantly confined to Caricaceae and pathotype W is predominantly confined to Cucurbitaceae (40). Regional variations occur in the host range of isolates.

In nature, several species of aphids transmit the papaya ringspot virus in a non-persistent manner. The most important vector is *Myzus persicae*. Others are *Aphis malvae*, *A. gossypii*, *A. medicaginis*, *A. rumicis*, *Microsiphum solonifolii*, *Rhopalosiphum maidis* and *Micromyzus formosanus*. No seed or dodder transmission is reported. In *Rhopalosiphum maidis* pre-acquisition starvation increases the transmission efficiency. For acquisition and transmission of the virus 15 min of feeding each is optimum (11). Infection increases with the number of aphids per plant, maximum being 5 apterous aphids, Adults are more efficient vectors than nymphs (11). Vectors lose the virus after feeding on the first healthy plant after acquisition.

Specific control measures are not known. Destruction of the affected plants should be undertaken as far as possible. Sprays of insecticides to keep away the vectors can reduce the spread. Resistance has not been found in any good variety of papaya under cultivation. In addition to papaya (*Carica papaya*), *C. microcarpa*, *C. cundinamarcensis* and *C. guoodontis* are also susceptible although *C. cundinamarcensis* is resistant to true papaya mosaic virus in Venezuela. In India, *C. cauliflora* is resistant (9).

■ LEAF CURL OF PAPAYA

Leaf curl of papaya is not as common as the papaya ringspot but in some areas papaya cultivation had to be stopped due to its prevalence. The disease is characterized by severe curling, crinkling and distortion of the leaves accompanied by vein clearing and reduction in size of the leaves. Leaves of the infected plants become leathery and brittle and the interveinal areas are raised on the upper surface due to hypertrophy which produces rugocity. The curled leaves have thickened veins. The petioles are twisted in a zig-zag manner. Sometimes all the leaves at the top of the plant are affected by these symptoms. The diseased plants fail to flower or bear any fruit. In advanced stage of the disease, defoliation takes place and growth of the plant is arrested.

The virus of papaya leaf curl disease is not mechanically transmitted (32). In nature, the most important agent of its transmission is the whitefly, *Bemisia tabaci*. These insects are common visitors of papaya plants. The virus can be transmitted from papaya to tobacco and tomato producing symptoms of leaf curl in these hosts and the Tobacco leaf curl virus can produce the disease in papaya when inoculated on it. On this basis it has been suggested that the papaya leaf curl is caused by the Tobacco leaf curl virus (geminivirus). Host range of the virus includes papaya, tomato, tobacco, zinnia, holyhock, *Althea*, and many other plant species (5). The transmission of the leaf curl virus in tomato has been extensively studied. The virus-vector relationship and the effect of granular insecticides on the incidence of leaf curl in tomato is also reported (38, 39). In tomato, dodder trasmission is reported (37).

The diseased papaya plants never recover. If allowed to stand in the plantation they simply help in the spread of the disease to healthy plants. Therefore, as soon as the diseased plants are detected they must be uprooted and destroyed. None of the good varieties of papaya are resistant to leaf curl. Sprays of insecticides are recommended to reduce the population of the vector.

Fig. 31. Leaf curl of papaya.
Courtesy: H.B. Singh

■ REFERENCES

1. Alvarez, A.M., J.W. Hylin and J.N. Ogota. 1977. Post-harvest diseases of papaya reduced by biweekly orchard sprays. *Plant Dis. Rep.* **61**: 731.
2. Ann, P.J. and W.H. Ko. 1988. Induction of oospore germination of *Phytophthora parasitica*. *Phytopathology* **78**: 335.
3. Aragaki, M. and R.B. Hine. 1963. Effect of radiation on sporangial production of *Phytophthora parasitica* on artificial media and detached papaya fruit. *Phytopathology* **53**: 854.

4. Aragaki, M., R.D. Mobley and R. B. Hine. 1967. Sporangial germination of *Phytophthora* from papaya. *Mycologia* **59**: 93.

5. Bock, K.R. 1982. Geminivirus diseases in tropical crops. *Plant Dis.* **66**: 266.

6. Bolkan, H.A., F.P. Cupertino, J. C. Dianese and A. Takatsu. 1976. Fungi associated with pre- and post-harvest rots of papaya and their control in central Brazil. *Plant Dis. Rep.* **60**: 605.

7. Capoor, S.P. and P.M. Varma. 1948. A mosaic disease of papaya in the Bombay Province. *Curr. Sci.* **17**: 265.

8. Capoor, S.P. and P.M. Varma. 1958. A mosaic disease of papaya in Bombay. *Indian J. Agric. Sci.* **28**: 225.

9. Capoor, S.P. and P.M. Varma. 1961. Immunity to papaya mosaic virus in the genus *Carica*. *Indian Phytopath.* **14**: 96.

10. Chau, K.F. and A.M. Alvarez. 1983. A histological study of anthracnose of *Carica papaya* fruit. *Phytopathology* **73**: 1113.

11. Cheema, S.S. and R.S. Reddy. 1985. Studies on the transmission of papaya mosaic virus by *Rhopalosiphum maidis*. *Indian J. Virol.* **1**: 49.

12. Dickman, M.B. and A.M. Alvarez. 1983. Latent infection of papaya caused by *Colletotrichum gloeosporioides*. *Plant Dis.* **67**: 748.

13. Dickman, M.B. and S.S. Patil. 1986. Cutinase deficient mutants of *Colletotrichum gloeosporioides* are non-pathogenic to papaya fruit. *Physiol. Mol. Plant Pathol.* **28**: 235.

14. Dickman, M.B. and S.S. Patil. 1988. The role of cutinase from *Colletotrichum gloeosporioides* in the penetration of papaya, pp. 175-182. In: *Experimental and Conceptual Plant Pathology*. R.S. Singh, U.S. Singh, W.M. Hess and D.J. Weber (eds.). Oxford and IBH Publishing Co., New Delhi.

15. Dickman, M.B., S.S. Patil and P.E. Kolattukudy. 1982. Purification, characterization and role in infection of an extracellular cutinolytic enzyme from *Colletotrichum gloeosporioides* in *Carica papaya*. *Physiol. Plant Pathol.* **20**: 333.

16. Ho, H.H. 1981. Synoptic keys to the species of *Phytophthora*. *Mycologia* **73**: 705.

17. Huang, T.H., D.W. Chen and L.S. Leu. 1976. Phytophthora fruit and root rot of papaya in Taiwan. *Plant Prot. Bull. (Taiwan)* **18**: 293.

18. Hunter, J.E. and I.W. Buddenhagen. 1969. Field biology and control of *Phytophthora parasitica* on papaya in Hawaii. *Ann. Appl. Biol.* **63**: 55.

19. Hunter, J.E. and I.W. Buddenhagen. 1972. Incidence, epidemiology and control of fruit diseases of papaya in Hawaii. *Trop. Agric. (Trinidad)* **49**: 61.

20. Hunter, J.E. and R.K. Kunimoto. 1974. Dispersal of *Phytophthora palmivora* sporangia by wind blown rain. *Phytopathology* **64**: 202.

21. Kadooka,, J.Y. and W.H. Ko. 1973. Production of chlamydospores by *Phytophthora palmivora* in culture media. *Phytopathology* **63**: 559.

22. Klienjunas, J.T. and W.H. Ko 1974. Effect of motility of *Phytophthora palmivora* zoospores on disease severity in papaya seedlings and substrate colonization in soil. *Phytopathology* **64**: 426.

23. Ko, W.H. 1971. Biological control of seedling rot of papaya caused by *Phytophthora palmivora*. *Phytopathology* **61**: 780.

24. Ko, W.H. 1978. Heterothallic *Phytophthora*: Evidence for hormonal regulation of sexual reproduction. *J. Gen. Microbiol.* **107**: 15.

25. Ko, W.H. 1987. Biological control of Phytophthora root rot of papaya with virgin soil. *Plant Dis.* **66**: 446.

26 Ko, W.H. 1992. Phytophthora root rot of papaya, pp. 300-307. In: *Plant Diseases of International Importance*. Vol. III. *Diseases of Fruit Crops*. J. Kumar *et al.* (eds.). Prentice-Hall, New Jersey.

27. Ko, W.H. and L.L. Chase. 1973. Aggregation of zoospores of *Phytophthora palmivora*. *J. Gen. Microbiol.* **78**: 79.

28. Ko, W.H. and M.J. Chan. 1973. Infection and colonization potential of sporangia, zoospores, and chlamydospores of *Phytophthora palmivora* in soil. *Phytopathology* **63:** 1307.

29. Luna, L.V. and R.B. Hine. 1964. Factors influencing saprophytic growth of *Pythium aphanidermatum* in soil. *Phytopathology* **54:** 955.

30. Marathe, T.S. and A.S. Summanwar. 1984. Detection of papaya mosaic virus in different plant parts of papaya. *Indian Phytopath.* **26:** 502.

31. Mishra, J.N. and A. Jha. 1955. Mosaic of papaya in Bihar. *Proc. Bihar Acad. Agric. Sci.* **4:** 102.

32. Nariani, T.K. 1956. Leaf curl of papaya. *Indian Phytopath.* **9:** 151.

33. Nishijima, K.A., C.K. Miura, J.W. Armstrong, S.A. Brown and B.K.S. Hsu. 1992. Effect of forced hot air treatment of papaya fruit on fruit quality and incidence of post-harvest diseases. *Plant Dis.* **76:** 723.

34. Parris, G.K. 1942. *Phytophthora parasitica* on papaya *(Carica papaya)* in Hawaii. *Phytopathology* **32:** 314.

35. Pathak, V.N. 1997. Post-harvest fruit pathology: Present status and Future possibilities. *Indian Phytopath.* **50:** 161.

36. Ramirez, B.N. and D.J. Mitchell. 1975. Relationship of density of chlamydospores and zoospores of *Phytophthora palmivora* in soil to infection of papaya. *Phytopathology* **65:** 780.

37. Reddy, K.S. and R.C. Yaraguntaiah. 1979. Transmission of tomato leaf curl virus through dodder. *Indian Phytopath.* **32:** 653.

38. Reddy, K.S. and R.C. Yaraguntaiah. 1981a. Virus-vector relationship in leaf curl disease of tomato. *Indian Phytopath.* **34:** 310.

39. Reddy, K.S. and R.C. Yaraguntaiah. 1981b. Effect of granular insecticides on the tomato leaf curl virus disease. *Indian Phytopath.* **34:** 291.

40. Roy, G., R.K. Jain, A.I. Bhat and A. Varma. 1999. Comparative host range and serological studies of papaya ringspot potyvirus isolates. *Indian Phytopath.* **52(1):** 14- 17.

41. Singh, R. S. 1997. *Plant Diseases*, 7th Ed. Oxford and IBH Publishing Co., New Dellhi.

42. Stanghellini, M.E. and M. Aragaki. 1966. Relation of periderm formation and callose deposition to anthracnose resistance in papaya fruit. *Phytopathology* **56:** 444.

43. Tandon, I.N. 1959. Studies in the control of damping off of papaya. *Hort. Adv.* **3:** 115.

44. Tandon, I.N. 1963. Control of post-emergence damping off of papaya with vapam. *Indian Phytopath.* **16:** 44.

45. Teakle, D.S. 1957. Papaya root rot caused by *Phytophthora palmivora*. *Queensland J. Agr. Sci.* **14:** 81.

46. Trujillo, E.E. 1965. The role of Pythiaceous fungi in the papaya replant problem in Hawaii. *Phytopathology* **55:** 126.

47. Trujillo, E.E. and R.B. Hine. 1965. The role of papaya residue in papaya root rot caused by *Pythium aphanidermatum* and *Phytophthora parasitica*. *Phytopathology* **55:** 1293.

48. Verma, A.K. 1997. Status of viral diseases of papaya *(Carica papaya)* in the northern plains of India, p. 103. In: *Int. Symp. Zonal Meeting IPS, Kanpur,* Dec.19-20, 1997.

49. Zentmyer, G.A., D.J. Mitchell, L. Jefferson, J. Roheim and D. Carnes. 1973. Distribution of mating types of *Phytophthora palmivora*. *Phytopathology* **63:** 663.

⇘ **8**

Strawberry Diseases

■ ANTHRACNOSE

Anthracnose is a major cultural problem in strawberry cultivation and one or more of its different phases cause severe losses throughout the world (28, 29). The term "strawberry anthracnose" is used as a general term to identify all diseases of strawberry caused by the species of *Colletotrichum*. These include anthracnose, crown rot, bud rot, flower blight, fruit rot, black leaf spots, and irregular leaf spots (21).

Symptoms

Anthracnose
Lesions on stolons and petioles are typical symptoms of anthracnose. These lesions are sunken, firm, dark, dry, and with a sharp line of demarcation between the diseased and symptomless tissue. When a lesion girdles a runner, the uprooted daughter plants beyond the lesion wilt and die. The lesions may continue to elongate until the entire runner or petiole is infected. When a petiole is girdled, the leaf dies and turns brown. The lesions often form on the underside of the petiole. The petiole then bends sharply at this point and the leaf hangs down. Such leaves may continue to live in this position and remain green for an extended period of time.

Crown rot
The crowns of plants in the nursery become infected when the fungus grows into the crown from the runners or petioles or when sufficient number of spores germinate in the central bud. The spores are splashed or washed into the central bud from runner or petiole lesions. The crown rot affected plants may die in the nursery or after being transplanted in the main field. Such plants may grow normally for some time but then suddenly wilt and die. When the infected crown is cut lengthwise a reddish brown firm rot or reddish brown streaking in the interior is seen. The reddish tint is a diagnostic feature (18).

Bud rot

Bud rot develops within a few days after transplanting or later after the plants have become established and have formed multiple crowns. In either case a damp, firm rot develops and the infected bud turns dark brown to black. At the time of transplanting, the plant has a single bud and if this single bud decays soon after transplanting the entire plant is lost. In multicrowned plants, as the rot progresses the entire plant may die or only the infected crown dies and the rest continue growing. Pink masses of *Colletotrichum* conidia are formed on the decaying tissue. A longitudinal cut through the infected bud and the crown exposes a sharp line of demarcation between the dark infected bud tissue and the crown tissue below the bud. The interior portions of the crowns of these plants appear white and healthy (31).

Flower blight

In some varieties of strawberry, flowers and flower buds are highly susceptible to anthracnose and may become infected at any time after the bud begins to emerge from the crown until the flower is fully open. When the newly emerging buds are infected the sepals dry and turn dark brown to nearly black. As the stem elongates, the entire bud becomes infected, dries and turns light brown. When the flower bud is infected after the stem has begun to elongate, a dark lesion forms on the calyx or on the stem just below the calyx. The infection spreads throughout the bud which dries and becomes light brown before or after the flower opens. When an open flower is infected, the pistil and ovules turn black and the calyx and a portion of the stem become tan or light brown and dry. Sap exudation in the form of a sticky, gelatinous mass may occur on the stems of the flowers and the flower buds. This mass may contain conidia of the fungus.

Fruit rot

Round, firm, sunken spots develop on the ripening fruits. They usually become black but sometimes remain tan for a few days, especially in wet weather (41). The spots may enlarge until the entire fruit is affected. Then the fruits dry and are mummified. In severe attacks in the field the green fruits in all stages of development are also affected. Lesions on green fruits are dark brown to black and hard. Some seeds in anthracnose-affected fruits have minute crusty masses of spores.

Black leaf spot

These spots are round and range in diameter from 0.5 to 1.5 mm, occasionally up to 3 mm. Lesions are usually black but may remain light grey. Leaflets can have numerous spots without dying. These spots develp before symptoms on the runners and petioles develop and, thus, can serve as warning for impending disease development.

Irregular leaf spot
These dry lesions are dark brown to nearly black and occur on the margins and tips of the leaves. They tend to elongate along the margin up to 13 mm with irregular margins.

The Causal Organisms

Colletotrichum fragariae Brooks, *C. acutatum* Simmonds, *C. gloeosporioides* Penz, *C. dematium* (Pers.) Grove, and *Gloeosporium* sp. are reported to be associated with strawberry anthracnose in different regions (5, 6, 21). *C. fragariae* and *C. acutatum* are most common. *C. fragariae* fits within the group species *C. gloeosporioides* and Von Arx (46) had included it in this group. However, many later reports suggest it a separate species, mainly because it has not yet been found to produce ascigerous stage in culture, although many isolates of *C. gloeosporioides* on different hosts also do not produce the *Glomerella cingulata* perithecia in culture.

Smith and Black (43) conducted a detailed study of 24 isolates of *Colletotrichum* from strawberry to compare their conidia, appressoria, cultural characteristics and setae. Thirteen isolates were identified as *Colletotrichum fragariae*. They produced cylindrical conidia. Two isolates, identified as *Glomerella cingulata* (*C. gloeosporioides*), developed cylindrical conidia but formed perithecia in culture. The remaining 9 isolates were identified as *C. acutatum*. They produced fusiform conidia but did not form perithecia in culture. The cylindrical conidia of *C. gloeosporioides* and *C. fragariae* isolates were pointed on one end and measured 12.9–16.1×4.4–5.4 µm in *C. gloeosporioides* and 12.4–15.0×4.4–5.2 µm in *C. fragariae*. Generally, the conidial appressoria of both these species were similar and slightly more lobed and clavate than those of *C. acutatum*. The fusiform conidia of *C. acutatum* tapered to a point at both ends and measured 12.3–14.7×4.6–5.3 µm. Setae were not formed by *C. acutatum* and *C. gloeosporioides* in culture and on the host. All isolates of *C. fragariae* produced setae in culture and on the host. These setae were 1- or 2-septate and varied in length from 45 to 107 µm. In plate cultures at 32°C, the radial growth was 69, 63, and 13 mm, respectively, for *C. fragariae*, *C. gloeosporioides* and *C. acutatum*. In artificial inoculations the *C. fragariae* isolates caused leaf and fruit lesions while *C. acutatum* caused fruit rot but no leaf lesions. All isolates of *C. fragariae*, four of the five isolates of *C. acutatum* and one of the two isolates of *C. gloeosporioides* caused a crown rot in certain strawberry cultivars. The most severe petiole and crown symptoms were caused by *C. fragariae* isolates, less by *C. acutatum* isolates and least by *C. gloeosporioides* isolates. However, cultivar response differed in regard to these symptoms. There is strong evidence for existence of races in *C. fragariae* (10).

Disease Cycle

Survival of anthracnose fungi in the absence of the host is not clearly understood. All the three major species have some non-strawberry hosts on which they can survive. *C. fragariae* does not survive in soil (19). *C. acutatum* is reported to survive in soil for more than 7 months as appressoria, chlamydospore-like hyphal cells and conidia (11). The level of survival was 100% during winters (November to March) but quickly declined in the spring and summer. The survival is highest in cool (10°C) dry soils (12). The mummified fruits on soil surface or 5-8 cm below soil surface are a source of survival of this species during the winter but later the recovery declines (29). The decay of mummies in soil leads to reduction of inoculum source. Obviously, where there is a long gap between two successive strawberry crops and temperatures are high the soil survival does not play much important role. Since strawberry is propagated through runners and crown, these infected parts carry the fungus to new plantings (19). Conidia formed on these plants are splashed by rain to other plants or parts of the same plant and thus cause spread of the disease (26, 50). Latent period (time of first sporulation) varies from 2-3 days at 25°C to 6-17 days at 5°C. *C. acutatum* has a shorter latent period than other species at 5°C and 10°C. It also produces more conidia at these temperatures than other species (23). The pathogens are favoured by temperatures above 24°C and prolonged wetness of host surfaces (48). The optimum temperature range for sporulation is 22°–26°C. The temperature of 32°-35°C is good for the infection of seedlings (42). Two to four weeks old strawberry seedlings (age after transplanting) are more susceptible than 14-18 weeks old seedlings (44).

Management

The best combination of measures for control of strawberry anthracnose is the use of disease-free planting stock (29), removal and destruction of mummies, fumigation of the nursery and main field soil with a mixture of methyl bromide and chloropicrin, and regular fungicide sprays in the nursery. Among the fungicides, captan, captafol, folpet, prochloraz-Zn, prochloraz-Mn, combination of prochloras-Zn and folpet, propiconazole, difenoconazole and benomyl have been used for spray (9, 14, 20). Resistance to benomyl and dicarboximides in *C. fragariae* has been reported especially where these fungicides have been extensively used. Benzimidazole resistance is more common than dicarboximide resistance (25). Heat treatment (5 min at 49° C) is also reported to be as effective as some of the best fungicides.

Strawberry cultivars vary in susceptibility to anthracnose of petioles, stolons, crowns, leaves, flowers, and fruits. The cultivar resistant to one fungus is generally resistant to the other two also. Resistance to one phase of the disease makes the cultivar resistant to other phases also. However, no cultivar is resistant to all forms or races of *C. fragariae*, *C. acutatum* and *Glomerella cingulata*.

■ MYCOSPHAERELLA LEAF SPOTS

Leaf spot diseases are found in all strawberry growing regions of the world. They cause loss in yield and poor quality of fruits. Several parasitic fungi are reported to cause leaf spots. These include *Mycosphaerella fragariae*, *Dendrophoma obscurans*, *Hainesia lythri* (*Pezizella lythri*) and *Macrophomina phaseolina*. Mycosphaerella and Dendrophoma leaf spots are more common and sometimes destructive in certain areas.

Symptoms

The typical leaf lesions caused by *Mycosphaerella* appear as small, purplish, circular spots on the upper surface of young leaflets. When the lesions enlarge (2.5-3.0 mm diameter) the central portion becomes grey to white and is surrounded by a broad, reddish purple margin. If the infection is light, the spots are scattered on the lamina surface but in severe infections large number of spots coalesce and give blighted appearance to the leaf. In Dendrophoma blight, the central portion is /dark grey, surrounded by a light brown area and finally there is the purple margin. The central portion of the spots shows the presence of pin-point black pycnidia of the fungus. The Mycosphaerella spots may occur on petioles, fruit pedicels, suckers and calyx. The fruits may also be attacked and develop black spots.

The Causal Organism

Mycoospphaerella leaf spots are caused by *Mycosphaerella fragariae* (Tul,) Lind. (anamorph *Ramularia tulasneii* Sacc.). In the conidial stage the fungus produces conidia on short conidiophores in pycnidia formed on the central necrotic area of the lesions on leaves, petioles, pedicels and calices. The conidia are oblong, slender, hyaline, 0-4 septate and measure

20-40 × 3-4 μm. In the *Mycosphaerella* stage the loculate ascostromata vary considerably in size (88-180 μm). They are black, partly embedded, erumpent, globose, without ostiole, and mostly epiphyllous. The asci are cylindrical to clavate, fasciculate, borne on short stalks and measure 30-40 × 10-15 μm. The ascospores are hyaline, 1-septate and 12-15 × 3-4 μm in size. A spermogonial stage is also reported. During off-season the fungus may produce sclerotia which resemble ascostromata in shape and size but have no locules.

Disease Cycle

In areas where strawberry remains in the field all round the year, conidia on the diseased plants serve as primary inoculum. At other places, *M. fragariae* survives in soil through sclerotia and ascostromata (2, 38). Spores formed by these structures cause primary infection in the new plantings. The secondary spread is by conidia dispersed by rains or by contact between the healthy and diseased leaves. Ascospores are air-borne. The infection occurs on both surfaces of the leaves. Penetration is either direct or through stomata.

The progress of lesions is a slow and steady process and is influenced by environmental variables like high temperature and high humidity. Saharan and Badiyala (39) recorded maximum increase in the number of lesions on different cultivars when the maximum temperature was 25.5°C, minimum 19.3°C, mean temperature 23.4°C and relative humidity 87.4%, with rainfall of 39.9 mm in 7 days of rains. Fall in minimum temperature, mean relative humidity and number of rainy days reduced the increase in the size of lesions.

Management

Destruction of crop refuse in the field reduces the soil-borne inoculum. Spacing between plants and proper drainage prevent high humidity. One or two sprays of a suitable fungicide during spring provide satisfactory control of the disease. Although in the past copper based fungicides had been used, they were not very effective and left spots on the fruits. Later, 0.2% captan was used. Carbendazim (0.1%), benomyl (0.1%), captafol (0.2%) and ziram (0.2%) are currently recommended for control of strawberry leaf spots. Saharan and Badiyala (39) had reported that no cultivar is free from leaf spots in Punjab (India). However, varieties Elista and Tioga are highly resistant while Florida-90, Senga Sengana, Howard-17 and Blackmore are highly susceptible. The degree of resistance was judged from lesion numbers as well as lesion size. Earlier, Sindhan and Roy (40) had reported varieties and selections Albritton, Jem, Macherauch,

Jacunda, Premier, Dilpasand, Cavalier, Pusa Early Dwarf, Howard-17, CH-III-12, and CH-III-40 as resistant in Kumaon Hills of India. Albritton, Premier, Cavalier and Dilpasand were resistant to both Mycosphaerella spots and Dendrophoma blight.

■ RED STELE OF STRAWBERRY

Red stele or red core, brown core, black stele and Phytophthora disease is a major factor limiting strawberry fruit production in cooler regions around the world. It was first investigated in Scotland in 1926 but is now known to occur in all countries of the world (33). Once the field gets infested with the pathogen it becomes unfit for strawberry cultivation for many years. The damage to strawberry during the first 2-3 years of soil infestation is not very heavy but later there are severe losses. Initially, the infections start on plants on the ill-drained portion of the field but soon they spread to the well-drained area also.

Symptoms

The diseased plants remain stunted and in ill-drained soil they may be killed. Older leaves dry and wither while new leaves remain small sized. The petioles are short. Leaves on the stunted plants look bluish green while the older leaves are yellowish, reddish or brown tinged. At high temperatures the entire plant or its leaves wilt. The badly affected plants either do not produce fruits or if fruits are formed they are undersized and dry before ripening.

The pathogen first destroys the feeder roots or rootlets. If the diseased plants are uprooted, the main root is found smooth without laterals. After entering the main root the pathogen grows upward through the central cylinder. When the stele is invaded its tissue appears reddish brown which can be seen in a longitudinal split of the main root. This discoloured portion can be easily distinguished from the surrounding white tissue of the cortex. The cortical tissues also later decay either due to the same fungus or due to attack of secondary saprophytic organisms. The ends of the roots turn brown or black and die from the tip backward.

The Causal Organism

Red stele of strawberry is caused by *Phytophthora fragariae* Hickman (3, 16). Sporulation of the fungus occurs on the roots. The sporangia are inverted pear-shaped, without papilla, and measure 32-90 × 22-52 μm. Mostly the sporangia germinate by producing zoospores and sometimes by a germ tube. The zoospores are of irregular shape or ellipsoid, biflagellate and 12 μm in dia in the encysted stage. The oospores are found in vascular bundles of the roots. Oogonia are terminal or lateral and mostly spherical. The base is funnel-shaped. These oogonia are golden brown in colour and measure 28-44 μm in dia. Antheridia are mostly terminal, sometimes intercalar, paragynous, and 16-30 × 12-23 μm in size. The oospores are mostly spherical and 22-44 μm in dia. They are hyaline. The fungus grows best at 20°C. At 10°C there is good growth and at 4°C there is very slow growth. The fungus ceases to grow at 30°C.

Atleast 7 races of the fungus are known (8, 30). These races are separated by their ability to infect and produce oospores in roots of differential strawberry cultivars (32). The level of host susceptibility to a given race is also quantified by the number of oospores produced in infected roots.

Disease Cycle

The fungus survives through its oospores in roots and crown of infected plants which are kept for new planting. The pathogen is introduced into new areas by planting infected suckers (17). Sometimes, suckers with no visible symptoms carry the fungus to new plantings. Zoospores in sporangia produced by infections from oospores spread the disease through movement of zoospores in soil water. The zoospores cause infection through root hair (15).

Management

Freedom of the planting stock from visible infection or presence of propagules with soil is a necessary precaution. Hence, the planting stock should be taken from a field where the pathogen is not present. Good drainage should be ensured by proper levelling of the field. Fumigation of the soil with a mixture of methyl bromide and chloropicrin is also recommended. This ensures protection from many other diseases of strawberry. The fungus attacks only the strawberry and a rotation of 3-4 years considerably reduces the inoculum load in the soil.

Excellent control of root and collar diseases caused by *Phytophthora* has been achieved through several systemic fungicides such as metalaxyl, fosetyl-Al, ethazol and propamocarb. Dipping of planting stock in any of these fungicides can eliminate the pathogen. However, the danger of resistance development in the fungus against systemic fungicides limits their use.

■ BLACK ROOT ROT COMPLEX

In many strawberry growing regions of the world black root rot is a serious disease which has, so far, defied the normal control measures because of the multiplicity of the causal agents. Root necrosis is the typical symptom of the disease. It starts with blackening and necrosis of cortical tissues and death of feeder rootlets. Eventually, there is partial or total rot of the root system. The plant growth is stunted. The infected adult plants remain dwarfed. The dwarfing or stunting depends on the extent of the root decay. If young plants are attacked they are not dwarfed but lack vigour and in hot weather they droop and finally die. Such plants do not bear fruits. If fruits are formed they dry before ripening. In early stages, the main roots show brown lesions and laterals are dead and black. In the final stage the entire root system turns black and dies. The affected roots become corky.

The black root rot of strawberry has been attributed not to one causal agent but to a complex of several fungi and a nematode. The common fungi isolated from diseased roots include *Cylindrocarpon destructans*, *Pythium ultimum*, *Pythium irregulare*, *Rhizoctonia solani*, binucleate *Rhizoctonia* spp., *Phoma* sp., and *Alternaria* spp. No vascular wilt organism has been detected (51). Among these fungi *Pythium ultimum* (47) and binucleate species of *Rhizoctonia* (27) have been found pathogenic on strawberry but they alone do not cause the black root rot. Soil treatment with metham sodium (Vapam) or metalaxyl (effective against *Pythium*) fails to reduce the disease or improve root development and yield although it modifies the microbial population densities in soil and rhizosphere (51). Thus, other pathogens, in addition to *Rhizoctonia* and *Pythium* are also involved. Binucleate species of *Rhizoctonia* are frequent on roots of old strawberry plants. The lesion nematode, *Pratylenchus penetrans*, also causes root rot of strawberry (7) and increases severity of root rot caused by binucleate *Rhizoctonia* spp. (24). The disease is common in severe cold and some reports suggest that the cold also is a contributing cause. Management of black root rot involves crop rotation, selection of healthy planting stock, and proper drainage. Incomplete and temporary reduction in disease incidence by application of metalaxyl or vapam is reported. The best control has been achieved by soil fumigation with a mixture of methyl bromide and chloropicrin. Although these fumigants are declared as restricted use pesticides because of possible health and environmental hazards, there is no alternative (51).

■ FUNGAL ROTS OF STRAWBERRY FRUITS

Strawberry fruits are very succulent and highly perishable. In the field as well as in the containers or in the market they are liable to attack of many fungi that induce post-harvest fruit rots. Fruit rot is the most destructive disease of strawberry. If the disease occurs in the field it creates problems in picking and packing and carries the infection for post-harvest rot during transit and in the market. In countries lacking proper systems of picking, packing, and marketing, fruit rot is especially a serious problem. In a season of heavy and continuous rains 50-75% losses may occur.

Grey Mold Rot

The grey mold rot caused by *Botrytis cinerea* is very similar to the bunch rot of grapes. It is also known as Botrytis rot, brown rot of strawberry, and dry rot and is common in cool climates. The symptoms of the disease start from that part of the fruit which is in contact with soil, dead and fallen diseased leaves, and rotting fruits in the packing containers during transportation. The decaying tissue turns light brown and soft but there is no leakage of fluid from the fruits. The rot rapidly spreads on the entire fruit which dries and becomes hard. Greenish or grey fungal growth develops on such fruits. This growth consists of conidia of the fungus. The infection can occur any time from the growth to the ripening of the fruit. Infection also occurs on leaves, calices, pedicels and flowers. The symptoms are similar to those on grapes (1, 2, 38).

Sexual stage of *Botrytis cinerea* is not common but sclerotia are formed on dried strawberry fruits and other parts of the host. Dormant mycelium and sclerotia present on the affected plant parts and in crop residue left in the field are sources of primary inoculum. Conidia are dispersed by wind, rains and by hands of pickers. In cool (18°-23°C), humid weather, especially in continuously cloudy weather, the conidia germinate and the germ tubes directly penetrate the host tissue. Mycelium can also grow and cause infection. Anthers are the main route of entry into the developing receptacle of the fruit. In strawberry cultivars that lack anthers, the amount of infection by *B. cinerea* is considerably reduced. When there is good sunshine and atmospheric humidity is low the chances of Botrytis grey mold rot are also low.

Field sanitation practices such as removal of crop debris, proper spacing between plants to permit good air circulation and sunshine, mulching of the open space between rows reduce grey mold rot in the field. Mulching with straw avoids contact between the fruits and the soil. Chemical control of grey mold rot in the field has been only partially successful, especially when the weather is favourable for the pathogen. Sprays or dusts of captan (0.2%), thiram or benomy are recommended (22). The first spray of captan is given when the flowers are opening. Two more sprays are given until the fruits develop colour. The contact fungicides iprodione and vinclozolin are excellent fungicides against Botrytis rot. The fungus has a tendency to develop resistance to benomyl, iprodione, or even captan .(25, 34, 37). Benzimidazole resistance is more common than resistance to carboximodes. Therefore, different fungicides or fungicide combinations are used in different sprays.

In biological control, *Gliocladium roseum* is reported as a versatile adversary of *Botrytis cinerea*. It suppresses the pathogen by more than 98% on leaves, petals and stamens, better than even captan. In field plots weekly sprays are required. The suppression is through competition for nutrients (45). Extracts of *Allium* and *Capsicum* are highly effective against spore germination of *B. cinerea* (49). Oil of palmarosa (*Cymbopogon martinii*), cinnamon leaf and clove buds are also very highly antifungal against the pathogen.

Botrytis rot of strawberry fruit after the harvest may arise from field infections. This is not readily controlled by chemical treatments. Well-timed fungicide sprays in the field are, therefore, necessary. Post-harvest dip of fruits in 0.1 - 1.0% captan and pre-harvest sprays of captan in the field are recommended chemical control measures for post-harvest decay of strawberry. Bicarbonates have been used against many post-harvest decay fungi and *in vitro* studies have shown that bicarbonates inhibit the colony growth of *B. cinerea* also (36). Exposure of fruits to hot air at 43° C (98% relative humidity) for 30 min controls the grey mold rot as well as Rhizopus rot (4).

Leak or Rhizopus Rot

Leak is a transit and market disease of strawberry and many other fruits and vegetables. It may occur in the field also. Rhizopus rot can cause serious losses in areas where quick and hygienic means of transport are not present and relatively warm temperatures exist. When conditions are favourable, the disease spreads rapidly throughout the containers and losses can be very high within a short period of time.

The symptoms of strawberry leak differ from other rots. The fruits do not show change of colour in early stages but later they change to light brown and become soft and watery. The skin easily ruptures during picking, transportation and other handling operations and the fruit juice leaks out. In containers or packing cases the rotting fruits are covered with a hairy, white fungal growth on which black sporangial heads are formed turning the growth black and whisker-like. The affected tissues at first give off a mildly pleasant smell but soon yeasts and bacteria move in and a sour odour develops.

Strawberry leak is caused by *Rhizopus stolonifer* (Fr.) Lind. The aseptate mycelium has prominent aerial branches (stolons) which arise from the assimilative hyphae in the substrate. The stolons are large enough to be seen with naked eyes. As these stolons reach a height of 1-2 cm they bend down to form a tuft of branched rhizoids at the point of contact with the substrate. Opposite to these rhizoids, branches arise vertically in fascicles and become the sporangiophores. At the tip of these branches a terminal sporangium is formed except one branch that bends down to repeat the process. The sporangia release innumerable, non-motile spores which are dispersed into the atmosphere. Ripe spores are mostly oval with striate wall. These spores float in the atmosphere throughout the year. The spores germinate by a germ tube. Sexual reproduction is heterothallic and results in the formation of zygospores.

The fungus is a saprophyte and exists on debris of the diseased plants on soil and also on other dead organic substrates to produce spores that are carried by wind in to the atmosphere. These floating spores land on the wounds of fleshy fruits caused during handling operations and grow on the dead cells of the wounds. The hyphae thus produced secrete pectinolytic enzymes which break down and dissolve the pectic substances of the middle lamella and separate the cells. The action of cellulolytic enzymes dissolves the cell walls and cell contents are released for use by the fungus which reaches the site after these enzymic actions have occurred. The hyphae of the fungus, thus, never come in contact with the living host cells and existence of the fungus in or on the fruit is saprophytic.

The initiation of infection and the invasion of the host tissue by the fungus are influenced by temperature, humidity and the stage of ripening of the fruit. Unfavourable combination of temperature and humidity or insufficient maturity of the fruit slow down the growth and activity of the fungus.

Since the spores of the fungus are present in the atmosphere and strawberry fruits are easily bruised during handling, the chances of infection of fruits are high. Wounds on the fruits should be avoided during picking and transportation as much as possible. If wounds can be detected on the fruits, the latter should be discarded. Such fruits as well as over-ripe and soft fruits should not be packed with healthy fruits. The field should be kept clean and no decomposable organic matter should be allowed near the fruits. Chilling of harvested fruits at 7°-10°C before transportation at 10°C has been found to minimize the rot even if the fruits have contacted infection. The hot air treatment as listed for grey mold rot above is also effective. Edney (13) could check the rot caused by *Rhizopus nigricans* and *Mucor* sp. by dipping fruits in a solution of the antifungal antibiotic mycostatin. The fungicide Botran (DCNA) is highly effective against *Rhizopus stolonifer* but not against other species such as *R. arrhizus* and *R. oryzae* while iprodione (Rovral) is effective against other *Rhizopus* spp. (35).

■ REFERENCES

1. Agrios, G.N. 1988. *Plant Pathology*. 3rd Ed. pp. 321-329; 403-407. Academic Press.
2. Anderson, H.W. 1956. *Diseases of Fruit Crops*. McGraw Hill.
3. Bain, H.F. and J.B. Demaree. 1945. Red stele root disease of the strawberry caused by *Phytophthora fragariae*. *J. Agr. Res.* **70**: 11.
4. Barkai-Golan, R. and D.J. Phillips. 1991. Postharvest heat treatment of fresh fruits and vegetables for decay control. *Plant Dis.* **75**: 1085.
5. Beraha, L. and W.R. Wright. 1973. A new anthracnose of strawberry caused by *Colletotrichum dematium*. *Plant Dis. Rep.* **57**: 445.
6. Brooks, A.N. 1931. Anthracnose of strawberry caused by *Colletotrichum fragariae* n. sp. *Phytopathology* **21**: 739.
7. Chen, T.-A. and A.E. Rich. 1962. The role of *Pratylenchus penetrans* in the development of strawberry black root rot. *Plant Dis. Rep.* **46**: 839.
8. Converse, R.H. and D.H. Scott. 1962. Physiologic specialization in *Phytophthora fragariae*. *Phytopathology* **52**: 802.
9. Delp, B.R. and R.D. Milholland. 1980. Control of strawberry anthracnose with captafol. *Plant Dis.* **64**: 1013.
10. Delp, B.R. and R.D. Milholland. 1981. Susceptibility of strawberry cultivars and related species to *Colletotrichum fragariae*. *Plant Dis.* **65**: 421.
11. Eastburn, D.M. and W.D. Gubler. 1990. Strawberry anthracnose: Detection and survival of *Colletotrichum acutatum* in soil. *Plant Dis.* **74**: 161.
12. Eastburn, D.M. and W.D. Gubler. 1992. Effect of soil moisture and temperature on the survival of *Colletotrichum acutatum*. *Plant Dis.* **76**: 841.
13. Edney, K.L. 1964. Postharvest rotting of strawberries. *Plant Pathology* **13**: 87.
14. Freeman, S., Y. Nizani, S. Dolan, S. Even and T. Sando. 1997. Control of *Colletotrichum acutatum* in strawberry under laboratory, greenhouse and field conditions. *Plant Dis.* **81**: 749.
15. Goode, P.M. 1956. Infection of strawberry roots by zoospores of *Phytophthora fragariae*. *Trans. Brit. Mycol. Soc.* **39**: 367.
16. Hickman, C.J. 1940. The red core root disease of the strawberry caused by *Phytophthora fragariae* n. sp. *J. Pomol. Hortic. Res.* **21**: 83.
17. Hickman, C.J. and M.P. English. 1951. Factors influencing the development of red core in strawberries. *Trans. Brit. Mycol. Soc.* **34**: 223.
18. Horn, N.L. and R.G. Carver. 1963. A new crown rot of strawberry plants caused by *Colletotrichum fragariae*. *Phytopathology* **53**: 768.
19. Horn, N.L. and R.D. Carver. 1968. Overwintering of *Colletotrichum fragariae* in strawberry crowns. *Phytopathology* **58**: 540.
20. Howard, C.M. 1971. Control of strawberry anthracnose with benomyl. *Plant Dis. Rep.* **55**: 139.
21. Howard, C.M., J.L. Mass, C.K. Chandler and E.E. Albregts. 1992. Anthracnose of strawberry caused by the *Colletotrichum* complex in Florida. *Plant Dis.* **76**: 976.
22. Jordan, V.W.L. 1978. Control of fruit rot: *Botrytis cinerea* on strawberry. *Pflanzenschutz-Nachr.* **31**: 1.
23. King, W.T., L.V. Madden, M.A. Ellis and L.L. Wilson. 1997. Effect of temperature on sporulation and latent period of *Colletotrichum* spp. infecting strawberry. *Plant Dis.* **81**: 77.

24. La Mondia, J.A. and S.B. Martin. 1989. The influence of *Pratylenchus penetrans* and temperature on black root rot of strawberry by binucleate *Rhizoctonia* spp. *Plant Dis.* **73**: 107.

25. La Mondia, J.A. and S.M. Douglas. 1997. Sensitivity of *Botrytis cinerea* from Connecticut greenhouses to benzimidazole and dicarboximide fungicides. *Plant Dis.* **81**: 729

26. Madden, L.V., L.L. Wilson and M.A. Ellis. 1993. Field spread of anthracnose fruit rot of strawberry in relation to ground cover and ambient weather conditions. *Plant Dis.* **77**: 861.

27. Martin, S.B. 1988. Identification, isolation frequency and pathogenicity of anastomosis groups of binucleate *Rhizoctonia* spp. from strawberry roots. *Phytopathology* **78**: 379.

28. Mass, J.L. (ed.). 1984. *Compendium of strawberry diseases.* Am. Phytopath. Soc. Press.

29. McInnes, T.B., L.L. Black and J.M. Gatti Jr. 1992. Disease free plants for management of strawberry anthracnose crown rot. *Plant Dis.* **76**: 260.

30. McKeen, W.F. 1958. Races of and resistance to *Phytophthora fragariae*. *Plant Dis. Rep.* **42**: 768.

31. Milholland, R.D. 1982. Histopathology of strawberry infected with *Colletotrichum fragariae*. *Phytopathology* **72**: 1434.

32. Milholland, R.D., W.O. Cline and M.E. Daykin. 1989. Criteria for identifying pathogenic races of *Phytophthora fragariae* on selected strawberry genotypes. *Phytopathology* **79**: 535.

33. Montgomerie, I.G. 1977. *Red Core Disease of Strawberry*. Commonwealth Agric. Bureaux (CAB), U.K.

34. Northover, J. and J.A. Matteoni. 1986. Resistance of *Botrytis cinerea* to benomyl and iprodione in vineyards and greenhouses after exposure to the fungicide alone or mixed with captan. *Plant Dis.* **70**: 398.

35. Ogawa, J.M., R.M. Sonoda and H. English. 1992. Postharvest diseases of tree fruit, pp. 405-422. In: *Plant Diseases of International Importance*. Vol. III. *Diseases of Fruit Crops*. J. Kumar *et al.* (eds.). Prentice-Hall New Jersey.

36. Palmer, C.L., R.K. Horst and R.W. Langhans. 1997. Use of bicarbonates to inhibit in vitro colony growth of *Botrytis cinerea*. *Plant Dis.* **81**: 1432.

37. Pepin, H.S. and E.A. MacPherson. 1982. Strains of *Botrytis cinerea* resistant to benomyl and captan in the field. *Plant Dis.* **66**: 404.

38. Plakidas, A.G. 1964. *Strawberry Diseases*. Louissiana Univ. Press, Baton Rouge.

39. Saharan, G.S. and S.D. Badiyala. 1985. Progress of Mycosphaerella leaf spot on strawberry cultivars in relation to environment. *Indian Phytopath.* **38**: 139.

40. Sindhan, G.S. and A.J. Roy. 1981. Screening of different varieties and selections of strawberry against leaf diseases and their control. *Indian Phytopath.* **34**: 304.

41. Singh, S.J. 1974. A ripe fruit rot of strawberry caused by *Colletotrichum fragariae*. *Indian Phytopath.* **27**: 433.

42. Smith, B.J. and L.L. Black. 1987. Resistance of strawberry plants to *Colletotrichum fragariae* affected by environment conditions. *Plant Dis.* **66**: 559.

43. Smith, B.J. and L.L. Black. 1990. Morphological, cultural and pathogenic variation among *Colletotrichum* species isolated from strawberry. *Plant Dis.* **74**: 69.

44. Smith, B.J., L.L. Black and G.J. Galletta. 1990. Resistance to *Colletotrichum fragariae* in strawberry affected by seedling age and inoculation method. *Plant Dis.* **74**: 1016.

45. Sutton, J.C., De-Wei Li, G. Peng, H. Yu, P. Zhang and R.M. Valdebenito-Sanhueza. 1997. *Gliocladium roseum*: A versatile adversary of *Botrytis cinerea* in crops. *Plant Dis.* **81**: 316.

46. Von Arx, J.A. 1970. A revision of the fungi classified as *Gloeosporium*. *Bibl. Mycol.* **24**: 1.

47. Wilhelm, S. 1965. *Pythium ultimum* and the soil fumigation growth response. *Phytopathology* **55**: 1016.

48. Wilson, L.L., L.V. Madden and M.A. Ellis. 1990. Influence of temperature and wetness duration on infection of immature and mature strawberry fruit by *Colletotrichum acutatum*. *Phytopathology* **80**: 111.

49. Wilson, L.L., J.M. Solar, A. El Ghouth and M.E. Wisniewski. 1997. Rapid evaluation of plant extracts and essential oils for antifungal activity against *Botrytis cinerea*. *Plant Dis.* **81**: 205.

50. Yang, X., L.L. Wilson, L.V. Madden and M.A. Ellis. 1990. Rain splash dispersal of *Colletotrichum acutatum* from infected strawberry fruit. *Phytopathology* **80**: 590.

51. Yuen, G.Y., M.N. Schroth, A.R. Weinhold and J.G. Hancock. 1991. Effect of soil fumigation with methyl bromide and chloropicrin on root health and yield of strawberry. *Plant Dis.* **75**: 416.

9

Diseases of
Coconut Palm

■ **BUD ROT OF PALM**

Coconut palm (*Cocos nucifera*) and toddy palm (*Borassus flabellifer*) suffer badly from bud rot disease caused by *Phytophthora palmivora* (2, 6, 9). In the past this disease had been a limiting factor in palm cultivation but due to drastic measures, including cutting down of diseased trees, the disease has been brought under control to a great extent but even now 30 - 40% affected plants may be found in a single garden (*cf*. 9).

The young leaves and shoots are affected by the disease. Discoloured spots are seen on the leaves and leaf bases. The central expanding leaves turn yellow and dry. In the early stages the disease can be prevented and the trees saved, but when the rot has spread to the growing point and the bud rot phase has set in, the tree is doomed to die. Palms of all ages are susceptible but the disease is more severe on 15 - 45 years old trees.

The causal organism, *Phytophthora palmivora* Butler, has a very wide host range which includes various palms, cacao, rubber, cinchona, castor, safflower and species of *Citrus*. The mycelium of the fungus is intercellular producing haustoria in the host cells. The hyphae are large and often swollen at regular intervals. They are upto 7 μm in diameter. The hyphae ramify between the cells of parenchyma. Sporangiophores are simple or branched with inverted pear-shaped, rarely round, and always terminal sporangia. The sporangia measure 38 - 72 × 33 - 42 (average 50 × 35) μm and germinate in water giving rise to large zoospores. The resting zoospores measure 8-10 μm in diameter. Oospores are spherical and measure 35-45 μm in diameter, the wall being up to 4 μm thick. They germinate readily giving rise to a short branch on which a secondary sporangium is formed early. A number of morphological and pathological strains of the species are involved in the disease.

The pathogen perpetuates through mycelium in diseased tissues in leaf axils and crown and through oospores, where formed, lying in dead plant parts. The same species causing wilt of black pepper (*Piper nigrum*) in south India (16) is reported to survive through oospores in decaying leaves and twigs throughout the dry season (January to June).

After initiation of the disease in the beginning of rainy season, the production of sporangia and zoospores on tree tops carries the infection from tree to tree and to new spots on the same tree. Tappers may help in dissemination of inoculum. The rhinoceros beetle (*Orycles rhinoceros*) and rains are other agents of inoculum dispersal and disease spread. The infection goes back and forth between coconut palm and palmyra palm. The disease is closely related with humid atmospheric conditions. Microclimatic conditions of temperature and moisture at the leaf axil level of palm trees exert considerable influence on the disease incidence. A temperature range of 21°-24°C and relative humidity of 98-100% in leaf axil of the palm tree are highly conducive for the development of the disease. It takes a minimum of 5 weeks for the pathogen to express the symptoms after infection (*cf.* 9).

The control of bud rot was earlier largely confined to cutting down the diseased trees and burning them. Later, sprays of Bordeaux mixture were found very effective. As soon as the disease makes its first appearance the tree should be thoroughly sprayed with 5:5:50 Bordeaux mixture. The trees around an infected tree should also be given a protective spray. Copper oxychloride fungicides were then used as substitute of Bordeaux mixture. Under experimental conditions, trunk injection of Ridomil (metalaxyl) or Aliette (fosetyl-Al) has been found effective against the bud rot. A tree requires 3 g a.i./injection and two treatments are required per year, in March and May (9).

A destructive fruit and heart (bud) rot of coconut palms, caused by a *Phytophthora* resembling *P. katsurae* has been reported from the Hawaii Islands by Uchida *et al.* (19). The early symptoms are dark fruit rots and premature loss of young nuts. The fruit rot starts as brown to black elliptical lesion with a grey centre, mostly from the stem end. The trees producing such fruits show gradual decline. Initially, the youngest or the spear leaf dies. This is followed by other younger leaves and then finally the older leaves. Eventually, only the leafless trunk is left. A cross section through the apical portion of the trunk shows severe rancid heart rot or bud rot.

Sporangia of the fungus measure 30.8- 49.0 × 22.8 - 41.8 µm. The species is homothallic and produces abundant oospores in the infected tissues and in the culture. The oogonia have tapered base and measure 22.1 - 30.8 µm in diameter. The oospores are 22.2 - 25.8 µm in diameter.

■ STEM-BLEEDING DISEASE

This disease of coconut was first reported in Sri Lanka in 1906. Later, its presence was reported from India, Malaysia, Philippines and Trinidad also. The disease is quite common in south India.

On the lower part of the trunk, about 2-3 cm above the soil line, deep reddish brown ooze is seen coming out through the cracks. On the surface of the stem this ooze dries and becomes black. The tissue in the affected area rots and turns yellow (10). In early stages, the infection is localized but since many infections take place at different points of the stem, they coalesce and affect a large area. In newly established plants, the infection spreads very rapidly and there is extensive decay of internal tissues. This creates a cavity in the pith which is filled with a thin, yellow fluid. When this cavity is opened, the fluid comes out with some force. Sometimes, the decay develops from inside without any external symptom. Plants looking healthy from outside have been found with cavities 10×15 cm in dimensions.

The disease is caused by *Ceratocystis paradoxa* (Dade) C. Moreau [syn. *Ceratostomella paradoxa* Dade; *Ophiostoma paradoxa* (Dade) Mannfeldt]. The fungus is a member of Ophiostomataceae of Sphaeriales in Pyrenomycetes (Ascomycotina). The fungus causes soft rot of sugarcane seed setts (pineapple disease) and base rot, leaf rot and fruit rot of pineapple also. Mycelium of the fungus is hyaline or light brown with 3.5-7 μm wide hyphae. Two types of conidia are produced: macrospores or chlamydospores and microspores or microconidia. The macrospores are produced on short lateral conidiophores which are 20-80 μm long and bear spores in short chains. These chlamydospores are obovate to oval, thick-walled, and brown. They measure $10\text{-}25 \times 7.5\text{-}20$ μm. The microconidia are at first hyaline but later turn brown. They are thin-walled, smooth, cylindrical to oval when mature, and measure 6-24 μm in length (average 13 μm) and 2-5.5 (mostly 3.5-5) μm in width. These spores are produced endogenously in the conidiophores and pushed out in a chain. The conidiophores may measure up to 100 μm, sometimes 200 μm, in length with short cells at the base and long apical cells. Perithecia are partly or completely immersed, light brown, globose, 190-350 μm in diameter, and are ornamented with numerous stellate or corraloid appendages. The long, narrow beak of the perithecium is black, pale brown towards the tip, tapering, and up to 1400 μm in length. Ostiole at the tip of the beak contains hyaline, erect or slightly divergent periphyses. Ascospores are ellipsoid, often with unequally curved sides, 1-celled, smooth, and $7\text{-}10 \times 2.5\text{-}4$ μm in size.

The fungus is a wound parasite and enters the host through the cracks or the wounds. The infection can also occur if the plant is weakened by some stress. Plants with low fertilizer application generally show rapid spread of invasion. The fungus survives in the cracks and cavities in the stem in the form of perithecia and chlamydospores. As perithecia, it can survive in soil also. The spores washed down on soil and moved by various means cause infection of the stem at the soil line. Insects also disperse the conidia from the ooze.

As soon as the disease is detected it can be managed to some extent by removing the affected area and cleaning the wound with some disinfectant or by using flame. Hot tarcoal or Bordeaux paste have also been found effective. In old trees, after draining the fluid the cavities can be stuffed with a mixture of tarcoal and sawdust. Since the pathogen attacks the trunk near the soil line, the practice of covering the lower part of the trunk with tarcoal or Bordeaux paste can protect the trees.

- **ROOT (WILT) DISEASE OF COCONUT**

The root (wilt) disease of coconut had become conspicuous in central Kerala, the main coconut growing state of India, after a severe flood in 1882 when the land was water-logged for a long period. Now, it has spread to over 30% of the coconut growing area of Kerala and parts of Tamil Nadu. It is estimated that about 15 million palm trees are affected by the disease and the annual loss amounts to Rs. 300 million*(9). The characteristic symptom of the disease is a slow wilting of the foliage. The earliest symptoms are flaccidity and ribbing of leaflets accompanied by an abnormal bending of the petiole. Later, yellowing of the outer whorl of leaves and marginal necrosis of leaflets occurs. In rare cases, the yellowing and drying of intermittent leaves and abnormal shedding of buttons and immature nuts precedes the flaccidity of leaves. Sterility of pollen and necrosis of spadix also occurs. Rotting of roots is an important feature of the disease. The quality and quantity of coconuts are adversely affected by the disease.

In the root rot, the major portion of the root system is destroyed. Many of the main and lateral roots start drying from the tip backward. In older parts of the roots cracks and lesions may develop. The cortex turns brown and gradually dries. Sometimes the roots look healthy but their absorbing region is covered by hypodermis which interferes with root functions. It has been proved that these root symptoms are secondary and in the diseased plants production of new roots is more than in the healthy plants. A very high percentage of root decay is common in very old trees and loss of 50% root system is not uncommon even in healthy plants (11).

---

*almost U.S $ 6.9 million.

Development of the symptoms is very slow. The affected plants may survive for 15 years after the first appearance of the symptoms. Pre-bearing and early-bearing stage of plants (6-15 years old) is more susceptible than 15-50 years old plants. The young plants show rapid development of symptoms and succumb within 3-4 years. However, the infection of plants younger than 4 yr is rare.

Etiology of the disease is still not very clear (7). Earlier the malady was attributed to the lack of nutrients, water-logging and other soil conditions. However, investigations conducted on soil and nutritional aspects have failed to indicate involvement of these factors in disease initiation. Since 1954 onward, a sap transmissible virus was suspected as the cause of coconut root wilt. Summanwar *et al.* (18) had reported that the disease is caused by a virus, probably a strain of Tobacco mosaic virus. The particles of the virus were seen under electron microscope in root and leaf extracts but not in coconut water. Transmissibility of the causal agent by transmission of sap through insects and grafting was also reported (8). The possible vector was supposed to be *Stephanitis typicus*. Cowpea was considered an indicator host of the virus (17). According to Summanwar *et al.* (18) the virus had *in vitro* longevity of 1 year. It could be inactivated at 90° C. Fungi, bacteria, and nematodes were also implicated with coconut wilt. *Rhizoctonia solani* and *R. bataticola* are often found in decaying roots. But their involvement in full disease syndrome could not be proved. There is evidence that *R. solani* grows better in the root zone of virus affected plants. However, later studies revealed that the rod-shaped particles seen and extracted were of host-origin and a virus is not responsible for the disease (*cf.* 9). The more recent view is the mycoplasmal etiology of the disease. Electron microscopic examinations of tissue samples from the diseased plants have revealed MLOs in the phloem elements of apical meristem, petioles of very young leaves, and root tips.

The disease is more common in water-logged, poorly aerated, heavy soils. In sandy and sandy loam soils also the disease is common if the water table is high. Water-logging causes accumulation of soluble iron, manganese and ammonia nitrogen in the root zone which interferes with uptake of major nutrients. The rate of loss of water from the leaf surface is about 25 % more in the diseased than in the healthy plants (11).

No effective control measures have yet been developed. The destruction of affected palms in early stages in areas where the incidence of the disease is low has been recommended as one of the measures.

■ CADANG-CADANG DISEASE OF COCONUT PALM

Cadang-cadang, meaning "dying" in local dialect of Philippines where it was first detected, is considered a disease of international importance

(15) and is one of the diseases that will cause greater damage in the future (1). The disease has so far killed more than 30 million coconut palms (4). It is caused by a viroid (Coconut cadang-cadang viroid, CCCVd). A related viroid (CTiVd) causes a similar disease of coconut in Guam in the Pacific.

The disease develops slowly in palms and passes through three well-defined stages. In the early stage, lasting 2-4 years, the nuts become rounded, with characteristic equatorial scarifications, and the first non-necrotic, translucent, bright yellow leaf spots appear. In the second stage of symptom expression, lasting 2 years, the inflorescences become necrotic, nut production ceases, new frond production slows down and the leaf spots become larger and more frequent so that from a distance the leaves look chlorotic. In the last stage, lasting about 5 years preceding the death of the plant, the leaf spots are almost confluent, the whole crown is distinctly yellowish or bronze-coloured and very much reduced in size and number of leaves. The overall time from the first symptoms to the death of the tree is about 8 years for 22 years old palms and about 16 years for 44 years old palms (4). Palms before the age of flowering are rarely infected. If infected they remain stunted and do not produce inflorescences. In artificial inoculation experiments a variation of the symptoms of cadang-cadang has been observed. It is a more severe type of lamina reduction to the extent that there is only midrib of the leaf.

Oil palm (*Elaeis quinnensis* Jacq.), naturally infected or inoculated with CCCVd in the Philippines, develops bright orange leaf spots that are larger and more numerous on the older fronds. Palm species successfully inoculated with CCCVd include oil palm, buri palm (*Corypha elata* Roxb.), betelnut palm (*Areca catechu* L.), golden cane palm (*Chrysalidocarpus lutescens*), and date palm (*Phoenix dactylifera*), royal palm and Manila palm. Preliminary observations have suggested that several monocot species growing around cadang-cadang infected coconut palms occasionally contain viroid-like molecules resembling CCCVd (4). Such plants have no apparent symptoms.

Association of viroid-like molecules with cadang-cadang was reported in 1975 (12). The viroid RNA is of low molecular weight, single stranded, predominantly circular, and infectious (5, 13, 14, 15). The disease is insensitive to tetracyclines. The viroid is present in the husk of the nut, in the embryo and in the pollen. It is seed transmitted at a low rate of 1 in 300. No insect vectors are known but certain coleopterous insects are suspected to transmit it through feeding wounds. No specific control measures are known.

■ **LEAF ROT OF COCONUT**

This disease was first reported in 1908 in Kerala. Nearly one third of the coconut palms in central Kerala are affected by the disease. It is of common occurrence in nearly 20 % of the root (wilt) disease affected palms. The loss due to leaf rot, calculated on the basis of reduced yield of nuts, exceeds Rs. 5.6 million* per year (10).

The first visible symptom of leaf rot is the blackening and shrivelling of the tips of leaflets in some of the inner whorls of leaves. The affected portion is broken and blown away by wind giving the leaves a fan-like appearance which is characteristic of the disease. The central spindle also gets infected. Reddish brown spots appear on tender leaves and these penetrate the tissue to cause soft rot of leaves. When the infection is severe the leaves fail to unfold. The leaf rot does not kill the plant outright but it progresses slowly and steadily until the eventual death of the plant. The disease does not attack seedlings in nursery and is severe on palms below 25 years of age. It is most severe during the peak of mon~ when the atmospheric humidity is very high.

Helminthosporium halodes, Gloeosporium sp. and *Gliocladium roseum* ι been found associated with leaf rot and in more than 80% artificial inoculations they have produced the disease. However, it is also reported that none of these fungi is the primary cause of the disease. Imbalance of boron is reported to play an important role in the development of the disease symptoms.

Successful control of leaf rot is observed when the leaves are sprayed quarterly with 1 % Bordeaux mixture. Basic arsenate of copper and copper oxychloride are also equally effective. Sprays of 0.5 % Bordeaux mixture bimonthly are more effective than quarterly sprays of 1% Bordeaux mixture. Preliminary experiments have suggested the possibility of reducing the disease by soil and foliar application of boric acid.

■ **RED RING DISEASE OF COCONUT PALM**

In tropical America, the red ring disease of coconut palm, caused by the bud and leaf nematode, *Rhadinaphelenchus cocophilus (Aphelenchus cocophilus)*, is common and causes severe losses. The nematode attacks various palms including the oil and coconut palm. Coconut palms are susceptible for approximately 2 years before and after they come into bearing. Three years old plants are most affected. The seedling tissue may not be suitable for the development of the nematode (20).

Usually the first visual symptom is yellowing of the lower leaves. The yellowing starts at the tip and progresses toward the base. The yellowing

*about U.S. $ 130,000.

is followed by browning and the eventual death of the leaves. Discolouration spreads to the outer leaves with same results and the tree dies approximately 3 months after appearance of the first symptom.

Examination of the infected trees shows an orange-red ring and yellow ring of discolouration, approximately 2.5 - 4 cm wide, located 2.5 - 5 cm below the stem surface. A large number of larvae of the nematode are present in the discoloured area of the stem and petiolar tissue. Adults are present only in the extremities of the ring. Root cortex assumes a reddish brown colour and becomes spongy. A small number of nematodes are seen in this area also.

The breakdown of tissues due to nematode infection releases a thermostable toxin that adversely affects the efficiency of leaves and may cause wilting of young plants of some small palms and of tomato. The infection also causes dysfunction of the vascular system in the affected area although the nematode does not enter the vascular bundles.

The nematode can penetrate the undamaged surface of all types of coconut tissue and the disease can be initiated by exposure of the crown or the roots of the tree to the infection. The infection of the crown is usually initiated by the palm weevil (*Rhynchophorus palmarum*). The nematode has been recovered from the body, including digestive tract and body cavity, of the weevils. The nematode lives for 2-3 days in the weevil when picked up in fragments of diseased tissue and for 6-7 days when picked up from suspensions. Nematodes within the alimentary canal and body cavity can survive for 10 days (3). In addition to insects, the nematode can also spread by contact between the roots of diseased and healthy trees (20).

No effective control measures are known. But the incidence can be reduced. The most important is phytosanitation. If an infected tree is allowed to stand it is usually attacked by weevils. The eggs laid in the plant produce new generations which are heavily contaminated with the nematode. These migrate to other trees and spread the disease. In the past it had been suggested that the infected trees be cut down and burnt as soon as they show red-ring symptoms. However, this may not be a very effective step. If root to root transfer of nematodes occurs it is not feasible to remove all the roots which spread to several metres around the tree. The falling chips during cutting may also spread the nematodes. Further, the process is very costly and time consuming especially when large number of trees are affected. Injection of sodium arsenate into the trunk has been found to reduce the incidence by preventing further spread of nematodes within the tree. It does not affect the nematodes in the colonized area because no chemical can penetrate the discoloured area. Control of the weevils is, thus, the only step that can possibly check the disease. Monthly treatment of leaf axils, where the weevils live during day time, with 1 % Endrin plus a sticker reduce the incidence by one third. Agrocide dust containing 0.65 % gammexane also gives similar result. Less frequent treatments are not effective.

■ REFERENCES

1. Agrios, G.N. 1988. *Plant Pathology*, 3rd Ed. p. 22. Academic Press.
2. Butler, E.J. 1907. Bud rot of palms. *Mem. Dep. Agric. India, Bot. Ser.* **3**:1.
3. Fenwick, D.W. 1968. Red ring disease of the coconut palm, pp. 38-48. In: *Tropical Nematology*. G.C. Smart and V.G. Perry (eds.). Univ. Florida Press, Gainesville.
4. Hanold, D. and J.W. Randles. 1991. Coconut cadang-cadang disease and its viroid agent. *Plant Dis.* **75**: 330.
5. Haseloff, D., N.A. Mohamed and R.H. Symons. 1982. Viroid RNAs of cadang-cadang disease of coconut. *Nature* **299**: 316.
6. Joseph, T. and K. Radha. 1975. Role of *Phytophthora palmivora* in bud rot of coconut palm. *Plant Dis. Rep.* **59**: 1014.
7. Lal, S.B., K. Radha and P. Shanta. 1970. Etiology of the root (wilt) disease of coconut palm, pp. 662-669. In: *Plant Disease Problems*. S.P. Raychaudhuri *et al*. (eds.). Indian Phytopathological Soc., New Delhi.
8. Nagaraj, A.N. and K.P.V. Menon. 1956. Note on the etiology of the wilt (root) disease of coconut palms in Travancore-Cochin. *Indian Coconut J.* **9**:161.
9. Peethambaran, C.K. 1989. Diseases of arecanut and coconut palms, pp. 227-239. In: *Perspectives in Plant Pathology*. V.P. Agnihotri, *et al*. (eds.). Today and Tomorrow Printers and Publishers, New Delhi.
10. Radhakrishnan, T.C. 1987. Symptom expression in stem bleeding disease of coconut in relation to season. *Indian Phytopath.* **40**: 100.
11. Ramadasan, A. 1970. On the nature of wilt in the root (wilt) disease of coconut palm, pp. 670-675. In: *Plant Disease Problems*. S.P.Raychaudhuri *et al*. (eds.). Indian Phytopathological Society, New Delhi.
12. Randles, J.W. 1975. Association of the ribonucleic acid species with cadang-cadang disease of coconut palms. Phytopathology **65**: 163.
13. Randles, J.W. 1987. Coconut cadang-cadang, pp. 265-277. In: *The Viroids* T.O. Diener (ed.). Plenum Press, New York.
14. Randles, J.W. and J.S. Imperial. 1984. Cadang-cadang viroid. *Description of Plant Viruses*, No. 287. CMI, Kew, U.K.
15. Randles, J.W., D. Hanold, E.P. Pecumbaba and M.J.B. Rodriguez. 1991. Cadang-cadang disease of coconut palm. In: *Plant Diseases of International Importance*. A.N. Mukhopadhyay *et al*. (eds.). Prentice Hall.
16. Sastry, M.L.N. and R.K. Hegde. 1988. Survival of *Phytophthora palmivora*. *Indian Phytopath.* **41**: 118.
17. Shanta, P. and K.P.V. Menon. 1960. Cowpea (*Vigna sinensis*), an indicator plant for the coconut wilt virus. *Virology* **12**: 309.
18. Summanwar, A.S., S.P. Raychaudhuri and K. Jagdish Chandra. 1971. Further studies on coconut root wilt disease, pp. 128-129. *Second Int. Symp. Plant Pathol*. Indian Phytopath. Soc. New Delhi.
19. Uchida, J.Y., M. Aragaki, J.J. Ooka and N.M. Nagata. 1992. Phytophthora fruit and heart rot of coconut in Hawaii. *Plant Dis.* **76**: 925.
20. Warvick, D.R.N. and A.P.T. Bezzera. 1992. Possible root transmission of the red ring nematode (*Rhadinaphelenchus cocophilus*) to coconut palms. Plant Dis. **76**: 809.

10

Miscellaneous Tropical Fruits

- **PINEAPPLE**

Ananas comosus

Heart or Stem Rot and Wilt Disease

Two Phytophthora diseases of pineapple caused by the same species (*Phytophthora parasitica* Dastur) have been separately described in India (4, 5). Heart or stem rot is described as a serious disease in other countries also. Although sporadic in the nature of its occurrence in Assam (India), the disease caused 7-25% mortality of plants in the affected areas. In some localities 75% of the newly set plants collapsed due to heart rot within 6 weeks of planting.

Heart rot (4) is generally found in newly planted pineapples and occasionally in developing plants. In the initial stages, the leaves turn yellowish green with brown tips. The central leaves easily detach when pulled gently. The leaf blades show yellowish white decaying areas which have brownish margins merging into the green colour of the unaffected area. The secondary rot causing bacteria enter the affected area and a foul smell is emitted. From the central leaf of the crown, the disease progresses to the stem (edible part) producing a cheesy soft rot. The sudden death of plants after heavy rains is a characteristic of the disease. Roots are not affected in the heart and stem rot phase of the attack of *Phytophthora*. The plants that escape death may later grow normally.

The wilt (5) is common during the periods of heavy rains and occurs in 1-2 yr old plantings. The growth of plants ceases. Leaves look almond-coloured, flaccid and drooping. These are initial visual symptoms which are followed by wilt. The roots are also affected even before the leaf symptoms appear and this is the cause of above-ground symptoms. The growth of immature edible stem (fruit) ceases and its colour changes before ripening. The affected fruits are spongy and salty in taste.

These diseases are caused by *Phytophthora parasitica* described under citrus diseases. The disease cycle is similar to root and collar rot of citrus. Heart or stem rot and wilt appear to be different phases of the same disease complex. The infection starting from leaves is due to the infection resulting from inoculum splashed from the infested soil onto the crown while the wilt phase is probably due to the root infection by inoculum already present in the soil at the planting site.

Proper drainage, disease-free planting material treated with suitable fungicides, and care to avoid contamination of aerial parts with soil are precautionary measures. In the past, dip of planting material in and spray of Bordeaux mixture was recommended. Among newer fungicides fosetyl-Al has been found highly effective against *Phytophthora* in pineapple. It can be used as set-dip, soil drench and spray material. From the foliage the fungicide moves into the roots. A single application of fosetyl-Al as a pre-plant dip protects the plants against heart rot for 18 months (*cf.* 6). Development of resistance in the fungus against fosetyl-Al can be avoided by using it in combination with mancozeb.

■ BASAL ROT, LEAF ROT AND FRUIT ROT

The problem of basal rot in pineapple arises when the planting material has not been properly dried before setting or when it is stored under poorly aerated conditions. The disease is common in late planted crop in cool, wet soils. In the standing crop when there are wounds in the stem during cultural operations there may be initiation of the basal rot. The leaf rot is characterized by yellow spots or broad areas on the leaves. Fruit rot is common in prolonged storage. The decay is of black rot type in which the black colour is mainly due to spores of the fungus.

The basal, leaf, or fruit rots (3) are caused by *Ceratocystis paradoxa*. The fungus is described as the cause of stem bleeding disease of coconut. It is a wound parasite and produces disease only when it gets entry into the host through some wound or cut. In pineapple plantations, the fungus is soil-borne. Infection of the planting material usually occurs through the cuts. In a standing crop infection occurs through the wounds resulting from hoeing and weeding operations. In stored fruit the infection is also through the cut end of the stem. High-wind rains cause injuries on leaves or during the movement of workers through the fields leaves may be scratched. The infection of leaves may occur through these wounds by spores dispersed by wind. Generally, the leaf rot is seen after the weather has been stormy. The fruit rot is common during the winter soon after crop harvest.

Precautions during cultural operations, harvesting and transportation to avoid unnecessary wounds on stems and leaves are essential. The planting stock should be properly dry at the time of planting. Application of benzoic acid to the cut ends of fruits is reported to reduce fruit rot. In Assam, *Xanthium* leaf extract in ethyl alcohol is reported to be a strong antifungal plant extract that prevents *Ceratocystis* infection of pineapple when sprayed on them. Post-harvest gamma irradiation of fruits is highly effective against the fruit rot but is not practical because of the cost involved and nonavailability of the facility everywhere.

■ POMEGRANATE

Punica granatum

Bacterial Leaf Spots

The disease appears as few to numerous, small, deep red spots of 2-5 mm diameter with indefinite margins on the leaf blade. The leaves may be distorted or malformed. Severely infected young leaves shed. In fruit infections, generally, the bacteria cause slightly raised, dark brown lesions of indefinite margins on the surface. The bacteria enter the host through wounds and natural openings and first produce water-soaked lesions within 2-3 days. Later, these lesions take the shape of deep red spots.

Hingorani and Mehta (7) had first reported the disease and Hingorani and Singh (8) identified the causal agent as *Xanthomonas punicae* sp. nov. The name was later changed to *Xanthomonas campestris* pv. *punicae* (Hingorani and Singh) Dye and finally to *X. axonopodis* pv. *punicae* by Vauterin. The bacterial cells are capable of surviving in soil for more than 120 days. They are dispersed by the wind and lodged on leaves where they cause infection through wounds and stomata. During March-July (in India) the disease spreads fast due to high temperature and low humidity. Sprays of copper fungicides provide some control of the disease.

■ ASPERGILLUS ROT OF POMEGRANATE FRUITS

Aspergillus rot is a common rot of pomegranate fruits affected by fruit flies and usually occurs in the market. Brown discolouration of the fruit surface is the beginning of the rot. The affected area gradually turns black and sticky. Fungal growth with numerous greenish conidia covers the entire fruit which soon shows soft rot and gives a fermented smell. This happens within 8-10 days of appearance of initial symptoms.

Several species of *Aspergillus* may be involved (22). Common species are *A. niger, A. flavus*, and *A. fumigatus*. Injury to the rind of the fruit, such as punctures caused by fruit flies, provide avenues for entry of the fungi. Control of fruit flies is, therefore, one of the most important steps in management of the rot.

Soft rot of pomegranate fruits is also caused by *Rhizopus arrhizus* Fischer (10). In this rot the fruit surface is covered by a black hairy growth of the fungus.

■ LITCHI OR LEECHEE

Nephelium litchi, Litchi sinensis

Leaf Spots

Leaf spots are common in the major litchi growing areas of India. The loss depends on the intensity of spots and extent of defoliation before flowering and fruiting. Many of the leaf spot fungi cause post-harvest fruit rot also. The fungi commonly found associated with leaf spots are *Pestalotia pauciseta, Botriodiplodia theobromae* and *Colletotrichum gloeosporioides* (13, 14). Mixed infections are most common. Rarely the spots yield a single fungus.

Initially the spots are small but later many of them coalesce and form large necrotic areas on the leaf surface. The leaf spots are common during the off season (July- December) but the pathogens may remain on the twigs to cause infection of fruit pedicels and fruits. Survival of the pathogens is through twig cankers, spots on leaves present on the trees and also through the fallen leaves.

Spore germination occurs best in litchi leaf extract. Better germination occurs in dew than in plain water. A temperature of 20°C is optimum for spore germination of *Pestalotia pauciseta* and *Colletotrichum gloeosporioides* while for *Botryodiplodia theobromae* the optimum is 30°C. The spores of this species can germinate even at 40°C. At 5°C there is no spore germination in any of these species and at 10°C also there is only nominal germination. Spores of *P. pauciseta* are killed at 52°C by 2.5 min exposure. Conidia of *B. theobromae* are killed by 5- and 2-min exposure to 55° and 57°C, respectively. Conidia of *C. gloeosporioides* are killed at 51°C (5 min exposure) or 54°C (2 min exposure).

The upper surface of the leaf is not infected without the presence of an injury. Young leaves are more susceptible than the old leaves and leaves on the lower portion of the tree get infected first. Control measures include frequent sprays of fungicides but they are not cost effective.

■ FRUIT ROT OF LITCHI

Fruit rot of litchi occurs when ripe fruits are transported over long distances under unhygienic conditions. A large number of fungi are reported on litchi fruits (15, 16, 17). These include *Aspergillus flavus, A. niger, A. nidulans* and some other species of *Aspergillus, Colletotrichum gloeosporioides, Cylindrocarpon tonkinense, Botryodiplodia theobromae, Pestalotia* sp. and *Rhizopus* spp. Infection during transit and storage occurs through the wounds. In some cases there is latent infection of fruits from the tree. The losses from the post-harvest decay can be reduced through precautions during picking and storage. One or two pre-harvest sprays of 0.2% captan or 0.1% copper oxychloride are reported to decrease the fruit rot after harvest. A 2-min dip of fruits in hot water at 52° C is reported to control the rot caused by *Aspergillus* and *Rhizopus* (1).

■ LOQUAT

Eriobotrya japonica

Die-back, Twig blight and Canker Diseases

Cytospora eriobotryae causes a die-back disease in loquat (21). The disease was first reported in Italy in 1927 and is known to occur in India since 1965. In the absence of proper control the disease may prove fatal for the tree.

The disease starts from the tip of the twigs. The bark becomes soft and wrinkled. The infected area progresses backward affecting a major portion or the entire twig which is defoliated and dries. The bark turns papery and ruptures exposing orange-coloured spore tendrils. From twigs the disease moves to the main branches which are girdled by the discolouration. This stage may lead to complete defoliation and the eventual death of the whole branch.

Cystospora ciliata also causes a twig blight and canker of loquat. The symptoms are similar to the die-back disease (12). However, drying of the affected portion is irregular and embossed pimples are found all over the affected area. The fungus attacks the bark and cambium resulting in weakening and loosening of the bark. Amber yellow fluid containing the spores oozes out from the ruptured bark. In dry weather, this ooze hardens and becomes horn-like. Presence of spherical, tuberculate or stromatic masses consisting of hyphal aggregates on the twigs is the characteristic feature of this disease.

Cystospora is a member of Deuteromycotina. The septate branched mycelium is hyaline to pale brown. Fructifications are stromatic, dark brown, convoluted and multilocular with a single ostiole. Conidiophores are hyaline, septate, branched irregularly at the base and above and formed in the locules. Conidia are embedded in distinct variously coloured spore drops or tendrils (cirrhus). They are thin-walled, 1-celled, hyaline and allantoid.

These species survive as dormant mycelium on the wood of the branches and through stromata on twigs and branches fallen on the ground. In rainy weather the spore tendrils are broken and conidia are splashed to the healthy twigs to cause infection.

A die-back of loquat caused by *Colletotrichum gloeosporioides* is also reported (11). Defoliation and withering of twigs from the tip are characteristic symptoms. The spots on leaves start as light green areas changing to a brown colour. These spots are more conspicuous on the tips and margins of the leaf. Sometimes, they are found around the midrib. In wet weather acervuli of the fungus appear as pin-point black bodies on the spots and produce pink masses of conidia.

Tree sanitation is an economical and effective method of control. The affected branches should be pruned 30 cm behind the affected area and cut ends protected by a fungicidal paste consisting of 800 g red lead oxide, 800 g copper carbonate in 1000 ml of linseed oil or lanolin.

■ CASHEWNUT

Anacardium occidentalis

Decline of Cashewnut Trees

This disease can affect young as well as old trees. The symptoms appear during the summer months. Twigs on the affected trees start drying and defoliation takes place. Gradually, in 2-3 yr, the entire tree is destroyed. Roots of the affected trees are brown and root hair are rotted. Fungus mycelium and oospores are found in the rotting tissue (19).

Pythium spinosum has been identified as the incitant of cashewnut decline to some extent. Hyphae of the coenocytic, hyaline mycelium are 3-5 μm thick. Sporangia are intercalary, spherical or sub-spherical, broadly ellipsoidal or broadly ovoid and 27-42 μm long. Zoospores are rarely formed and sporangia germinate directly by 1-3 germ tubes. In sexual reproduction, the spherical, echinulate oogonia are important characters of the species. They are terminal, rarely intercalary, and produced in abundance. The spines on the wall are few on intercalary oogonia. The spines are conical and obtusely tipped. The average size is 4.5 μm. The oogonia measure 14-28 μm in diameter. The antheridia are both mono-

and diclinous and do not arise very close to the oogonium. The oospores are yellowish, spherical, smooth and 16 μm in diameter. In moist conditions the oospores germinate by a germ tube in 24-48 hours.

The fungus survives in soil through oospores. At the end of the rainy season, the germ tubes attack root hairs and destroy them. After 3-4 weeks of root infection the symptoms appear on the aerial parts.

Proper care of the plantation with suitable cultural practices and application of manures can keep the trees vigorous and less susceptible to damage by *Pythium*. Since the fungus is present in tissues of roots, it is difficult to eradicate it from soil. The application of systemic fungicides to soil is very costly. In the past soil drenching around roots with Cheshunt compound (2 parts copper sulphate and 11 parts ammonium carbonate) at the rate of 28.3 g per 10 lit water was recommended. After soil treatment, application of 40 kg compost, 1 kg ammonium sulphate and 0.75 kg super phosphate per tree was recommended. This stimulated growth of the tree and provided tolerance to *Pythium*. In addition, compost enhanced the growth of *Trichoderma* spp. which are antagonists of *Pythium*.

Anthracnose of Cashew

Anthracnose of cashew was reported in epidemic form during 1966 in the states of Kerala and Tamil Nadu. It is caused by *Colletotrichum gloeosporioides (Glomerella cingulata)*. The symptoms are similar to those of mango anthracnose. The disease occurs on new leaves (as leaf spots), twigs, flowers, flower stalks and also on fruits. The initial symptoms are the appearance of bright reddish brown, water-soaked lesions on leaves. Similar lesions appear on twigs and branches and cover a major portion of the branches. This causes drying of the branches. Gum exudation may also occur on the affected woody parts. The crumpling of affected young leaves and fruit mummies are characteristics of the disease. Unfertilized flowers turn black and shed. The drying of apical portion of twigs and branches eventually may cause death of the tree (18, 20).

The survival and dissemination of the fungus are similar to those in mango anthracnose. Inoculum surviving on dead twigs and branches or on fallen plant parts initiates the disease by infecting the young leaves. Sanitation and chemical sprays recommended for mango anthracnose may reduce the disease incidence satisfactorily.

Leaf spots of cashew are also caused by *Xanthomonas campestris* pv. *mangiferae-indicae* that causes leaf spots and canker of mango. Floral and foliage blight of cashew may result from powdery mildew caused by *Oidium anacardii*. Nutritionally weak trees are often attacked by a species of *Pestalotia*.

■ SAPODILLA PLUM (Sapota, Cheeku)

Achras zapota

Pestalotia and other Leaf Spots

The Pestalotia leaf spot disease of sapodilla is characterized by the appearance of small, reddish spots, irregularly scattered on the leaf blade. Gradually, these spots increase in size and become roundish. Fully developed spots have a grey centre surrounded by a red or dark brown area. Minute black dots, representing acervuli of the fungus, develop in the grey portion. The fungus may cause fruit rot also.

The species causing these spots is *Pestalotiopsis versicolor* (Speg.) Stey (syn. *Pestalotia sapotae*, *Pestalotia versicolor* and *Pestalotia podocarpi*). The fungus is similar to *Pestalotiopsis* described for mango and guava. On sapodilla the acervuli are 80-175 μm wide. Conidia are 5-celled, clavate-fusiform, erect or somewhat curved, 22-27 × 7.5-9.5 μm in size and bear 3, rarely 4, apical, flexuous, widely divergent setullae which are 17-27 μm long. The fungus survives on fallen leaves and fruits around the trees. Copper oxychloride at 0.3% has been recommended as spray for control of the disease.

Pestalotiopsis mangiferae, *Botryosphaeria dothidea* (9) and *Gloeosporium rubi* (23) are also reported to cause leaf spots of sapodilla. The symptoms of *G. rubi* are almost similar as described above. Conidia of the species are oblong and measure 10-16 × 3-4 μm (average 12.2 × 3.8 μm). For control of this leaf spot zineb has also been recommended in addition to copper oxychloride. Three to four sprays during the year are recommended.

Fusicoccum sapoticola causes a leaf blight disease of sapodilla (2). Minute, pin-point, brown spots appear on the leaf blade, mostly on the margins. These spots coalesce to form bigger spots. Eventually the leaf is blighted. Pycnidia in the form of black pin-point bodies may appear in these spots. The pycnidia are unilocular, ostiolate, variously shaped, and measure 95-100 × 38-114 μm. Conidiophores are hyaline, clavate, and 3.6-5.4 × 2 μm in size. The conidia measure 5.5-9.2 × 1.8-3.6 μm Certain species of *Fusicoccum* have their perithecial stage in *Botryosphaeria*. The mode of survival and infection in this disease is probably same as for the above pathogens.

■ REFERENCES

1. Barkai-Golan, R. and D.J. Phillips. 1991. Post-harvest heat treatment of fresh fruits and vegetables. *Plant Dis.* **75**: 1085.
2. Chinnappa, B. and V.G. Rao. 1970. A new species of *Fusicoccum* on *Achras sapota*. *Sci. & Cult.* **36**: 295.

3. Chowdhury, S. 1945. Ceratostomella disease of pineapple. *Indian J. Agric. Sci.* **15**: 135.
4. Chowdhury, S. 1945. Heart or stem rot of pineapple. *Indian J. Agric. Sci.* **15**: 139.
5. Chowdhury, S. 1946. Wilt of pineapple. *Curr. Sci.* **15**: 82.
6. Cohen, Y. and M.D. Coffey. 1986. Systemic fungicides and control of Oomycetes. *Annu. Rev. Phytopathol.* **24**: 311.
7. Hingorani, M.K. and P.P. Mehta. 1952. Bacterial leaf spot of pomegranate. *Indian Phytopath.* **5**: 55.
8. Hingorani, M.K. and N.J. Singh. 1959. *Xanthomonas punicae* sp. nov. on *Punica granatum. Indian J. Agric. Sci.* **29**: 45.
9. Jain, S.K., A.K. Saxena and S.B. Saxena. 1981. Two new fruit rot diseases of *Achras sapota. Indian Phytopath.* **34**: 403.
10. Kanwar, Z.S., D.P. Thakur and O.P. Kadian. 1973. A note on the effect of temperature and relative humidity on the development of soft rot of pomegranate fruits due to *Rhizopus arrhizus. Indian Phytopath.* **26**: 742.
11. Lele, V.C. and Asha Ram. 1969. Two new diseases of loquat in India. *Indian Phytopath.* **22**: 502.
12. Mandahar, C.L. 1971. A new disease of loquat (*Eriobotrya japonica*). *Sci. & Cult.* **37**: 112.
13. Prasad, S.S. 1963. Two new leaf spot diseases of *Nephelium litchi* Camb. *Curr. Sci.* **31**: 93.
14. Prasad, S.S. 1967. Leaf spot diseases of *Nephelium litchi. Indian Phytopath.* **20**: 530.
15. Prasad, S.S. and R.S. Bilgrami. 1969. Investigation on diseases of litchi. I. Phyllosphere mycoflora of *Litchi sinensis* in relation to fruit rot. *Indian Phytopath* **22**: 507.
16. Prasad, S.S. and R.S. Bilgrami. 1973. Investigations on diseases of litchi. II. Influence of temperature and humidity on decay of fruits caused by nine virulent pathogens. *Indian Phytopath.* **26**: 517.
17. Prasad, S.S. and R.S. Bilgrami. 1973. Investigations on diseases of litchi. III. Fruit rots and their control by post-harvest treatments. *Indian Phytopath.* **26**: 523.
18. Ramachandran, A. 1969. Pests and diseases of cashew. *J. Mysore Hortic. Soc.* **14**: 9.
19. Ramakrishnan, T.S. 1955. Decline in cashewnut. *Indian Phytopath.* **8**: 58.
20. Singh, S., H.S. Sehgal, P.C. Pandey and B.K. Bakshi. 1967. Anthracnose disease of cashew (*Anacardium occidentalis*). Its cause, epidemiology and control. *Indian Forester* **93**: 374.
21. Singh, S.B., J. Upadhyay and B. Prasad. 1989. Die-back of loquat caused by *Cytospora eriobotryae* Curzi and Barbaini. *Indian Phytopath.* **22**: 525.
22. Srivastava, M.P. and R.N. Tandon. 1971. Aspergillus rot of pomegranate. *Indian Phytopath.* **24**: 172.
23. Wilson, K.I., S. Balakrishnan and N.G. Nair. 1970. Leaf spot of *Achras sapota* in Kerala. *Sci. & Cult* **36**: 109.

Index

Acremonium brevae, biocontrol of fruit
   decay 27, 121
Agrimycin 36
Agrocide dust 289
Agrocin 135
Agrobacterium 132
   radiobacter 132-134
   rhizogenes 132
   tumefaciens 132-134
Aldehyde fumigation of fruits 120
Aldicarb (Temik) 8, 54
Algatol paint 135
Aliette (fosetyl-Al)   7-9, 103, 282
Almond
   brown rot 106
   crown gall 131, 135
Alternaria 15, 18, 19, 27
   apple 119
   citrus 18, 19, 24, 27
Alternaria
   alternata 18, 20, 118, 120
   citri 18, 24, 26, 27, 257
   tenuis 18, 24
Ammonia for fruit paper wraps 23
Ampelomyces quisqualis
   biocontrol of powdery mildew 96, 225,
   226
Amylovorin 127
Anilopyrimidines 87, 224
Anival 224
Anthracnose of
   banana 193
   citrus 9
   grapes 233
   guava 24
   mango 159
   papaya 257
   strawberry 265

Aphids vectors of
   Tristeza virus 43
   papaya ringspot virus 261
Apple
   Armillaria root rot 12
   basal rot 99
   bitter rot 115, 118
   black mold 118
   black rot 113
   blossom blight 109, 125, 126,
   blue mold 118
   Botryosphaeria rot 118
   brown rot 106, 118, 119
   canker 116, 127, 128
   collar rot 99
   crown gall 131
   crown rot 12, 99, 103
   fire blight 125
   foot rot 99
   grey mold rot 118, 228
   Mucor rot 122
   pink mold rot 121
   post-harvest fruit decay 115
   powdery mildew 88, 91
   root rot 103, 111
   scab 75
   stem brown 111
   twig canker 113
   white root rot 103
   white rot 111
Apricot
   bud blast 136
   brown rot 106
   crown gall 131
Armillaria mellea 12, 13
Armillaria root rot 12
Ascochyta caricae 257

Aspergillus 15, 119, 227
  awamori 245
  flavus 196, 293, 294
  fumigatus 168, 170, 196, 293
  nidulans 168, 170
  niger 17, 118, 168, 170, 196, 293, 294
  terreus 168
Athelia bombacina
  biocontrol of apple scab 91
Aureofungin 86, 105
Azoxystrobin 88, 217

Bacillus subtilis in biocontrol of
  apple collar rot 103
  citrus canker 36, 37
  citrus fruit decay 27
  pome and stone fruit decay 110, 121
Bacterial cankers of
  citrus 28
  mango 153
  stone fruits 139
Bacteriocin 135
Banana
  anthracnose fruit rot 193
  bacterial wilt (Moko) 197
  bunchy top 201
  Cercospora leaf spots 189
  crown rot 195
  Fusarium wilt (Panama disease) 183, 197
  post-harvest fruit decay 193
  root and rhizome rot 204
  Sigatoka 189
  stem-end rot 195
Banana bunchy top virus (BBTV) 203
Bavistin (carbendazim) 11, 25, 39, 85-87, 94, 95, 98, 224, 245
Baycor (bitertanol) 87, 94, 95, 98, 110
Bayleton (triadimefon) 95, 98, 224, 225
Bayton (triadimenol) 224
Benlate (benomyl) 25, 85, 87, 94, 110, 193, 225
Benomyl (Benlate) 11, 23, 25, 26, 85, 87, 94, 110, 115, 119, 162, 229, 235, 245, 271
Benzimidazoles 23, 25, 85, 86, 227, 245
Biological control of
  apple powdery mildew 96
  apple scab 91
  Armillaria root rot 13
  banana bacterial wilt 201
  citrus canker 36, 37

citrus fruit rots 27
crown gall 135
fire blight 131
grape downy mildew 218, 219
grape powdery mildew 225, 226
grey mold of
  grapes 230
  strawberry 276
guava fruit decay 245
mango bacterial blight 156
pome and stone fruit decay 245
mango bacterial blight 156
pome and stone fruit decay 110, 121
strawberry fruit rots 276
Tristeza virus 45
Biphenyl (diphenyl) 23, 24
Bitertanol (Baycor) 87, 88, 94, 95, 110
Bitter rot of apple 115, 118, 120
Black mold rot of
  citrus 17
  mango 168-170
Black root rot complex of strawberry 274
Black rot of grapes 230
Blitox 216
Blossom blight of
  apple 109
  cashewnut 296
  citrus 11
  grapes 230
  mango 157
  stone fruits 135
Blue mold rots 15, 27, 119, 120
Borax for fruit treatment 23, 24
Bordeaux mixture 11, 24, 36, 98, 130, 139, 162, 192, 193, 212, 216, 244, 254, 282, 287
Bordeaux paste 8, 284
Boron deficiency 59, in mango 172
Botran (dichloran) 119, 120, 277
Botryodiplodia 165
  theobromae 15, 17, 23, 27, 160, 163-166, 169, 195, 245, 257, 293, 294
Botryosphaeria 111, 165
  dothidea 111, 165, 297
  obtusa 113, 115
  rhodina 169
  ribis 111, 112, 160, 163-166
  stevensii 118
Botrytis cinerea 118-121, 227-229, 275, 276
Brassicol 105
Bravo 98, 110

Brown rot of
  citrus fruits 1, 2
  pome and stone fruits 118-120
Brown rot gummosis of citrus 1
Bunch rot of grapes 227
Bunchy top of banana 201
Bupirimate 94
Burrowing nematode 55, 56, 204

Cachexia of citrus 46
Cadang-cadang viroid 286
Calcium deficiency 58
Calcium prevents fruit rots 120, 121
Calixin (tridemorph) 94, 98, 224
Candida guilliermondii for biological control of fruit decay 27, 121
Canopy management in grapes 229, 236
Canteconidia furcellata 156
Captafol (Difolatan) 11, 85, 86, 98, 166, 229, 269, 271
Captan 9, 23, 24, 85, 86, 109, 117, 162, 216, 229, 235, 254, 269, 271
Carbon dioxide for fruit decay control 120
Carbon disulphide 14, 105
Carpophilus hemipterus dissiminates *Mucor piriformis* 124
Cashewnut
  anthracnose 296
  decline 295
  leaf spots 296
Ceratocystis
  fimbriata 160, 163, 195
  paradoxa 195, 196, 283, 294
Ceratostomella paradoxa 283
Cephaleuros
  mycoides 167
  viriscence 257
Cercobin 85
Cercospora musae 191
Chaetomium globosum
  biocontrol of apple scab 91
Chemical control of
  apple
    black rot and twig canker 114
    Botryosphaeria fruit rot 115
    collar rot 103
    powdery mildew 94, 95
    fire blight 130
    fruit decay 119
    scab 85-90
  citrus
    anthracnose 11, 12

brown rot 9
canker 36
fruit decay 23-27
greening 39
gummosis and root rot 7, 8
root nematode 54
coconut palm
bud rot 282
banana
sigatoka 192, 193
grapes
anthracnose 235
black rot 233
bunch rot 229, 230
downy mildew 216-218
powdery mildew 223-225
guava
anthracnose 235
fruit decay 245
mango
anthracnose 160-162
bacterial blight 156
powdery mildew 158
papaya
stem, foot and root rot 254, 255
fruit rots 259
strawberry
anthracnose 269, 270
leaf spots 271
red stele 273
Chemical control of
  canker and gummosis of stone fruits 139
  post harvest decay of
    citrus fruits 21-27
    pome and stone fruits 119
Chilling for control of fruit decay 22
Chlorine for
  fruit treatment 9, 24, 36, 244
  soil disinfestation 8
Chloropicrin 269
Chlorothalonil 11, 85, 98, 110, 229, 235, 236
Citrus
  Alternaria black rots 18
  anthracnose 9
  black mold rot 17
  blue green mold rot 15
  brown rot 1, 2, 15
  burrowing nematode 56
  cachexia 46
  canker 28

collar rot 1
crown rot 1
damping off 1
decline 2, 9, 12, 41
die-back 9, 11, 15, 28
exocortis viroid 49
fibrous root rot 1, 2
foot rot 1
fruit drop 1, 10
greening 11, 37
gummosis 1-9
melanose 15
mold rots 15
mushroom root rot 12
shoe-string root rot 12
nutritional disorders 56
post bloom fruit drop 10, 11
post-harvest fruit decay 15, 21-27
psorosis 47, 48
root nematode 50
root rot 1, 2
scaly bark 47
shoe-string root rot 12
sour rot 20, 24
spreading decline 55
stem-end fruit rot 10, 17
Tristeza virus 39
wither tip 9
xyloporosis 46
Citrus psylla,
vector of greening bacteria 38
Citrusnin A 34
Cladosporium 124
herbarum 118
Cladosporium
antagonist of apple scab fungus 85
Coconut Palm
bud rot 281
cadang cadang 285
fruit rot 282
heart rot 282
leaf rot 287
red ring disease 287
root (wilt) disease 284
stem bleeding 283
Collar rot of
apple 99
citrus 1
Colletotrichum 23
acutatum 268-270
dematium 268
fragariae 268, 269

gloeosporioides 10, 11, 15, 110, 116, 117,
159, 160, 167, 194, 243, 245, 257, 259,
268, 269, 293-296
musae 194, 196
papayae 257
psidii 243
Compost for apple scab control 91
Copper deficiency 60
Copper oxychloride 11, 98, 156, 162, 192,
216, 244, 245, 282, 287
Cotoneaster aitchinsonii
C. bacillaris, hosts of apple scab fungus
74
Cross protection against Tristeza 45, 46
Crown gall 131
Cultural practices, disease control
banana 188, 192, 196, 203, 206
citrus 7, 11, 14, 21, 35, 39, 44, 49, 54
grapes 216, 223, 229, 232, 236
guava 249
mango 152, 171
papaya 251, 257
pome and stone fruit trees 84, 102,
105, 110, 117
strawberry 269, 271, 273
Cuman 85, 86
Cuprasol 245
Cupric hydroxide 94, 98, 230, 244
Curvularia lunata 96
Cuscuta, reflexa 43
subinclusa 43
Cylindrocarpon
destructans 274
tonkinense 294
Cymoxanil 217
Cyprex (Dodine) 85
Cyproconazole 224
Cyprodinil 87, 224
Cytospora
ciliata 299
eriobotryae 294

Daconil 85, 86, 98
Dasanit 54
Debaromyces hansenii (see *Candida
guilliermondii*)
Deficiency diseases 56
Deightonella torulosa 195, 196
Delan 86
Dematophora necatrix 104, 105
Dendrophoma obsurans 270

Dendrophthoae 176
  falcata 177
Dexon (diazoben) 103
Diaporthe citri 17, 18
Diazinon 39
Diaphorina citri, vector of greening
  bacterium 38
Dicarboximides 230, 269
Dichlofluanid (Euparen) 235, 236
Dichlone (Phygon) 110
Dichloran (Botran) 23, 110, 230
Die-back of
  citrus 9, 11, 15
  loquat 294
  mango 159
Difenoconazole 86-88, 94, 98, 99, 235, 269
Difolatan (captafol) 11, 85, 86, 98, 103,
  120, 235, 244, 254
Dikar 85, 105
Dinocap (Karathane) 94, 95, 224
Diphenyl 23
Dimethoate 54
Dimethomorph 217
Diplodia 23, 27, 163, 165, 169
  natalensis 15, 17, 24, 160, 163, 169
Disease cycle of
  Apple and stone fruits
  bacterial canker 137
  bitter rot 116
  Botryosphaeria rots 112, 114
  brown rot 109
  collar rot 100
  crown gall 133
  fire blight 127
  powdery mildew 93
  scab 79
  white root rot 105
  Banana
  bacterial wilt 199
  Fusarium wilt 187
  root and rhizome rot 205
  Sigatoka 191
  Citrus
  anthrancnose 10
  Armillaria root rot 13
  canker 32
  Grape
  anthracnose 235
  black rot 232
  bunch rot 228
  downy mildew 214

powdery mildew 221
Guava
anthracnose 243
wilt 249
Mango
anthracnose 161
bacterial canker 156
twig and leaf blight 165
Papaya
Phytophthora root rot   256
Peach
leaf curl 97
Strawberry
anthracnose 269
Mycosphaerella leaf spots 271
red stele 272
Disease forecasting 82-84, 115, 131, 217,
  224
Dithane M-45 86, 98
Dithiocarbamates 85
Dodder, transmission of
  citrus greening 39
  exocortis viroid 50
  Tristeza virus 43
Dodine (Syllit) 85, 86, 88
Dothiorella 112
Downy mildew of grapes 89
Dolomite lime, suppression of *Venturia*
  inaequalis 85
Drosophila melanogaster
  disseminates fruit molds 124
  Mucor piriformis 124

Eagle 85
Elsinoe ampelina 234
Endrin 39, 289
Enterobacter aerogenes
  biocontrol of apple collar rot 103
Enterobacter cloacae
  biocontrol of fruit decay 27, 121
Environmental factors, effect on
  apple Botryosphaeria rot 112, 114
  apple collar rot 101, 102
  apple powdery mildew 93
  apple scab 79-83
  banana Sigatoka 192
  banana wilt 187-188
  bud rot of plam 282
  fire blight 128
  grape black rot 232
  grape bunch rot 228

grape downy mildew   214-216
grape powdery mildew   221-223
guava anthracnose   243
mango bacterial cankers   156
mango blight   166
mango powdery mildew   158
papaya stem and root rot   256
peach leaf curl   98
strawberry anthracnose   269
strawberry leaf spot   271
Enzone   7, 87
Epiphytic survival of
    Erwinia amylovora   128
    Pseudomonas syringae   137
Erwinia, amylovora   127
    carotovora   168
Erwinia herbicola, biocontrol of,
    citrus canker   37
    fire blight   131
Erysiphe   88, 158
Etaconazole (Vangard)   94, 110, 224
Ethanol   110
Ethaprophos   54
Ethylene in fruit rots   17, 34
Eupoecilia ambiguella, promotes grape
    bunch rot   229
Euparen   235
Exocortis viroid   49
Extracellular polysachharides (EPS)   32,
    34, 127

Fenarimol (Rubigan)   85-88, 94, 224, 225
Fensulfothion   54
Fire blight of apple and pear   125
Flooding for control of banana wilt   188,
    189
Fluazinam (Shirlan)   235, 236
Fluorescent pseudomonads, biocontrol of
    P. parasitica   6
Fluquinconazole   88, 94
Flusilazole   87, 94, 99, 155, 224
Flutriazole (Impact)   224
Fosetyl-Al (Aliette)   7-9, 103, 217, 257, 273,
    282, 291
Fumigation of fruits   23, 24, 259
Fusarium moniliforme   175
Fusarium oxysporum   175
    f. *cubense*   186, 187, 197
    f. *psidii*   247
Fusarium proliferatum
    antagonist of *P. viticola*   218

Fusarium roseum   195-196
Fusarium solani   51, 52, 247
Fusicladium dendriticum   77
Fusicoccom
    aesculi   111, 165
    sapoticola   297

Gamaxane   289
Gamma irradiation   22, 245, 292
Genetically engineered protection   46
Geotrichum candidum   20, 21, 24, 118, 196
Gliocladium
    biocontrol of apple collar rot   103
Gliocladium roseum
    antagonist of *B. cinerea*   276
    on palms   287
Gloeosporium
    ampelophagum   234
    foliicolum   10
    fructigenum   116
    limetticolum   10
    musarum   194, 196
    papayae   257
    psidii   243
    rubi   297
Glomerella cingulata   116, 118, 120, 121,
    160, 161, 194, 243, 268, 270, 296
Glyodin   85
Glyodex   85
Grafting, role in tree diseases   6, 102
Grape berry moth promotes bunch rot
    229
Grapefruit, hot water treatment   9, 23
Grapes
    anthracnose   233
    Armillaria root rot   12
    black rot   230
    bunch rot   227
    crown gall   131, 132
    crown rot   12
    downy midew   212
    powdery mildew   88, 89, 96, 219
Green mold rot of citrus   15, 26, 27
Grey mold rots   118-120, 228
Guava
    anthracnose   242
    anthracnose rot   245
    Aspergillus rot   245
    Botryodiplodia rot   245
    die-back   243
    grey blight   245

Pestalotiopsis rot 245
Phomopsis rot 245
post-harvest decay 243-245
Rhizopus rot 245
scab 245
wilt 246
zinc deficiency 249
Guignardia bidwelllii 231

Hainesia lythri 270
Heat removal for fruit decay control 22
Heat treatment of
citrus fruits 22
mango fruits 163
orchard floor 85
peach fruits 110, 125
planting stock 49
soil 14
stone fruits 139
Helminthosporium halodes 287
Hemicriconemoides mangiferae 163
Hendersonula toruloidea 160, 163
Hexacap (captan) 98
Hexaconazole 87, 94, 224
Hot air treatment of fruits 117, 163, 259, 276, 277
Hot water treatment of fruits 9, 23, 110, 125, 162, 163, 245
Hydrocarbons against crown gall 135
Hydrocooling of fruits 120

Imazalil (Fungaflor) 23, 26, 245
Impact 224
Incipient infection 195, 243
Insect vectors of
banana bunchy top 203
citrus canker 33
citrus greening bacterium 38
citrus tristeza virus 43
fruit rot pathogens 124
papaya ringspot virus 261
papaya leaf curl virus 262
red ring of palm 288
Iprodione (Rovral) 23, 110, 119, 120, 229, 276, 278
Iron deficiency 61
Irradiation for fruit decay control 22, 245, 292

Karathane (dinocap) 85, 94, 95, 158, 224
Kloeckera apiculata
biocontrol of fruit rot 121

Lasiodiplodia theobromae 15, 17, 18, 27, 166, 169, 245, 257, 259
Latent infection of fruits 15, 161, 195
Leaf curl of
papaya 262
peach 96
Leaf miners disseminate citrus canker 33
Leaf removal, control of grape bunch rot 229
Ledermycin 39
Lemon, hot water treatment 23
Liberobacter asiaticum
cause of citrus greening 38
Liberobacter africanum 38
Lime sulphur 94
Litchi
(*Nephelium litchi*)
fruit rot 294
leaf spots 293
Lobesia botrana promotes grape bunch rot 229
Loquot, (*Eriobotrya japonica*), canker, die back, twig blight 294

Macrophoma 111, 165
mangiferae 160, 163-166
Macrophomina 165
phaseolina 248, 257, 270
Magnesium deficiency 59
Mancozeb 11, 85, 86, 98, 103, 115, 216, 217, 230
Maneb 11, 117, 216
Managanese deficiency 61
Mango
anthracnose 159
bacterial blight and canker 153
black mold rot 168, 170
black tip 172
blossom blight 157, 159
die-back 159, 163
fruit drop 153, 157
fruit rots 155, 163, 167
gummosis 164
grey blight 166
leaf blight 163
malformation 173
necrosis 172
powdery mildew 157
red rust 167
soft rot 168, 170
stem-end rot 163, 168
twig blight 163

Margos (see Neem)
Melanose of citrus 15
Metalaxyl (Ridomil) 7-9, 103, 217, 255, 257, 273, 282
Methyl bromide 11, 269
Mineral oils, control of
grape powdery mildew 225
banana Sigatoka 192, 193
Mistletoe on mango 176
MLO 38, 285
Mold rots
black 27, 118, 168
blue or green 27, 118
grey 118, 227, 230
pink 121, 122
Molybdenum deficiency 62
Monilia
cinerea 107
fructigena 107
laxa 107
Monilinia 22, 106, 108, 110, 118
fructicola 107-110, 120, 121, 124
fructigena 107, 108
laxa 1C. *.09, 120
Mucor piriformis 118, 122-124
Mucor rot of pome and stone fruits 122, 123
Myclobutanil (Enzone) 86-88, 94, 224, 225
Mycosphaerella
fijiensis 191, 183
fragariae 270, 271
musicola 191
Mycostatin 277
Myllocerus discolor transmits mango bacterial canker 156
Myxosporium psidii 248

Nanavirus 203
Nectarine
brown rot 106
Neem
oil for grape bunch rot control 230
products for powdery mildew control 226
seed cake for, citrus canker control 37
guava wilt control 249
Nemagon, (DBCP) 54
Nimrod (bupirimate) 94
Nitrogen deficiency 57
Nucellar seedlings 39, 45

Oidium anacardii 296
Ophiostoma paradoxa 283
Oranges
hot water treatment 9, 23
Organic amendments
citrus root nematode control 54
Ornalin (vinclozolin) 229
Orthega vadrusalia transmits mango canker bacterium 156
Orycles rhinoceros promotes bud rot of palms 282
Oxytetracycline 130
Oxythioquinox 94

Papaya
algal spots 257
anthracnose rot 257
damping off 253
foot rot 253
fruit rot 254
leaf curl 262
mosaic 161, 259
Phytophthora root rot 254
post-harvest fruit rots 257
ringspot virus 259, 261
stem rot 253
Parathion 39
Paraffin wax, fruit coat 120
Peach
brown rot 106, 110
canker 135
crown gall 131
leaf curl 96
Rhizopus soft rot 110
Pear
Armillaria root rot 12
blossom blight 125, 126
brown rot 106
bud blast 136
canker 135
crown gall 131
crown rot 12
fire blight 125
while root rot 103
Penconazole (Topas) 85, 87, 88, 94, 115, 224
Penicillium rots of, citrus 15, 16
pome and stone fruits 118-120
Penicillium 15, 16, 24, 119, 124, 227
digitatum 15, 17, 21, 24-27

expansum 15, 16, 23, 118-121
italicum 15, 24-26
vermosenii 248
Pentalonia nigronervosa
vector of banana bunchy top 203
Pestalotia
disseminata 245
mangiferae 166
pauciseta 293
podocarpii 245-297
psidii 246
sapotae 297
Pestalotiopsis
mangiferae 166, 246, 297
psidii 246
versicolor 245, 297
Pezizella lythri 270
Phaltan (folpet) 24
Phloem-restricted fastidious bacterium 38
Phoma 163, 165, 231
mali 118
Phomopsis 257
citri 15, 17, 18, 23-25, 27
psidii 245
Phoradendron serotinum 176
Phorate 54
Phosethyl-Al (see fosetyl-Al)
Phosphate spray for disease control 158, 225, 230
Phosphorus deficiency 57
Phyllocnistis citrellla disseminates citrus
canker 33
Phyllosticta 231
Physalospora 165
obtusa 113
rhodina 160, 163, 169
Phytoalexins 6, 34
Phytomycin 36
Phytophthora 1, 5, 23, 99, 103, 253
cactorum 99, 100, 118
cambivora 100
capsici 256
cinnamomi 100, 256
citricola 3, 100
citrophthora 2, 3
drechsleri 100
fragariae 272
hibernalis 2
katsurae 282
megasperma 100
nicotianae var. *parasitica* 2, 3, 245
palmivora 2, 3, 254, 255, 257, 281

parasitica 2, 3, 100, 254, 256, 257, 290, 291
syringae 100
Phytophthora diseases of
apple 99-101
citrus 1-7
guava 245
papaya 254-257
pineapple 290-291
strawberry 272-273
Phytophthora, susceptibility to antagonism 4, 6, 101
Pineapple
basal rot and leaf rot 291
heart and stem rot 290
wilt 290
Pink mold rot of apple 121
Plant extracts, in disease control 26, 276, 292
Plant growth regulators, in disease
control 26
Plant oils, in disease control 89, 95, 120, 225, 226, 230, 259, 276
Planofix 245
Plasmids 132, 133
Plasmopara viticola 89, 213
Plum
brown rot 106
crown gall 131
Podosphaera leucotricha 92
Pomegranate
Aspergillus rot 292
bacterial leaf spots 292
Poncirus trifoliata, immune to Tristeza 45
Post-harvest deay of fruits 10, 15, 17, 20, 24, 114, 115, 117, 167-170, 243, 245, 257
Potassium deficiency 58
promotes guava anthracnose 243
Powdery mildew of
apple 91
grapevines 89, 219
mango 157
Pratylenchus penetrans 264
Prochloraz 269
Propamocarb (Previcur) 273
Propiconazole (Tilt) 110, 269
Pseudomonas cepacia
biocontrol of fruit decay 27, 121
Pseudomonas fluorescens
anatgonist of citrus canker bacterium 36

Erwinia amylovora 131
Pseudomnas solanacearum 197, 199
Pseudomonas syringae, biocontrol of canker
   and gummosis of stone fruits 135
   citrus canker 36
   fruit decay 27, 121
Pseudomonas syringae
   pv. *morsprunorum* 137, 138
   pv. *persicae* 137
   pv. *syringae* 135, 136, 219
Pyrimethanil 88
Pythium 1, 99, 253
   aphanidermatum 254, 255
   debaryanum 100
   irregulare 274
   polytylum 100
   spinosum 295
   ultimum 100, 274
   vexans 100

Quarantine against
   citrus canker 35
   fire blight 129
   Sigatoka 192
Quiescent infection 195

Radopholus similis on
   banana 205
   citrus 55, 56
Ramularia tulasneii 270
Ralstonia solanacearum (see also
   Pseudomonas solanacearum) 197, 199,
   200
Resistance to
   antibiotics 130
   fungicides 23-26, 87, 89, 95, 98, 110,
   217, 218, 225, 229, 230, 269, 270, 273,
   276, 291
Rhadinaphelenchus cocophilus 287
Rhinoceros beetle dissimantes bud rot of
   palm 282
Rhizoctonia 1
   bataticola 285
   solani 274, 285
Rhizomorph 13, 14
Rhizopus
   arrhizus 168, 170, 171, 245, 278, 293
   nigricans 277
   oryzae 170, 278
   stolonifer 110, 118, 120, 121, 168, 170,
   171, 196, 257, 277, 278

Rhizopus rot of apple and peach 119-
   121, 123, 124
Rhynchophorus palmarum
   disseminates red ring of palm 288
Ridomil 7, 9, 103, 217, 282
Ringspot virus 261
Ronilan (vinclozolin) 229
Rootstock, role in disease 1, 7, 41, 45,
   102, 103
Rosellinia necatrix 104
Rovral (iprodione) 110, 114, 119, 229, 278
Rubigan (fenarimol) 86, 87, 94, 95, 224

Sanitation 8, 9, 11, 14, 21, 34, 44, 47, 49,
   50, 84, 85, 96, 98, 102, 110, 112, 124,
   129, 130, 134, 159, 162, 166, 171, 189,
   192, 195, 200, 203, 204, 223, 229, 232,
   244, 246, 249, 269, 271, 273, 275, 291,
   292
Sapodilla Plum
   Pestalotia spots 297
Saprol (triforine) 85, 86, 94, 95, 245
Sclerotinia 108
   americana 107
   fructicola 107
   fructigena 107
   fuckeliana 227
   laxa 107
Score 86-88, 94
Shirlan 235
Silicates, reduce grey mold and bunch rot
   230
Sodium bicarbonate, grape powdery mil-
   dew control 225
Sodium carbonate, in disease control 26,
   119, 226
Sodium orthophenyl phenate 23, 25, 36,
   88, 118, 120
Sodium metabisulphite 120, 124
Soil fumigation 14, 105, 254
Sour rot of citrus 20, 24
Sphaceloma ampelinum 234
Sphaeropsis 165
   malorum 113
Sphaerotheca 88
   fuliginea 226
Spilocaea pomi 77
Sporobolomyces roseus
   biocontrol of apple decay 110
Spray schedules for apple scab 89, 90
Stem-end rots 17, 163
Stem-pitting 41

Sterol inhibitors 25, 85, 87
Strawberry
  anthracnose 265
  black leaf spots 267
  black root rot complex 274
  Botrytis rot 275
  brown rot 275
  bud rot 267
  crown rot 265
  Dendrophoma leaf spots 270
  flower blight 267
  fruit rot 267
  grey mold 275
  leak 276
  Mycosphaerella leaf spots 270
  red stele 272
  Rhizopus rot 276
Streptocyline 156
Streptomycin 36, 130, 139
Strobilurin fungicides 88, 94, 193, 217, 224
Strobilurus tenacellus 88
Sulfex 158, 224
Sulphur deficiency 59
Sulphur dioxide fumigation of fruits 120, 230, 244
Sulphur dioxide injury 172
Sulphur fungicides 85, 86, 110, 158, 223, 225
Syllit 85, 86
Systhane 86, 94

Taphrina deformans 97, 98
Tebuconazole 88, 89, 110, 115
Tetraconazole 94
Thiabendazoles 23, 25, 26, 114, 119, 162, 245
Thimet (phorate) 54
Thiovit 224
Thiophanates 23, 25, 85, 94, 110, 119, 158, 166, 229, 235
Thosconyrsa citri disseminates citrus canker 33
Tilt (propiconazole) 110, 269
Ti-plasmid 133
Topas 88, 94
Topas-C 85, 89
Topsin-M 85, 87, 94, 98, 110, 158, 224, 229
Toxoptera, aurantii 43
  citircida 43

Tree eradication 14, 28, 35, 56
Triadimefon 95, 224, 225
Triadimenol (Bayton) 224
Triazoles 87, 224
Trichoderma, in biocontrol of
  apple collar rot 103
  fruit decay 27, 85, 230, 245
Trichoderma harzianum, biocontrol of
  Botrytis grey mold 230
Trichoderma viride
  antagonist of *Armillaria mellea* 14
  biocontrol of citrus fruit rots 27
Trichothesium roseum 118, 227
  pink mold rot 122
Tridemorph (Calixin) 94, 95
Triforine (Cela, Funginex) 23, 25, 85, 86, 94, 95, 110
Triflumazole 94
Tristeza virus 37, 42-46
Tylenchulus semipenetrans 8, 50, 51

Uncinula necator 88, 220, 224-226
Urea for suppression of apple scab 84, 85

Vangard (etaconazole) 94, 110, 224
Vapam 254
Venturia
  asparata 77
  inaequalis 77, 78, 89
  pyrina 74, 82
Verticillium theobromae 195, 196
Vesicular arbuscular mycorrhizae 6
Vinclozolin (Ornalin, Ronilan) 110, 229, 230, 276
Viroids 49
Viscum album 177
Vision 88
Vorlan 229

Warning systems 82-84, 115, 131, 224
Whiskers rot 118
Wilt of
  banana 183, 197
  guava 246
  coconut 284
  pineapple 290

Withertip of
  citrus 9
  mango 159

Xanthomonas axonopodis
  pv., *citri*   30-32
  punicae 292
Xanthomonas campestris
  pv., *citri*   30
  citrumelo 30
  mangiferae-indicae   156, 167, 296
  punicae 292

Xanthomonas
  citri   30, 36
  punicae 292
Xyloporosis of citrus 46

Yeasts, as biocontrol agents 27

Zinc deficiency 60, 249
Zineb 11, 85, 86, 103, 162, 244
Ziram 117, 235, 236, 271

About the Author

Dr. R.S. Singh had been engaged in teaching and research in plant pathology for more than 40 years. He spent 26 years at the G.B. Pant University of Agriculture and Technology, Pantnagar as Associate Professor, Professor and Head of Department and also as Dean Post Graduate Studies before retiring from active service. However, he has continued his association with plant pathology through his books. As a research scientist he published nearly 200 papers in national and international journals. His other widely read books are **Plant Diseases, Principles of Plant Pathology, Diseases of Vegetable Crops** and the **Plant Pathogens series (Fungi, Bacteria and Nematodes).** His latest contribution, in addition to the present volume, is **Plant Disease Management.** He was co-editor of **Conceptual and Experimental Plant Pathology** and **A Handbook of Economic Nematology.**

Dr. Singh had done post-doctoral research at the University of Wisconsin (Madison) in 1960 and had visited that country on a lecture tour in 1971. He was President of the Indian Phytopathological Society and India Society of Mycology and Plant Pathology.